Advanced Analytical and Numerical Methods with their Application to Industrial Problems

Advanced Analytical and Numerical Methods with their Application to Industrial Problems

Edited by

Dr. Ramakanta Meher
Department of Mathematics,
Sardar Vallabhbhai National Institute of Technology,
Surat, India

CWP

This edition has been published by Central West Publishing PTY LTD, (ABN 13 683 898 722) Australia

For more information about the books published by Central West Publishing PTY LTD, please visit https://centralwestpublishing.com

NATIONAL LIBRARY OF AUSTRALIA

A catalogue record for this book is available from the National Library of Australia

ISBN (print): 978-1-922617-67-5

About the Book

This book covers the models of different real-world problems that includes the models using Ordinary and partial differential equations and Dynamical systems, Uncertainty quantifications using Fuzzy systems, Computational methods for differential equations, Modern control theory and applications, Neural network and Neural computing, Computational heat and mass transfer and Computational fluid dynamics. It also includes various research problems on developing advanced analytical and numerical methods for solving real-world situations with its theoretical derivations and engineering and science applications.

About the Editor

Ramakanta Meher works as an Associate Professor in the Department of Mathematics, SVNIT, Surat. His current research interests include Differential and Integral equations, Fractional Differential Equations, Linear Algebra, Approximation Theory, Numerical Analysis, and Fuzzy theory. His main study focus is Fluid Dynamics. Collaboration Research is ongoing with researchers from India, China, Iran, Turkey, South Africa, and the United States. He has supervised 07 PhD students and 29 M.Sc. students and oversees 06 additional PhD students. He has published over 100 international research publications in various reputed international journals and has writ-ten and published 06 textbooks. He has received several internation-al travel grants such as the CIMPA-KAUST Travel Grant to visit King Abdullah University of Science and Technology, Thuwal, Saudi Arabia in 2012, CIMPA-UNESCO-MESR-MICINN-APSA-SOUTH AFRICA Travel Grant to visit African Institute of Mathematical Sciences, Muizenberg, Cape town, South Africa in 2011, CIM-PA-UNESCO-MICINN-THAILAND Travel Grant to visit Chulalongkorn University, Bangkok, Thailand in 2011 and CIM-PA-UNESCO-SCHOOL MONGOLIA Travel Grant to visit Mongolian State University, Mongolia in 2013. He has delivered invited talks at several scientific events in India, Thailand, Saudi Arabia, UAE, Mongolia, and South Africa. He has coordinated many prestigious programs such as Mathematics Training Talent Search Programme (MTTS)-2013 (Level - 0,1 and 2) with Prof. S. Kumresan by receiving a grant from the National Board of Higher Mathematics (NBHM), Two STTP by receiving grants from the Indian Society of Technical Education, New Delhi and some STTPs by receiving grants from TEQIP and SVNIT, Surat. He has organised over 10 Short-Term Training/Faculty Development Programs and delivered over 20 invited talks in various reputed institutes.

Preface

This book covers the models of different real-world problems, including various research problems on developing advanced analytical and numerical methods for solving various real-world situations. It also includes the modelling related to heat and mass transfer problems with its theoretical derivations and engineering and sci-ence applications.

There are a total of 17 chapters. Chapter 1 starts with an interactive, non-linear model investigating agricultural practices' impact on the bat population in a woodland environment. Chapter 2 uses the homotopy analytic approach to the effect of the transfer of heat generated by nanofluid flow on two parallel plates. In contrast, Chapter 3 includes the study of the Couple Stress on Micro-Polar Fluid Flow Saturating a Porous Medium with Hall Effect. Chapter 4 discusses a cryptographic algorithm using a scalar potential technique with ma-trix formation. Chapter 5 applies the Homotopy Perturbation Meth-od to solve general linear and non-linear fractional partial differential equations. Chapter 6 discusses the one-dimensional mathematical model of pollutant transport in rivers with different input sources, whereas a new computational technique for the FitzHugh-Nagumo model is covered in Chapter 7. Chapter 8 discusses the multiple coexisting attractors in Magneto convection. Chapter 9 secretly discusses the Share text using the system of equations and the Cayley Hamilton theorem. Chapter 10 discusses the study of fuzzy differential equations using a double-parametric homotopy approach. A nonconventional cryptographic approach based on PMF and CDF is discussed in Chapter 11, whereas Chapter 12 includes the Stability analysis for a Predator-prey Model with Fractional Order. Chapter 13 discusses the application of Pythagorean fuzzy soft linear space, whereas Chapter 14 discusses the strain-controlled magnetic domain wall motion in multi-ferroic devices induced by current and travelling waves. Chapter 15 discusses Some invariant solutions of (3+1)-dimensional Zakharov-Kuznetsov-Burger's equation, whereas Chapter 16 discusses the Parallel Computing of Beta, Incomplete Gamma and Meijer-G Functions using GPU Platform magnetic field. Chapter 17 discusses the Second Kind of Chebyshev-Collocation Scheme for Solving Non-linear Fredholm Integro-Differential Equations with Weakly Singular Kernels.

Table of Contents

Chapter 1

An Interactive Nonlinear Model to Investigate the Impact of Agricultural Practices on Bat Population in a Woodland Environment

Abhinav Tandon[1], Vaishnudebi Dutta[2]
[1]Department of Mathematics, Birla Institute of Technology Mesra, Ranchi - 835215, Jharkhand, India
[2] School of Engineering Mathematics and Technology, University of Bristol, University Walk, BS8 1TW Bristol, United Kingdom.

Abstract

Woodlands are heavily impacted by habitat loss and are enmeshed in an agricultural system. Because it provides numerous bat species with places to rest and forage, the woodland is one of their most crucial habitats. Based on this, a nonlinear interactive dynamical model is developed in terms of system of ordinary differential equations. The proposed model is predicated on the idea that the bat population entirely depends on woodland vegetation for their existence, but the use of woodlands for agricultural purposes leads to the loss of woodland vegetation and bat population. In the system, the presence and stability of equilibrium solutions are investigated. It is being found that a stable system may be created when woodland vegetation, bat population, and crop yield densities coexist under the criteria shown by the mathematical study. Quantitative outcomes obtained through numerical simulation also provide evidence in support of these intuitive assumptions. Both qualitative and quantitative studies indicate limiting the overuse of woodland systems for agricultural purposes since doing so might destabilize the system.

Keywords: Woodland vegetation, crop-yield, mathematical model, differential equation, stability analysis, numerical simulation

Mathematics Subject Classification (2000) 92D40 · 34C60

1.1 Introduction

A woodland is identical with wood, and it is a low-density forest that creates open environments with little to no shelter from the sun [1]. Such an area is characterized by the presence of towering trees, rich undergrowth, and a plethora of interconnected life forms. Woodland ecosystem is a diverse and vibrant community of organisms that thrive in the enchanting embrace of woodlands [2]. The majestic canopy formed by the trees provides shelter, food, and nesting sites for a wide array of species. Bats are one of the many species that depend on woodland vegetation with adequate roosting cavities, such as hollow trees or cracks in tree bark [3, 4].

Over 1300 species of bats are known to exist globally, making them the second-largest animal order [5]. They act as a significant contributor to seed dissemination, pollination, and insect pest control services [6, 7]. Woodland is considered to be one of the most crucial environments for bats as it provides important habitats for them due to the abundance of roosting and feeding locations [8, 9, 10]. Here, the vegetation structure is of great importance in determining habitat quality and availability for bats [11, 12, 13].

The development of agriculture has had a tremendous influence on the woodland ecology, which was formerly replete with lush vegetation and a vibrant web of life [9, 10]. As agricultural activities expand, woodlands are often cleared to make way for farmland, resulting in the destruction of bat roosting sites and foraging habitats [11, 12]. Bats are one of the many species that depend on old trees with adequate roosting cavities, such as hollow trees or cracks in tree bark, which are frequently lost during land clearance [14]. Consequently, the loss of these crucial roosting sites disrupts bat populations and can lead to declines in their numbers. Moreover, the conversion of forests into agricultural fields can fragment bat habitats. This fragmentation restricts the movement of bats and hinders their ability to find food and suitable roosting sites [15].

Agricultural practices often involve the use of pesticides and herbicides to control pests and weeds, which can have indirect effects on bats [16]. These chemicals can reduce insect populations, which form a crucial part of the bat diet. The decline in insect abundance can lead to food scarcity for bats and ultimately impact their survival and reproductive success [17, 18]. Insectivorous bats in woodlands are advantageous for maximizing agricultural productivity because they control pests, which eventually emphasizes the need to integrate bat conservation into integrated pest management programs [19, 20, 21, 22].

Given this perspective, it appears crucial to suggest a research endeavor that elucidates the interconnected relationships among three key elements: woodland vegetation, bat population, and agricultural product or crop yield within woodland ecosystems. To address this, the current study proposes a nonlinear mathematical model, employing a system of differential equations, that is widely acknowledged as a dependable and effective tool. Within the formulated model, the dynamics of woodland vegetation, bat population, and crop yield are regarded as the focal variables of interest. In this analysis, the model is examined to explore its qualitative characteristics, such as the presence of equilibrium points, stability, and bifurcations. These investigations aim to assess the long-term dynamical behavior of the system. The qualitative analysis reveals different scenarios in which the bat population survives in an agricultural setting of a woodland system. The qualitative results are supported through numerical simulations to portray the varied dynamic patterns that would emerge with change in parameters.

1.2 Mathematical Model Formulation

Both local and landscape-level factors have an impact on bat populations in wooded areas inside an agricultural environment. At the local level, patch layout is found to be less significant than the characteristics of the forest vegetation. Based on the species-specific habitat relationships, environmental variety appears to be crucial for meeting the foraging needs of various bat species [10]. The number of bats in areas where the agricultural activities occur is deemed to be the most pressing issue in a woodland system. Keeping the study of Fuentes et al. [10] in mind, the interactive dynamics that changes due to involvement of bats in agricultural productivity in wooded settings is modelled mathematically in this article.

In the presence of crop-yield density $C(t)$ at any time $t > 0$, the woodland system is assumed to be finite where $W(t)$ and $B(t)$

represent the vegetation and the population of bats, respectively, giving rise to the following system of differential equations:

$$\frac{dW}{dt} = pW(1 - \frac{W}{K}) - \alpha WB - \beta WC - \gamma W^2 C \quad (1.2.1a)$$

$$\frac{dB}{dt} = \theta\alpha BW - \nu B^2 - \mu B \quad (1.2.1b)$$

$$\frac{dC}{dt} = qC(1 - \frac{C}{L}) + \delta BC \quad (1.2.1c)$$

with $W(0) \geq 0, B(0) \geq 0, C(0) \geq 0$.

The parameters $p, K, \alpha, \beta, \gamma, \theta, \nu, \mu, \delta, q$ and L are all positive. However, the meanings of model parameters are summarized in Table 1.1.

Table 1.1 Model parameters description

Parameter	Definition
p	Intrinsic growth rate of woodland vegetation
K	Carrying capacity of woodland vegetation
α	Rate coefficient responsible for the depletion of woodland vegetation due to its use by bat population
β	Rate coefficient responsible for the depletion in the growth of woodland vegetation due to encroachment for crop production
γ	Rate coefficient responsible for the depletion in the carrying capacity of woodland vegetation due to encroachment for crop production
θ	Proportion of β utilised by the bat population to expand
ν	Intra-specific competition rate of bat population
μ	Natural death rate of bat population
q	Intrinsic growth rate of crop yield
L	Carrying capacity of crop yield
δ	Rate coefficient responsible for the increase in crop yield due to bat population

The following points are taken into consideration to formulate the system (1.2.1):

- Assuming that the woodland has a finite amount of space, its vegetation grows logistically, with an intrinsic growth rate of p and a carrying capacity of K. In the same system, farming is also practised and is represented by crop yield density, that exhibits logistic growth with intrinsic growth rate q and carrying capacity L.
- As already said, the woodland vegetation provides bats with shelters, food, and other necessities for survival. Bat population hence increases due to woodland vegetation. In exchange, they have a detrimental effect on the development of woodland vegetation. On the basis of this, it is assumed that the growth of woodland vegetation will be negatively impacted by bats at a rate coefficient of α. A portion of it $\theta\alpha$, is employed for the growth of bats from woodland. The growth of bats is intended to be dropping with rate coefficient ν, taking into account the intra-specific competition amongst bat population for scarce resources. The νB^2 term in the second equation (1.2.1b) of the model serves as its representation. The bat population is assumed to exhibit natural death rate μ.

- The natural growth rate of forest vegetation and its carrying capacity are both impacted by the usage of woodland systems for agriculture. Based on this, the two terms viz. βWC and $\gamma W^2 C$ are incorporated in the first equation related to the rate of change in woodland vegetation. The first term βWC exhibits the negative impact of crop-yield density on the natural growth of woodland vegetation at a rate of β, however, the second term $\gamma W^2 C$ represents the negative impact of crop-yield density on the carrying capacity of woodland vegetation at the rate γ.

- Numerous bat species act as natural pest killers, crop productivity may improve as a result of the bat population. Therefore, it is presumable that bat population boost crop yield at a rate of δ.

By adding the following dimensionless variables, the described model (1.2.1) may be made simpler and the qualitative properties can be more clearly understood:

$$\bar{t} = \mu t, \bar{W} = \frac{W}{K}, \bar{C} = \frac{C}{L}, \bar{B} = \frac{\alpha}{\mu} B$$

With these dimensionless variables, the following dimensionless parameters are obtained:

$$\bar{p} = \frac{p}{\mu}, \bar{\beta} = \frac{\beta L}{\mu}, \bar{\gamma} = \frac{\gamma KL}{\mu}, \bar{\theta} = \frac{\theta \alpha K}{\mu}, \bar{\nu} = \frac{\nu}{\alpha}, \bar{q} = \frac{q}{\mu}, \bar{\delta} = \frac{\delta}{\alpha}$$

The dimensionless model is produced as follows when the bars are removed:

$$\frac{dW}{dt} = pW(1 - W) - WB - \beta WC - \gamma W^2 C \quad (1.2.2a)$$

$$\frac{dB}{dt} = \theta BW - \nu B^2 - B \quad (1.2.2b)$$

$$\frac{dC}{dt} = qC(1 - C) + \delta BC \quad (1.2.2c)$$

The revised model, (1.2.2), contains fewer parameters and is devoid of units of measurement than the original, (1.2.1).

1.3 Qualitative Properties

It is necessary to elucidate the qualitative characteristics of the dimensionless model (1.2.2) in order to gain understanding of the system's long-term behaviour. These qualitative features are discussed in the present section.

1.3.1 Solution Positivity

The method outlined in [23] may be used to improve the effectiveness of system solutions. The dimensionless model (1.2.2) is transformed as:

$$\frac{dZ}{dt} = X(Z) \quad (1.3.1)$$

Consider initial values as $Z(0) = (W(0), B(0), C(0))^T = Z_0 \in \Re^3$, where $Z = (Z_1, Z_2, Z_3)^T = (W, B, C)^T \in \Re^3$ and T represents matrix transpose. The matrix $X(Z)$, defined as $X: \mathbb{Q}_+ \to \Re^3$ and $Z \in \mathbb{Q}^\infty(\Re^3)$, is

$$X = X(Z) = \begin{pmatrix} X_1(Z) \\ X_2(Z) \\ X_3(Z) \end{pmatrix} \quad (1.3.2)$$

where

$X_1(Z) = pW(1 - W) - WB - \beta WC - \gamma W^2C$

$X_2(Z) = \theta BW - \nu B^2 - B$

$X_3(Z) = qC(1 - C) + \delta BC$

It is simple to prove that for $i = 1,2,3,\ X_i(Z)|_{Z_i=0} \geq 0$ whenever $Z(0) \in \Re^3$ such that $Z_i = 0$. According to Nagumo's theorem [24], the solution to the system (1.3.2) with the initial condition $Z(0) = Z(0)$ is then guaranteed to be $Z(t)$ for all $t > 0$. Based on this, the following theorem may be stated for positive system solutions:

Theorem 1: The solutions of system (1.2.2) in $\Re^3$ are positive for all time $t > 0$ with initial values $W(0) \geq 0, B(0) \geq 0, C(0) \geq 0$.

1.3.2 Boundedness of Solutions

Considering that the system (1.2.2) has some solution $(W(t), B(t), C(t))$. Following that, the equation (1.2.2a), reduces as:

$$\frac{dW}{dt} \leq pW(1 - W) \tag{1.3.3}$$

Then, using the standard comparison theorem ([25]; [26]), the derived differential inequality yields:

$$\lim_{t\to\infty} sup W(t) \leq 1 \tag{1.3.4}$$

Again, the equation (1.3.3) gives the following differential inequality:

$$\frac{dB}{dt} \leq \theta W - \nu B - 1 \leq \theta - \nu B - 1$$

Now, the supremum of B is

$$\lim_{t\to\infty} sup B(t) \leq \frac{\theta - 1}{\nu} \tag{1.3.5}$$

The equation (1.3.4) reduces into the inequality as:

$$\frac{dC}{dt} \leq qC(1-C) + \delta BC \leq C[q(1-C) + \delta(\frac{\theta-1}{\nu})]$$

Hence

$$\lim_{t\to\infty} sup C(t) \leq \frac{q\nu + \delta(\theta-1)}{q\nu} \tag{1.3.6}$$

It shows the boundedness of all model solutions in a region of attraction ζ. Hence, the following theorem can be stated.

Theorem 2: The solutions of system (1.2.2) are bounded in a region of attraction ζ defined as:

$$\zeta = \{(W,B,C) \in \Re^3 : 0 \leq W \leq 1;\ 0 \leq B \leq \frac{\theta-1}{\nu};\ 0 \leq C \leq \frac{q\nu + \delta(\theta-1)}{q\nu}\}$$

provided the condition

$$\theta > 1 \tag{1.3.7}$$

remains valid.

1.3.3 Equilibrium Analysis

The identified equilibrium points of the model (1.2.2) are:

1. **Trivial - equilibrium $E_0(0,0,0)$:** It is always existent.

2. **Woodland vegetation and bat population - free equilibrium $E_1(0,0,1)$:** It is always existent.

3. **Bat population and crop yield - free equilibrium $E_2(1,0,0)$:** It is feasible as well.

4. **Bat population free equilibrium -** $\boldsymbol{E_3(\bar{\bar{W}}, 0, \bar{\bar{C}})}$**:** To locate the equilibrium E_3, the following equations are simultaneously solved

$$p - p\bar{\bar{W}} - \beta C - \gamma WC = 0 \tag{1.3.8}$$

$$q(1 - C) = 0 \tag{1.3.9}$$

From which it can be determined that:

$$\bar{\bar{W}} = \frac{p-\beta}{p+\gamma} \tag{1.3.10}$$

and,

$$\bar{\bar{C}} = 1 \tag{1.3.11}$$

So, for the existence of $E_3(\bar{\bar{W}}, 0, \bar{\bar{C}})$ the following condition must satisfy:

$$p > \beta \tag{1.3.12}$$

5. **Crop yield - free equilibrium** $\boldsymbol{E_4(\widehat{W}, \hat{B}, 0)}$**:** For the existence of E_4, it is required to solve the following equations simultaneously

$$p(1 - \widehat{W}) - \hat{B} = 0 \tag{1.3.13}$$

$$\theta\widehat{W} - \nu\hat{B} - 1 = 0 \tag{1.3.14}$$

Simultaneously solving the two equations, the following expressions are deduced:

$$\widehat{W} = \frac{p\nu+1}{p\nu+\theta} \tag{1.3.15}$$

and,

$$\hat{B} = \frac{p(\theta-1)}{p\nu+\theta} \tag{1.3.16}$$

which implies existence of the equilibrium $E_4(\widehat{W}, \hat{B}, 0)$, as $\theta > 1$ (condition (1.3.7)).

6. **Coexistence - equilibrium $E_5(\widetilde{W},\widetilde{B},\widetilde{C})$:** To locate the equilibrium E_5, the following equations are to be solved simultaneously:

$$p(1-\widetilde{W})-\tilde{B}-\beta\tilde{C}-\gamma\widetilde{W}\tilde{C}=0 \quad (1.3.17)$$
$$\theta\widetilde{W}-\nu\tilde{B}-1=0 \quad (1.3.18)$$

$$q(1-\tilde{C})+\delta\tilde{B}=0 \quad (1.3.19)$$

From equation (1.3.19),

$$\tilde{C}=\frac{q+\delta\tilde{B}}{q} \quad (1.3.20)$$

Now, from (1.3.18),

$$\tilde{B}=\frac{\theta\widetilde{W}-1}{\nu} \quad (1.3.21)$$

However, $\tilde{B}$ is positive under the following condition:

$$\theta\widetilde{W}>1 \quad (1.3.22)$$

From the derived $\tilde{C}$ and $\tilde{B}$, the quadratic in $\widetilde{W}$ is deduced to be:

$$\tilde{a}\widetilde{W}^2-\tilde{b}\widetilde{W}+\tilde{c}=0 \quad (1.3.23)$$

where,

$$\tilde{a}=\delta\gamma\theta \quad (1.3.24)$$

$$\tilde{b}=\gamma\delta-pq\nu-q\theta-\beta\delta\theta-q\gamma\nu \quad (1.3.25)$$

$$\tilde{c}=q\beta\nu-pq\nu-\beta\delta-q \quad (1.3.26)$$

$\widetilde{W}_1$ and $\widetilde{W}_2$ are the roots of the quadratic equation (1.3.23), defined as:

$$\widetilde{W}_1=\frac{\tilde{b}+\sqrt{\tilde{b}^2-4\tilde{a}\tilde{c}}}{2\tilde{a}} \quad (1.3.27)$$

$$\widetilde{W}_2 = \frac{\tilde{b}-\sqrt{\tilde{b}^2-4\tilde{a}\tilde{c}}}{2\tilde{a}} \tag{1.3.28}$$

Here, $\tilde{a}$ is positive. Now, for feasible values of $\widetilde{W}$, discriminant $\tilde{\Delta} = \tilde{b}^2 - 4\tilde{a}\tilde{c}$ should be greater than or equal to zero. But, whatever be the values of $\widetilde{W}$, it should satisfy condition (1.3.22). So, this feasibility condition should be checked for $\widetilde{W}_1$,$\widetilde{W}_2$ and also, when $\widetilde{W}_1 = \widetilde{W}_2 = \frac{\tilde{b}}{2\tilde{a}}$ $(\tilde{\Delta} = 0)$.

By checking the condition (1.3.22) for above values of $\widetilde{W}$, it is being obtained that the inequality would not hold true for $\widetilde{W}_2$ and also, for $\widetilde{W}_1 = \widetilde{W}_2 = \frac{\tilde{b}}{2\tilde{a}}$ $(\tilde{\Delta} = 0)$. It implies that there exists only unique feasible value of $\widetilde{W}$, that is, $\widetilde{W}_1$. The following inequality in parametric form is derived by plugging $\widetilde{W}_1$ in (1.3.17),

$$p > \frac{\theta\beta+\gamma}{\theta-1} \tag{1.3.29}$$

under the condition $\theta > 1$ given by equation (1.3.7). With obtained positive $\widetilde{W} = \widetilde{W}_1$, others $\tilde{B}$ and $\tilde{C}$ can be determined from equations (1.3.21) and (1.3.20) respectively. Hence, the unique coexistence equilibrium $E_5(\widetilde{W}, \tilde{B}, \tilde{C})$ exists under the condition (1.3.29).

1.3.4 Stability Analysis of Equilibrium States

Around different identified equilibrium states, the local stability analysis of the system (1.2.2) can be derived from the sign of the eigenvalues of the Jacobian matrix.

The general Jacobian matrix of the system (1.2.2) is

$$J(E_i) = \begin{pmatrix} f_{11} & -W & -\beta W - \gamma W^2 \\ \theta B & f_{22} & 0 \\ 0 & \delta C & f_{33} \end{pmatrix} \tag{1.3.30}$$

where

$$f_{11} = p - 2pW - B - \beta C - 2\gamma WC$$

$$f_{22} = \theta W - 2\nu B - 1$$

$f_{33} = q - 2qC + \delta B$

Below is a discussion of the local stability behaviour of possible equilibria:

1. At the equilibrium state $E_0(0,0,0)$, the Jacobian matrix is

$$J(E_0) = \begin{pmatrix} p & 0 & 0 \\ 0 & -1 & 0 \\ 0 & 0 & q \end{pmatrix} \tag{1.3.31}$$

The eigenvalues are p, -1 and q. Two positive eigenvalues result in an unstable manifold on the $W - C$ plane for equilibrium E_0.

2. At the equilibrium state $E_1(0,0,1)$, the Jacobian matrix is:

$$J(E_1) = \begin{pmatrix} p-\beta & 0 & 0 \\ 0 & -1 & 0 \\ 0 & \delta & -q \end{pmatrix} \tag{1.3.32}$$

The eigenvalues are $p - \beta$, -1 and $-q$, out of which one eigenvalue will become negative when the following condition is satisfied:

$$p < \beta \tag{1.3.33}$$

the other two eigenvalues are unconditionally negative. However, the above condition is just opposite of the condition (1.3.12), obtained for the existence of equilibrium E_3. Here, it is apparent that the equilibrium $E_1(0,0,1)$ demonstrates asymptotic stable behavior only when the equilibrium $E_3(\bar{\bar{W}}, 0, \bar{\bar{C}})$ is not feasible, otherwise whenever $E_3(\bar{\bar{W}}, 0, \bar{\bar{C}})$ is feasible, $E_1(0,0,1)$ is unstable. The connection of stability and instability of the equilibrium $E_1(0,0,1)$ with the feasibility and infeasibility of the equilibrium $E_3(\bar{\bar{W}}, 0, \bar{\bar{C}})$ indicates the occurrence of transcritical bifurcation between the two involved equilibria.

3. At the equilibrium state $E_2(1,0,0)$, the Jacobian matrix is:

$$J(E_2) = \begin{pmatrix} -p & -1 & -\beta - 2\gamma \\ 0 & \theta - 1 & 0 \\ 0 & 0 & q \end{pmatrix} \tag{1.3.34}$$

Here, the eigenvalues are $-p, \theta - 1$ and q. As $\theta > 1$ and $q > 0$, so two of the eigenvalues are always positive, so the equilibrium is unstable in the B and C direction.

4. At the equilibrium state $E_3(\bar{\bar{W}}, 0, \bar{\bar{C}})$, the Jacobian matrix is:

$$J(E_3) = \begin{pmatrix} -p\bar{\bar{W}} - \gamma\bar{\bar{W}}\bar{\bar{C}} & -\bar{\bar{W}} & -\beta\bar{\bar{W}} - 2\gamma\bar{\bar{W}} \\ 0 & \theta\bar{\bar{W}} - 1 & 0 \\ 0 & \delta\bar{\bar{C}} & -q\bar{\bar{C}} \end{pmatrix} \tag{1.3.35}$$

Here, the eigenvalues are $-p\bar{\bar{W}} - \gamma\bar{\bar{W}}\bar{\bar{C}}, \theta\bar{\bar{W}} - 1$ and $-q\bar{\bar{C}}$. Here, the stability behavior is decided from the eigenvalue $\theta\bar{\bar{W}} - 1$. If

$$\theta\bar{\bar{W}} < 1 \tag{1.3.36}$$

then the system exhibits asymptotic stable behavior around E_3. The above condition reduces into the following inequality with the value of $\bar{\bar{W}}$

$$p < \frac{\theta\beta+\gamma}{\theta-1} \tag{1.3.37}$$

It may be noted that the above condition is opposite of the condition (1.3.26), derived for the feasible coexistence equilibrium $E_5(\tilde{W}, \tilde{B}, \tilde{C})$. Analyzing both the conditions, it can be deduced that equilibrium E_3 shows unstable behavior whenever coexistence equilibrium E_5 is feasible. Such a situation gives rise to transcritical bifurcation between the two.

5. At the equilibrium state $E_4(\widehat{W}, \hat{B}, 0)$, the Jacobian matrix is:

$$J(E_4) = \begin{pmatrix} -p\widehat{W} & -\widehat{W} & -\beta\widehat{W} - \gamma\widehat{W}^2 \\ \theta\hat{B} & -\nu\hat{B} & 0 \\ 0 & 0 & q + \delta\hat{B} \end{pmatrix} \tag{1.3.38}$$

Here, the eigenvalues are $-p\widehat{W}, \nu\hat{B}$ and $q + \delta\hat{B}$ where the system is unstable in the C direction.

6. At equilibrium state $E_5(\widetilde{W},\tilde{B},\tilde{C})$, the Jacobian matrix is:

$$J(E_5)=\begin{pmatrix} -p\widetilde{W}-\gamma\widetilde{W}\tilde{C} & -\widetilde{W} & -\beta\widetilde{W}-\gamma\widetilde{W}^2 \\ \theta\tilde{B} & -\nu\tilde{B} & 0 \\ 0 & \delta\tilde{C} & -q\tilde{C} \end{pmatrix} \tag{1.3.39}$$

The following characteristic equation determines the eigenvalues:

$$\varphi^3+b_1\varphi^2+b_2\varphi+b_3=0 \tag{1.3.40}$$

where

$$b_1=q\tilde{C}+p\widetilde{W}+\gamma\tilde{C}\widetilde{W}+\nu\tilde{B} \tag{1.3.41}$$

$$b_2=pq\tilde{C}\widetilde{W}+q\gamma\tilde{C}^2\widetilde{W}+q\nu\tilde{B}\tilde{C}+(\theta+(p+\gamma\tilde{C})\nu)\tilde{B}\widetilde{W} \tag{1.3.42}$$

$$b_3=(\beta+\gamma\widetilde{W})\delta\theta\tilde{B}\tilde{C}\widetilde{W}+q(\theta+(p+\tilde{C}\gamma)\nu)\tilde{B}\tilde{C}\widetilde{W} \tag{1.3.43}$$

By establishing $b_1>0,\ b_3>0$ with $b_1b_2-b_3>0$, it is possible to conclude using the Routh-Hurwitz criterion that the coexistence equilibrium E_5 is locally asymptotically stable. Here, it is apparent that $b_1>0$ and $b_3>0$ for positive coexistence equilibrium. However, $b_1b_2-b_3$ is obtained to be:

$$\begin{aligned} b_1b_2-b_3 = {} & (2pq\nu-\beta\delta\theta)\widetilde{W}\tilde{B}\tilde{C}+pq^2\widetilde{W}\tilde{C}^2+p^2q\widetilde{W}^2\tilde{C} \\ & +q^2\gamma\widetilde{W}\tilde{C}^3+q\gamma^2\widetilde{W}^2\tilde{C}^3+2pq\gamma\widetilde{W}^2\tilde{C}^2 \\ & +p\theta\widetilde{W}^2\tilde{B}+\gamma(\theta-\delta\theta+2p\nu)\widetilde{W}^2\tilde{B}\tilde{C} \\ & +q^2\nu\tilde{B}\tilde{C}^2+p^2\nu\widetilde{W}^2\tilde{B} \\ & +2q\gamma\nu\widetilde{W}\tilde{B}\tilde{C}^2+\gamma^2\nu\widetilde{W}^2\tilde{B}\tilde{C}^2+\theta\nu\tilde{B}^2\widetilde{W}+\gamma^2\nu\widetilde{W}^2\tilde{B}\tilde{C}^2 \\ & +q\nu^2\tilde{B}^2\tilde{C}+p\nu^2\widetilde{W}\tilde{B}^2+\gamma\nu^2\widetilde{W}\tilde{B}^2\tilde{C} \end{aligned} \tag{1.3.44}$$

Hence, the coexistence equilibrium E_5 develops local asymptotic stable behavior if $b_1b_2-b_3>0$.

Theorem 3:
The model system (1.2.2) develops

1. Unstable behaviors without any condition around trivial equilibrium $E_0(0,0,0)$, bat population and crop yield - free equilibrium

$E_2(1,0,0)$, and crop yield - free equilibrium $E_4(\widehat{W}, \hat{B}, 0)$.

2. Around woodland vegetation and bat population - free equilibrium $E_1(0,0,1)$, local asymptotic stable behavior is exhibited if $p > \beta$, that is, when the bat population - free equilibrium $E_3(\bar{\bar{W}}, 0, \bar{\bar{C}})$ does not exist.

3. Around bat population - free equilibrium $E_3(\bar{\bar{W}}, 0, \bar{\bar{C}})$, local asymptotic stable behavior is exhibited if $p < \frac{\theta\beta+\gamma}{\theta-1}$, that is, when the coexistence equilibrium $E_5(\widetilde{W}, \tilde{B}, \tilde{C})$ is infeasible.

4. Around the feasible coexistence equilibrium $E_5(\widetilde{W}, \tilde{B}, \tilde{C})$, local asymptotic stable behavior is exhibited if $b_1 b_2 - b_3 > 0$.

1.3.5 Global Stability

The global stability of the model system (1.2.2) around it must be addressed since the coexistence equilibrium is of utmost relevance. In this instance, the global stability behaviour is established using Lyapunov's direct approach.

It is taken into account the following Lyapunov function:

$$Q = W - \widetilde{W} - \widetilde{W} \ln\frac{W}{\widetilde{W}} + l(B - \tilde{B} - \tilde{B} \ln\frac{B}{\tilde{B}}) + m(C - \tilde{C} - \tilde{C} \ln\frac{C}{\tilde{C}}) \tag{1.3.45}$$

where positive constants l and m are used. Now, the following result is obtained by differentiating the system (1.2.2) solutions along Q with regard to time t:

$$\frac{dQ}{dt} = (W - \widetilde{W})\frac{1}{W}\frac{dW}{dt} + l(B - \tilde{B})\frac{1}{B}\frac{dB}{dt} + m(C - \tilde{C})\frac{1}{C}\frac{dC}{dt} \tag{1.3.46}$$

Where, $\frac{dQ}{dt}$ can also be written as:

$$\begin{aligned} \frac{dQ}{dt} &= -(p + \gamma C)(W - \widetilde{W})^2 - lv(B - \tilde{B})^2 \\ &-mq(C - \tilde{C})^2 - (1 - l\theta)(B - \tilde{B})(W - \widetilde{W}) \\ &-(\beta + \gamma\widetilde{W})(W - \widetilde{W})(C - \tilde{C}) + m\delta(B - \tilde{B})(C - \tilde{C}) \end{aligned} \tag{1.3.47}$$

Choosing $l = \frac{1}{\theta}$ and $m = 1$, the above $\frac{dQ}{dt}$ reduces into:

$$\frac{dQ}{dt} = -(p+\gamma C)(W-\widetilde{W})^2 - lv(B-\tilde{B})^2 - mq(C-\tilde{C})^2 \\ -(\beta+\gamma\widetilde{W})(W-\widetilde{W})(C-\tilde{C}) + m\delta(B-\tilde{B})(C-\tilde{C}) \tag{1.3.48}$$

The $\frac{dQ}{dt}$ is negative-definite under the following inequalities:

$$\delta^2 < \frac{1}{2}\frac{vq}{\theta}$$

and

$$(\beta+\gamma)^2 < \frac{1}{2}pq$$

Consequently, the following statement can be stated on global asymptotic stability:

Theorem 4:
Under the following adequate conditions, the system (1.2.2) develops global asymptotic stable behaviour around the coexistence equilibrium $E_5(\widetilde{W},\tilde{B},\tilde{C})$:

$$\delta^2 < \frac{1}{2}\frac{vq}{\theta} \tag{1.3.49}$$

$$(\beta+\gamma)^2 < \frac{1}{2}pq \tag{1.3.50}$$

1.4 Numerical Simulation

The proposed model is numerically simulated with the following set of dimensionless parameters to support the above obtained qualitative results

$$p = 3.09, q = 0.95, \beta = 1.5, \gamma = 0.09, \theta = 5.8, v = 0.7, \delta = 1.07 \tag{1.4.1}$$

Performing the simulation with these parameters, it is being found that the coexistence equilibrium $E_5(\widetilde{W},\tilde{B},\tilde{C})$ is feasible with $\widetilde{W} = 0.2130$, $\tilde{B} = 0.3366$, and $\tilde{C} = 1.3791$. Besides this, it is also determined that the system settles down to the coexistence equilibrium $E_5(\widetilde{W},\tilde{B},\tilde{C})$ with these set of parameters. It shows the local asymptotic stable behavior of the system around the coexistence equilibrium only, depicted in Fig.1.1.

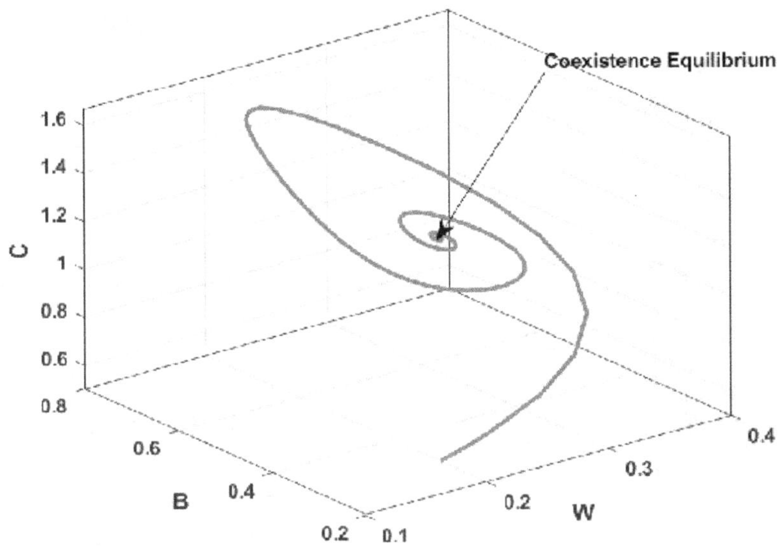

Figure 1.1 Phase Portrait in $W - B - C$ space.

The variations of different parameters may change the dynamical behavior of the system. With change in parameters, the system may also shift its stability from the coexistence equilibrium to another equilibrium which is without bats, that is, bat-free equilibrium. It indicates the appearance of transcritical bifurcation between the coexistence equilibrium E_5 to the bat-free equilibrium E_3, through the condition (1.3.37), that is obtained from the equation (1.3.36). In the condition (1.3.37), the parameters p, β, γ, and θ are involved. Based on this, these four can be considered to be key parameters of the model system and hence, it is quite crucial to consider their variations.

In Fig. 1.2, the time-series graphs are plotted for different values of p against each of the variables W, B, C. From the figure, it is clear that when the values of p increase, the periodic fluctuations in the system decrease and the solutions converge to the equilibrium. However, for very small value of p, bat population would not survive and collapse to zero at the transcritical bifurcation point $p_* = 1.83125$. For $p < p_*$, the coexistence equilbrium remains infeasible and therefore, bat population-free equilibrium becomes stable, as seen in the Fig. 1.2.

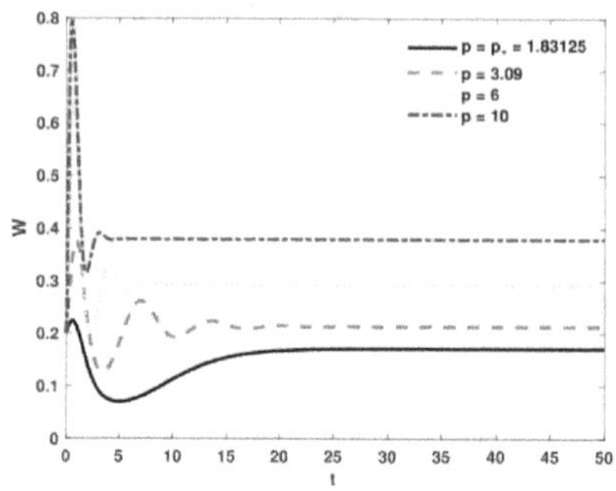

(a) W versus t

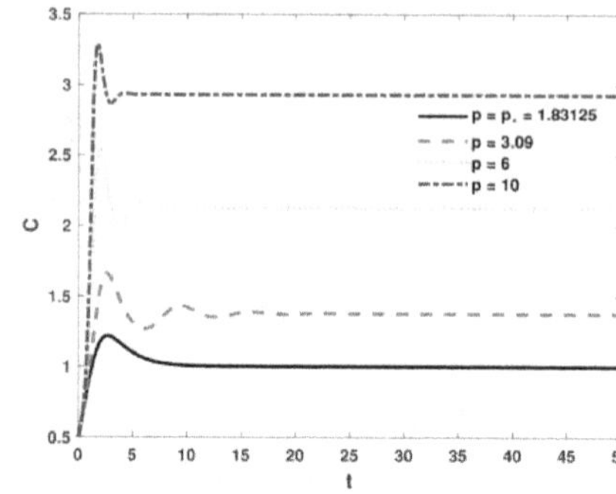

(b) B versus t

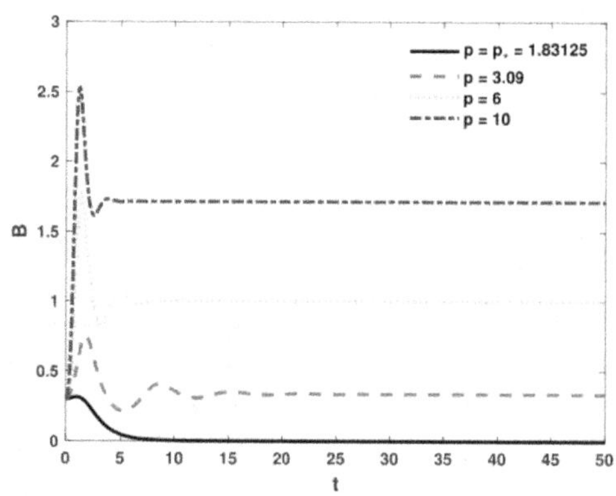

(c) C versus t

Figure 1.2 Time - series graphs with variations in p.

The time-series graphs of the variables W, B, C with variations in β are displayed in Fig. 1.3. Figure shows that the periodic fluctuations occur in the system, however, they decrease with increase in the value of β. It is also observed that the increase in the value of β has negative impact on both woodland and bat populations. Both of them decrease with increase in the values of β, however, at a certain threshold value $\beta = \beta_* = 2.54172$, transcritical bifurcation occurs when the bat population becomes zero. Beyond $\beta > \beta_*$, co-existence equilibrium does not exist and bat population free equilibrium becomes stable.

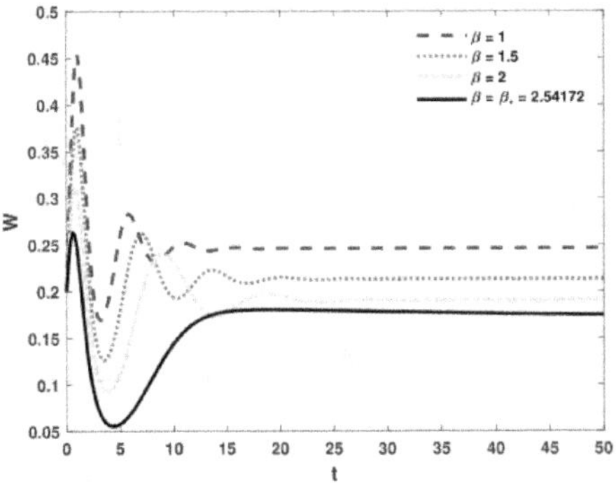

(a) W versus t

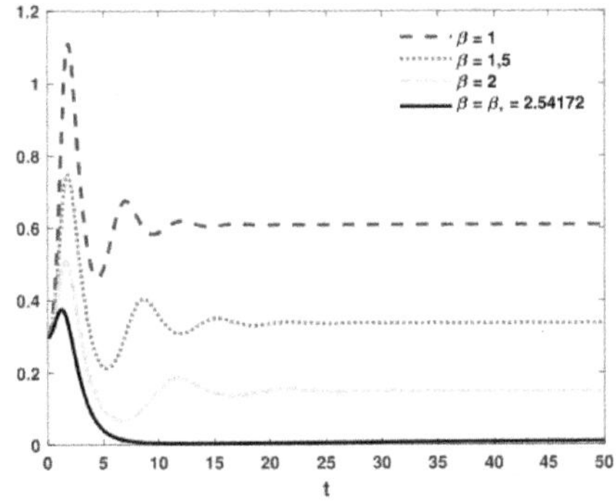

(b) B versus t

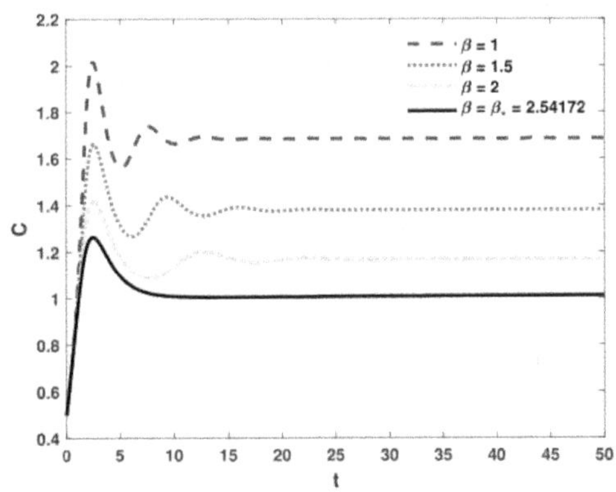

(c) C versus t

Figure 1.3 Time - series graph for change in β.

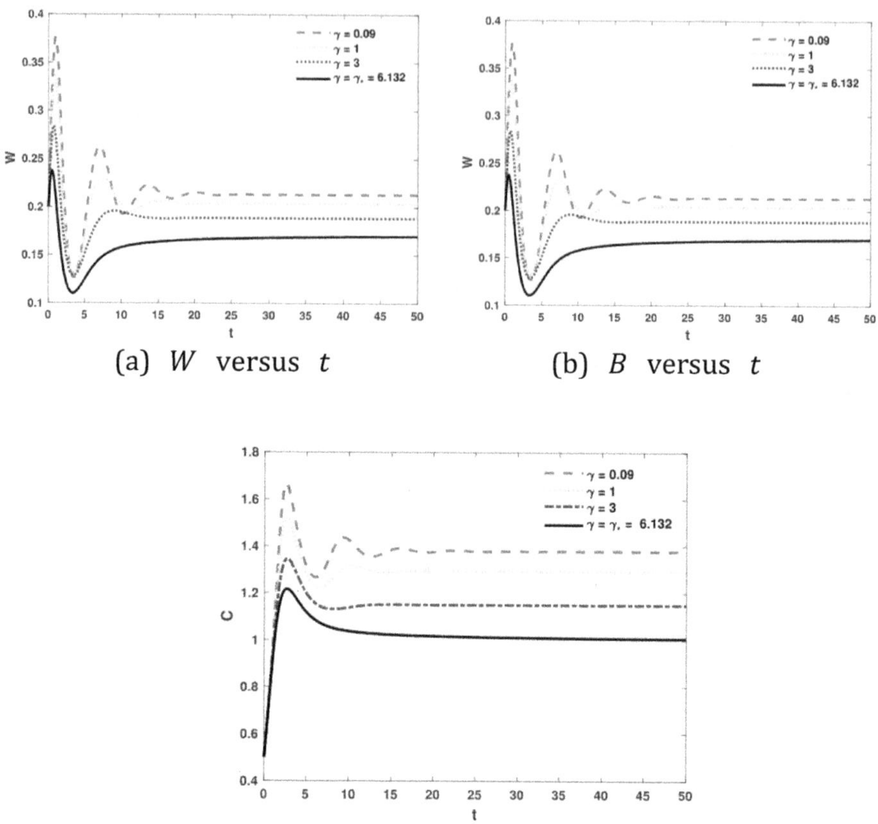

(a) W versus t (b) B versus t

(c) C versus t

Figure 1.4 Time - series graph for change in γ.

Just like β, similar behavior can also be noticed with the parameters γ and θ from the Fig. 1.4 and Fig. 1.5, respectively. With respect to γ and θ, transcritical bifurcation points can be obtained to be $\gamma = \gamma_* = 6.132$ and $\theta = \theta_* = 2$, respectively.

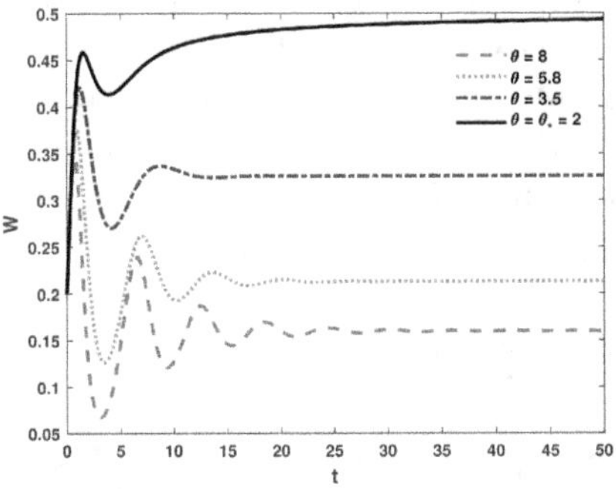

(a) W versus t

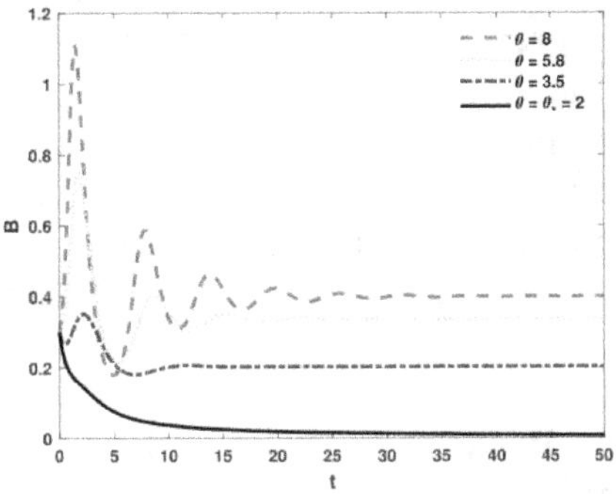

(b) B versus t

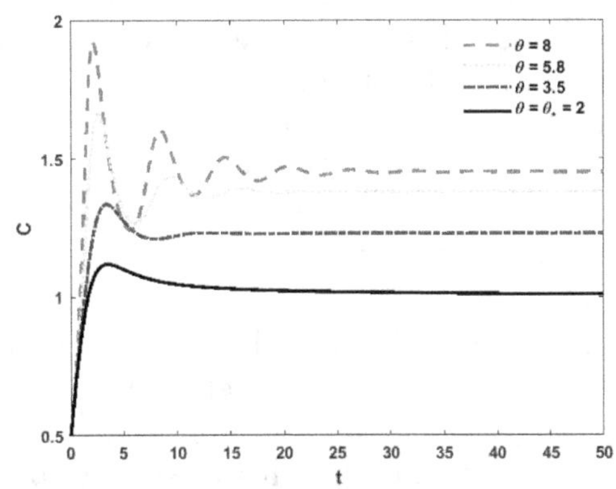

(c) C versus t

Figure 1.5 Time - series graph for change in θ.

1.5 Conclusion

For a woodland system, a nonlinear mathematical model has been proposed to demonstrate the changes that occur between interactive dynamics of woodland vegetation and bat population, when the woodland area gets excessively utilized for agricultural purposes. The constructed model underwent a qualitative analysis to examine the alterations in system behavior caused by different perturbations. Through this analysis, it was revealed that the system of equations developed in this study displays a diverse range of dynamic characteristics. Some notable aspects include:

1. The system exhibits six equilibrium states: a trivial equilibrium denoted as E_0, a woodland vegetation and bat population-free equilibrium known as E_1, a bat population and crop yield-free equilibrium denoted as E_2, a bat population-free equilibrium represented by E_3, a crop yield-free equilibrium referred to as E_4, and a coexistence equilibrium labeled as E_5. All these equilibria are feasible, except for the coexistence equilibrium and bat population-free equilibrium which have certain conditions to be met.

2. It has been examined how the system behaves locally around each equilibrium point. All other equilibria are inherently unstable, with the exception of the woodland vegetation and bat population-free, bat population-free and coexistence equilibria. The woodland vegetation and bat population-free equilibrium is stable provided bat population-free equilibrium does not exist. The bat population-free equilibrium is found to be stable when the coexistence equilibrium does not exist. Although it is subject to some parametric restrictions, the coexistence equilibrium has been shown to be both locally and globally stable.

To enhance understanding of the model's dynamic behavior, these results have been confirmed through numerical simulations.
The model results suggest conserving the bat population in a woodland environment with an agricultural setting for a stabilized system. If the woodland is continuously used for agricultural purposes, the system may settle down to a state devoid of vegetation and bat population. So, to mitigate the adverse effects of agricultural expansion on woodland bats, conservation efforts should focus on preserving and restoring suitable habitats. Implementing measures like maintaining patches of woodland within agricultural landscapes, creating wildlife corridors to connect fragmented habitats, and promoting sustainable farming practices that minimize chemical inputs can help protect and support woodland bat populations. Additionally, raising awareness about the importance of bats and their ecological roles can foster conservation actions and promote coexistence between agriculture and bat conservation.

References

[1] Peterken G. F., Woodland conservation and management. Springer Science & Business Media (1993).

[2] McElhinny C., Gibbons P., Brack C., Bauhus J., Forest and woodland stand structural complexity: its definition and measurement. Forest Ecology and Management, 218, no. 1-3: 1-24 (2005).
[3] Ovington, J. D., Quantitative ecology and the woodland ecosystem concept, In Advances in ecological research, vol. 1, pp. 103-192. Academic Press (1962).
[4] Kitching, R. L., An ecological study of water-filled tree-holes and their position in the woodland ecosystem, The Journal of Animal Ecology, 281-302 (1971).
[5] Simmons N. B., Order chiroptera, Mammal species of the world: a taxonomic and geographic reference 1 , 312-529 (2005).
[6] Wanger T.C., Darras K, Bumrungsri S, Tscharntke T, Klein A.M., Bat pest control contributes to food security in Thailand, Biological Conservation, 171, 220-223 (2014).
[7] Culina, A., Linton, D.M., Macdonald, D.W., Age, sex, and climate factors show different effects on survival of three different bat species in a woodland bat community, Global Ecology and Conservation, 12, 263-271 (2017).
[8] Groom, M. J., Carroll C. R., Meffe G. K., Meeting conservation challenges on the twenty-first century, Principles of Conservation Biology, third ed. Sinauer Associates, Inc., Massachusetts, 661-699 (2006).
[9] Boyles J. G., Cryan P.M., McCracken G.F., Kunz T.H., Economic importance of bats in agriculture, Science, 332, no. 6025, 41-42 (2011).
[10] Fuentes-Montemayor E., Goulson D., Cavin, L., Wallace J. M., Park K. J., Fragmented woodlands in agricultural landscapes: The influence of woodland character and landscape context on bats and their insect prey. Agriculture, ecosystems & environment, 172, 6-15 (2013).
[11] Bailey A. M., Ober H. K., Reichert B.E., McCleery R.A., Canopy cover shapes bat diversity across an urban and agricultural landscape mosaic, Environmental Conservation, 46, no. 3, 193-200 (2019).
[12] Froidevaux J.S., Laforge A., Larrieu L., Barbaro L., Park K., Fialas P.C., Jones G., Tree size, microhabitat diversity and landscape structure determine the value of isolated trees for bats in farmland, Biological Conservation, 267, 109476 (2022).
[13] Lacki M. J., Amelon S. K., Baker M.D., Foraging ecology of bats in forests, Bats in forests: conservation and management (MJ LACKI, JP HAYES, and A. KURTA, eds.), Johns Hopkins University Press, Bal-

timore, Maryland, 83-127 (2007).
[14] Carr A., Weatherall A., Zeale M.R., Rachwald A., Jones G. A., A method for rapid assessment of bat richness using woodland structure characteristics, Forest Ecology and Management, 529, 120680 (2023).
[15] Dietz M., Bögelsack K., Krannich A., Simon O., Woodland fragments in urban landscapes are important bat areas: an example of the endangered Bechstein's bat Myotis bechsteinii, Urban Ecosystems, 23, no. 6, 1359-1370 (2020).
[16] Geluso K. N., Altenbach J.S., Wilson D.S., Bat mortality: pesticide poisoning and migratory stress, Science, 194, no. 4261, 184-186 (1976).
[17] Torquetti C.G., Guimarães A.T., Soto-Blanco B., Exposure to pesticides in bats, Science of the Total Environment, 755, 142509 (2021).
[18] Slough B. G., Reid D.G., Schultz D.S., Leung M.C., Little brown bat activity patterns and conservation implications in agricultural landscapes in boreal Yukon, Canada, Ecosphere, 14, no. 3, e4446 (2023).
[19] Aizpurua, O., Budinski, I., Georgiakakis, P., Gopalakrishnan, S., Ibañez, C., Mata, V., Rebelo, H., Russo, D., Szodoray-Parádi, F., Zhelyazkova, V., Zrncic, V., Agriculture shapes the trophic niche of a bat preying on multiple pest arthropods across Europe: Evidence from DNA metabarcoding, Molecular ecology, 27, no. 3 , 815-825 (2018).
[20] Whitby M. D., Kieran T.J., Glenn T.C., Allen C., Agricultural pests consumed by common bat species in the United States corn belt: the importance of DNA primer choice, Agriculture, Ecosystems & Environment, 303 , 107105 (2020).
[21] Hughes M. J., de Torrez E.C., Ober H.K., Big bats binge bad bugs: Variation in crop pest consumption by common bat species, Agriculture, Ecosystems & Environment, 314, 107414 (2021).
[22] Ferreira, D. F., Gibb R., López-Baucells A., Nunes N.J., Jones K.E., Rocha R., Species-specific responses to land-use change in island insectivorous bats, Journal for Nature Conservation, 67, 126177 (2022).
[23] Gakkhar S., Singh A., Control of chaos due to additional predator in the Hastings–Powell food chain model. Journal of Mathematical Analysis and Applications, 385(1), 423-438 (2012).
[24] Nagumo, M.: ber die lage der integralkurven gewhnlicher differentialgleichungen, Proceedings of the Physico-Mathematical Society of Japan, 3rd Series, 24, 551-559 (1942).

[25] Hale, J.K., Ordinary differential equations, Wiley-Interscience, New York (1969).

[26] Freedman, H.I., So, J.W.H,, Global stability and persistence of simple food chains, Mathematical biosciences, 76, no. 1, 69-86 (1985).

Chapter 2

The Impact of the Transfer of Heat Generated by Nanofluid Flow on Parallel Plates Utilising the Homotopy Analytic Approach

Ajay Kumar, Ramakanta Meher
Department of Mathematics, Sardar Vallabhbhai National Institute of Technology, Surat - 395007, Gujarat, India

Abstract

In this chapter, the heat transmission in a flow with nanofluids squeezed between parallel plates is studied analytically utilising the homotopy analytic method (HAM). Copper nanoparticles have been considered, utilising water as the base fluid. Maxwell-Garnetts (MG) and Brinkman models are utilized in calculating the viscosity of a nanofluid and its effective thermal conductivity, respectively. This technique is found to be in excellent agreement when compared to other numerical ones. Nusselt number is studied as a function of the nanofluid volume fraction, δ, the squeeze number and Eckert number. The results reveal that when the plates are kept apart, the squeeze number, the nanoparticle volume fraction, the Eckert number, and δ are all closely connected to the Nusselt number, but the connection is reversed when the two plates are brought together.

Keywords: Heat transfer, nanofluid flow, parallel plates, homotopy analysis method

2.1 Introduction

Numerous sectors, including the energy, manufacturing, and transportation sectors, rely on efficient methods of heating and cooling fluids. Any high-energy equipment will require efficient cooling techniques. Although there are inefficient heat transfer media such as ethylene glycol, engine oil, and water counted as common fluids transferring heat. To develop a fluid-like heat transfer medium with the metal's thermal conductivity, combining the two substances is logical, as the metal's thermal conductivity can be up to three times higher than the fluid's. Choi [1] first coined the term "nanofluid" to describe manufactured colloids made up of nanoparticles and added to a base fluid. The enhancement in heat transmission caused in a differentially heated enclosure by the incorporation of nanoparticles was first studied numerically by Khanafer et al. [2]. They observed significant improvement in the heat transmission rate due to the dispersed nanoparticles at any given Grashof number. In a concentric annulus made up of a cold exterior square with hot internal circular cylinders, when there is a static radial magnetic field, the phenomenon of natural convection was studied by Sheikholeslami et al. [3] in which their findings are that whereas the Rayleigh number and the volume percentage of nanoparticles increase the average Nusselt number, the Hartmann number decreases it. Soleimani et al. [4] used Finite Element Method based on the Control Volume to study natural convection heat transfer in a nanofluid-filled semi-annulus enclosure in which findings suggest there is a considerable effect over isotherms, minimal and maximal values of the local Nusselt number, and streamlines by the angle of turn. Domairry et al.'s [5] analysis carried between two infinitely parallel vertical flat plates of a non-Newtonian copper-water nanofluid that convects naturally. They came to the conclusion that the thickness of the momentum and thermal boundary layer have increased and dropped, respectively as the volume percentage of nanoparticles raised. In a horizontally oriented magnetic field, Sheikholeslami et al. [6] carried out a computational investigation of the heat transfer within a hot inner sinusoidal circular cylinder containing a cold outer circular enclosure via natural convection of a Cu-water nanofluid. The study came to the conclusion that, the Rayleigh number increases as the enhancement ratio falls in the absence of a magnetic field, however this is not the case when one is present.

One of the most significant fields of research is the unsteady squeezing viscous flow between two parallel plates for studying the transfer characteristics of heat and mass, because it has so many uses in science and engineering, such as in hydrodynamical machines, chemical processing equipment, lubrication systems, polymer processing, food processing, crop damage from freezing, cooling towers and the formation and dispersion of fog. The initial research on a lubricating system's squeezing flow was published by Stefan [7]. Squeezed flow across a porous surface for transfer characteristics of heat investigated by Mahmood et al. [8].

The properties of mass and heat transmission when a viscous fluid squeezes between parallel plates were examined by Mustafa et al. [9] in which findings suggest that the Prandtl and Eckert numbers depend on the local Nusselt number to rise. Domairry and Aziz [10] studied the magnetohydrodynamic compression of a viscous fluid in a channel formed by two parallel discs. Many nonlinear engineering problems, including those involving heat transfer equations, necessitate either a numerical solution or one of various analytic methods, namely the perturbation method (PM), the differential transform method [16, 17], the homotopy perturbation method (HPM), the Adomian decomposition method [12, 13, 14, 15], the variational iteration method (VIM)[18, 19], and many more. In 1992, Liao [20, 21, 22] invented a new analytic method called the homotopy analytical method. Many non-linear problems in physical science and engineering [23, 24, 25, 26, 27, 28, 29, 30, 31] have significantly benefited from this method since Liao's 2004 publication of his book[32] on the subject. The HAM has demonstrated its ability to quickly, easily, and accurately approximate solutions to a wide variety of nonlinear problems, with those approximations quickly and accurately converging on the true solution. Most problems can be solved with minimal iteration after some initial exploration. Many studies in the engineering sciences use this approach. The main focus of this chapter is to use a semi-analytic approach called Homotopy Analysis Method to solve untenable squeezing nanofluid flow between plates placed parallelly to investigate heat transmission characteristics. Nusselt number is studied in terms of the nanofluid volume fraction, δ, the squeeze number, and the Eckert number.

2.2 Mathematical Formulation

The heat transfer analysis was considered for the nanofluid flow that is being squeezed between infinite parallel plates in two dimensions (Figure 2.1). The coordinate system for the placement of the two plates is: $z = \pm h(t) = \pm l(1-\alpha t)^{0.5}$. Both the plates touch at $t = 1/\alpha$, so they are squeezed until they touch for $\alpha > 0$, and the two plates are separated for $\alpha < 0$. When there is shear in the flow, friction generates heat, and this effect, known as viscous dissipation, is preserved. When the fluid is highly viscous or moving at high speeds, this impact becomes crucial. When the Eckert(>> 1) number is large, this phenomenon appears. Additionally, the symmetry of the movement is braced. Cu(copper) is suspended in a water-based nanofluid. Under the following conditions, the nanofluid can be viewed as a two-component mixture: incompressibility, negligible radiative heat transfer, negligible viscous dissipation, absence of chemical reaction, and thermal equilibrium between the base fluid and the nano-solid-particles, with no slippage between them. Table 2.1 lists the nanofluid's thermophysical properties.

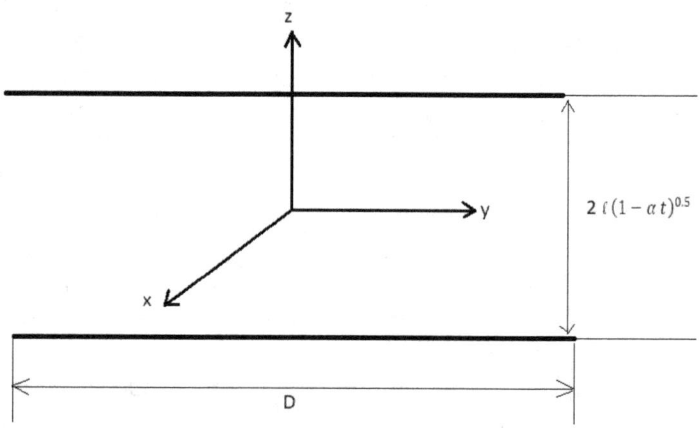

Figure 2.1 Geometrical presentation of the problem.

2D unsteady flow of a nanofluid's momentum and energy equations are given as:

$$\frac{\partial u}{\partial x}+\frac{\partial v}{\partial y}= 0 \tag{2.2.1}$$

$$\rho_{nf}\left(\frac{\partial u}{\partial t}+u\frac{\partial u}{\partial x}+v\frac{\partial u}{\partial y}\right)=-\frac{\partial p}{\partial x}+\mu_{nf}\left(\frac{\partial^2 u}{\partial x^2}+\frac{\partial^2 u}{\partial y^2}\right) \tag{2.2.2}$$

$$\rho_{nf}\left(\frac{\partial v}{\partial t}+u\frac{\partial v}{\partial x}+v\frac{\partial v}{\partial y}\right)=-\frac{\partial p}{\partial y}+\mu_{nf}\left(\frac{\partial^2 v}{\partial x^2}+\frac{\partial^2 v}{\partial y^2}\right) \tag{2.2.3}$$

$$\frac{\partial T}{\partial t}+u\frac{\partial T}{\partial x}+v\frac{\partial T}{\partial y}=\frac{k_{nf}}{(\rho C_p)_{nf}}\left(\frac{\partial^2 T}{\partial x^2}+\frac{\partial^2 T}{\partial y^2}\right)+\frac{\mu_{nf}}{(\rho C_p)_{nf}}(4(\frac{\partial u}{\partial x})^2+(\frac{\partial u}{\partial x}+$$

$$\frac{\partial u}{\partial y})^2) \tag{2.2.4}$$

Here, x and $y - axis$ directions velocities are u and v, respectively; T and P are temperature and pressure. For a nanofluid, we have the following expressions[3] for its effective density ρ_{nf}, the effective heat capacity $(\rho C_p)_{nf}$, the effective dynamic viscosity μ_{nf}, and the effective thermal conductivity k_{nf}, as follows:

$$\rho_{nf} = \phi \times \rho_s + (1-\phi) \times \rho_f ,$$
$$(\rho C_p)_{nf} = \phi \times (\rho C_p)_s + (1-\phi) \times (\rho C_p)_f ,$$
$$\mu_{nf} = \mu_f \times \frac{1}{(1-\phi)^{2.5}},$$
$$k_{nf} = k_f \times \frac{k_s + 2\times k_f - 2\phi(k_f - k_s)}{k_s + 2\times k_f + 2\phi(k_f - k_s)}. \tag{2.2.5}$$

The following are the conditions that pertain to the relevant boundaries:

$$At \quad y = h(t), \quad v = v_w = dh/dt, \quad T = T_H,$$
$$At \quad y = 0, \quad v = \partial u/\partial y = \partial T/\partial y = 0.$$

by introducing the following parameters:

$$\eta = \frac{y}{l(1-\alpha t)^{0.5}}, \qquad v = -\frac{\alpha l}{2(1-\alpha t)^{0.5}} f(\eta),$$

$$u = \frac{\alpha x}{2(1-\alpha t)} f'(\eta), \qquad \theta = \frac{T}{T_H}. \tag{2.2.6}$$

Using (2.2.6) together in (2.2.2), (2.2.3) and in (2.2.3), (2.2.4), respectively, gives following set of differential equations:

$$f^{iv} - SD_1(1-\phi)^{2.5}(\eta f''' + 3f'' - ff''' + f'f'') = 0 \tag{2.2.7}$$

$$\theta'' + PrS(\frac{D_2}{D_3})(f\theta' - \eta\theta') + \frac{PrEc}{D_3(1-\phi)^{2.5}}(f''^2 + 4\delta^2 f'^2) = 0 \tag{2.2.8}$$

where D_1, D_2, and D_3 are the constants given by:

$$D_1 = (1-\phi) + \frac{\rho_s}{\rho_f} \times \phi ,$$

$$D_2 = (1-\phi) + \frac{(\rho C_p)_s}{(\rho C_p)_s} \times \phi \ ,$$

$$D_3 = \frac{k_{nf}}{k_f} = \frac{2\times k_f + k_s - 2\phi(k_f - k_s)}{2\times k_f + k_s + 2\phi(k_f - k_s)} \ ,$$

with the boundary conditions as follows:

$$f(0)=0, \quad f''(0)=0, \quad \theta'(0)=0;$$

$$f(1)=1, \quad f'(1)=0, \quad \theta(1)=1.$$

where $Pr = \frac{\mu_f(\rho C_p)_f}{\rho_f k_f}$ is the Prandtl number, $S = \frac{\alpha l^2}{2v_f}$ is the squeeze number, $Ec = \frac{\rho_f}{(\rho C_p)_f}(\frac{\alpha x}{2(1-\alpha t)})^2$ is the Eckert number, and $\delta = \frac{l}{x}$.

Quantities of physical interest are as $C_f = \frac{\mu_{nf}(\frac{\partial u}{\partial y})_{y=h(t)}}{\rho_{nf} v_w^2}$ is the skin friction coefficient, and $Nu = \frac{-l k_{nf}(\frac{\partial T}{\partial y})_{y=h(t)}}{k T_H}$ is the the Nusselt number.

In view of Eq. (2.2.6), we have

$C_f^* = l^2/x^2(l-t\alpha)Re_x C_f = D_1(1-\phi)^{2.5} f''(1), \qquad Nu^* = \sqrt{1-t\alpha}Nu = -D_3\theta'(1).$ (2.2.9)

2.3 Application of HAM

In the current section, we will discuss the application of Homotopy approach [33, 34, 35] named HAM to acquire the findings of temperature and velocity profiles with various physical parameters for MHD nanofluid flow over parallel plates. The nonlinear operator using Eqs. (2.2.7) and (2.2.8) can be derived as following:

$$N_f = f^{iv}(\eta;q) - (1-\phi)^{2.5} \times SD_1(\eta f'''(\eta;q) + 3f''(\eta;q) + f'(\eta;q)f''(\eta;q) - f(\eta;q)f'''(\eta;q)) \qquad (2.3.1)$$

$$N_\theta = \theta''(\eta;q) + PrS(\frac{D_2}{D_3})(f(\eta;q)\theta'(\eta;q) - \eta\theta'(\eta;q)) + \frac{PrEc}{D_3(1-\phi)^{2.5}}(f''^2(\eta;q) + 4\delta^2 f'^2(\eta;q)) \quad (2.3.2)$$

with

$$f_j(\eta) = \frac{1}{j!}\frac{\partial^j f(\eta;q)}{\partial\eta^j}\bigg|_{q=0}, \qquad \theta_j(\eta) = \frac{1}{j!}\frac{\partial^j \theta(\eta;q)}{\partial\eta^j}\bigg|_{q=0} \quad (2.3.3)$$

where

$$f(\eta;q) = \sum_{j=0}^{\infty} f_j(\eta)q^j \quad and$$

$$\theta(\eta;q) = \sum_{j=0}^{\infty} \theta_j(\eta)q^j. \quad (2.3.4)$$

The initial approximation can be taken as

$$f_0(\eta) = \frac{1}{2}(3\eta - \eta^3) \ \text{and} \theta_0(\eta) = 1. \quad (2.3.5)$$

The linear operator can be presented as

$$\mathcal{L}^f = f^{iv}, \qquad \mathcal{L}^\theta = \theta''. \quad (2.3.6)$$

Our Linear operator $\mathcal{L}_f$ and $\mathcal{L}_\theta$ follows the conditions stated below:

$$\mathcal{L}^f(C_1\eta^3 + C_2\eta^2 + C_3\eta + C_4) = 0,$$

$$\mathcal{L}^\theta(C_5\eta + C_6) = 0. \quad (2.3.7)$$

Where, C_i's are constants. $(i = 1,2,3,4,5,6)$.

The jth- order deformation equations are given by

$$\mathcal{L}^f[f_j(\eta) - \chi_j f_{j-1}(\eta)] = \hbar_f R_j^f(\eta), \quad (2.3.8)$$

$$\mathcal{L}^\theta[\theta_j(\eta) - \chi_j \theta_{j-1}(\eta)] = \hbar_\theta R_j^\theta(\eta). \quad (2.3.9)$$

with boundary conditions

$$f(0) = 0, \quad \theta'(0) = 0, \quad f'(1) = 0,$$

$$f''(0) = 0, \quad f(1) = 0, \quad \theta(1) = 0. \tag{2.3.10}$$

Here, $R_j^f(\eta)$ and $R_j^\theta(\eta)$ can be expressed as

$$R_j^f(\eta) = f_{j-1}^{iv}(\eta) - (1-\phi)^{2.5} \times SD_1(\eta f'''_{j-1}(\eta) + 3f''_{j-1}(\eta) + \sum_{i=0}^{j-1} f'_i(\eta) f''_{j-1-i}(\eta) - \sum_{i=0}^{j-1} f_i(\eta) f'''_{j-1-i}(\eta)) \tag{2.3.11}$$

$$R_j^\theta(\eta) = \theta''_{j-1}(\eta) + Pr \times S \times \left(\frac{D_2}{D_3}\right)\left(\sum_{i=0}^{j-1} f_i(\eta)\theta'_{j-1-i}(\eta) - \eta\theta'_{j-1}(\eta)\right) + \frac{EcPr}{(1-\phi)^{2.5} \times D_3}(\sum_{i=0}^{j-1} f''_i(\eta) f''_{j-1-i}(\eta) + 4\delta^2 \sum_{i=0}^{j-1} f'_i(\eta) f'_{j-1-i}(\eta)) \tag{2.3.12}$$

and

$$\chi_j = \begin{cases} 0, & j \leq 1, \\ 1, & j > 1. \end{cases} \tag{2.3.13}$$

The general solution of Eqs. (2.3.18) and (2.3.19) can be written as

$$f_j(\eta) = f_j^*(\eta) + C_1\eta^3 + C_2\eta^2 + C_3\eta + C_4, \tag{2.3.14}$$

$$\theta_j(\eta) = \theta_j^*(\eta) + C_5\eta + C_6. \tag{2.3.15}$$

and the series solution can be written where, $f_j^*(\eta)$ and $\theta_j^*(\eta)$ are solutions obtained using solving the particular integral and constants C_i can be obtained using boundary conditions from Eq. (2.3.20).

And the series solution can be written as

$$f(\eta) = \sum_{j=0}^{\infty} f_j(\eta); \qquad \theta(\eta) = \sum_{j=0}^{\infty} \theta_j(\eta). \tag{2.3.16}$$

Using initial approximations from Eq. (2.3.15) in Eqs. (2.3.18) and (2.3.19), we have:

$$f_1(\eta) = SD_1(1-\phi)^{2.5} h_f(-0.18928571428571\eta^3 + 0.9285714285714\eta - 0.0035714285714\eta^7 + 0.1\eta^5) \tag{2.3.17}$$

$$\theta_1(\eta) = \frac{PrEc}{c(1-\phi)^{2.5}} h_\theta(-3.3\delta^2 - 0.75 + 0.3\delta^2\eta^6 - 1.5\delta^2\eta^4 + 0.75\eta^4 + 4.5\delta^2\eta^2) \quad (2.3.18)$$

The terms with $j \geq 2$, $f_j(\eta)$ and $\theta_j(\eta)$ can also be computed in a similar manner using Eqs.(2.3.8) and (2.3.9).

The final solution can be written as:

$$f(\eta) = \frac{1}{2}(3\eta - \eta^3) + (1-\phi)^{2.5} \times SD_1 \times h_f(0.1\eta^5 - 0.18928571428571\eta^3 + 0.9285714285714\eta - 0.0035714285714\eta^7) + \cdots \quad (2.3.19)$$

$$\theta(\eta) = 1 + \frac{PrEc}{c(1-\phi)^{2.5}} h_\theta(-3.3\delta^2 - 0.75 + 0.3\delta^2\eta^6 - 1.5\delta^2\eta^4 + 0.75\eta^4 + 4.5\delta^2\eta^2) + \cdots \quad (2.3.20)$$

Table 2.1 "Thermo physical properties of water and nanoparticles" [3]

	$C_p(J/kgK)$	$\rho(kg/m^3)$	$k(W/mK)$
Pure Water	4179	997.1	0.613
Copper (Cu)	385	8933	401

2.4 Results and discussion

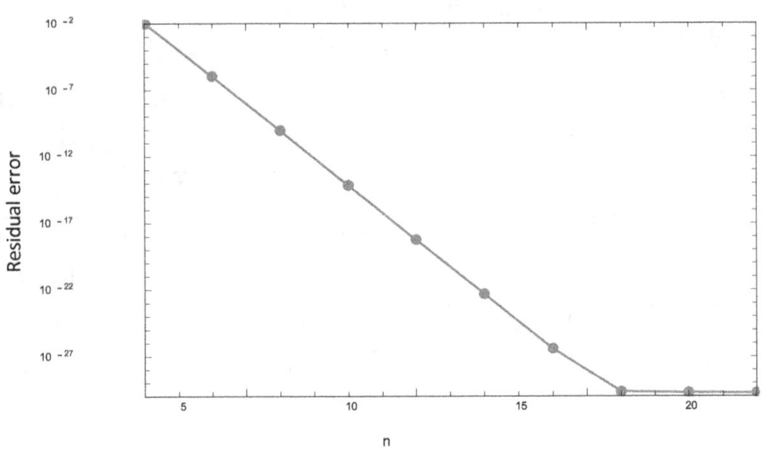

Figure 2.2 Residual error v/s order of approximation when $\delta = 1$, $\phi = 0.06$, $Ec = 0.1$, $S = 0.5$, and $Pr = 6.2$.

Table 2.2 Comparison of solutions at $\phi = 0.06$, $Ec = 0.1$, $S = 0.5$, $Pr = 6.2$, and $\delta = 1$.

η	$f(\eta)$		$\theta(\eta)$	
	RK-4[11]	HAM	RK-4[11]	HAM
0	0	0	2.955692	2.955692393
0.1	0.144419	0.144419188	2.931896	2.931896112
0.2	0.286441	0.286440854	2.861462	2.861461527
0.3	0.423607	0.423607011	2.747105	2.747105304
0.4	0.553338	0.553338282	2.592878	2.592878059
0.5	0.672872	0.672872111	2.403478	2.403477940
0.6	0.7792	0.779199726	2.18333	2.183330340
0.7	0.869002	0.869001637	1.935434	1.935433835
0.8	0.938582	0.938581546	1.659916	1.659915700
0.9	0.983799	0.983798825	1.352136	1.352136284
1	1	1	1	1

Table 2.3 Comparison of solutions at $\phi = 0.06$, $Pr = 6.2$, $Ec = 0.4$, $S = 1$, and $\delta = 1.5$.

η	$f(\eta)$		$\theta(\eta)$	
	RK-4[11]	HAM	RK-4[11]	HAM
0	0	0	14.92923	14.9292291712
0.1	0.140402	0.1404022149	14.72736	14.7273604654
0.2	0.278858	0.2788584097	14.13412	14.1341239069
0.3	0.413324	0.4133242745	13.18484	13.1848384796
0.4	0.541556	0.5415559341	11.93268	11.9326818602
0.5	0.661003	0.6610027134	10.44102	10.4410182111
0.6	0.768691	0.7686910011	8.773987	8.7739866037
0.7	0.861096	0.8610963564	6.985854	6.9858540046
0.8	0.934001	0.9340011784	5.108728	5.1087276696
0.9	0.982336	0.9823356118	3.136528	3.1365283322
1	1	1	1	1

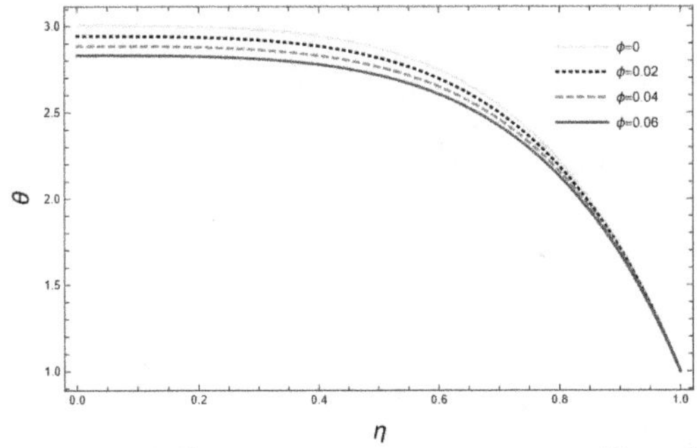

(a) Effect on temperature profile of nanoparticle volume fraction when S = 1.

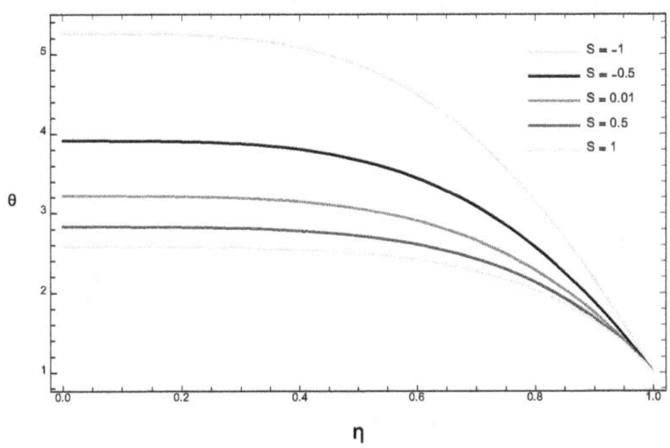

(b) Effect on temperature profile of squeeze number when ϕ = 0.06.

(c)

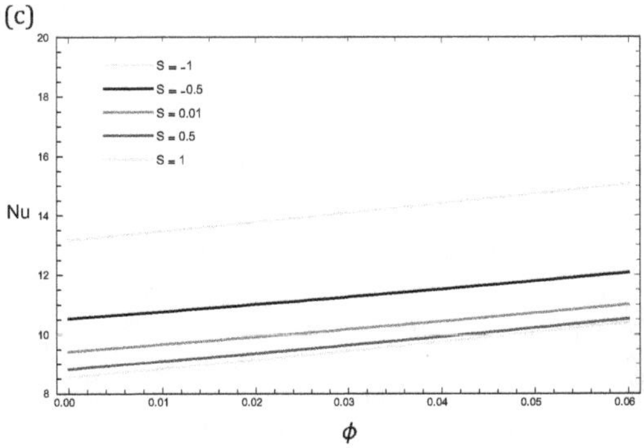

(d) Effect over Nusselt number of nanoparticle volume fraction and squeeze number.

Figure 2.3 Various effects when $\delta = 0.1$, $Pr = 6.2$, and $Ec = 0.5$

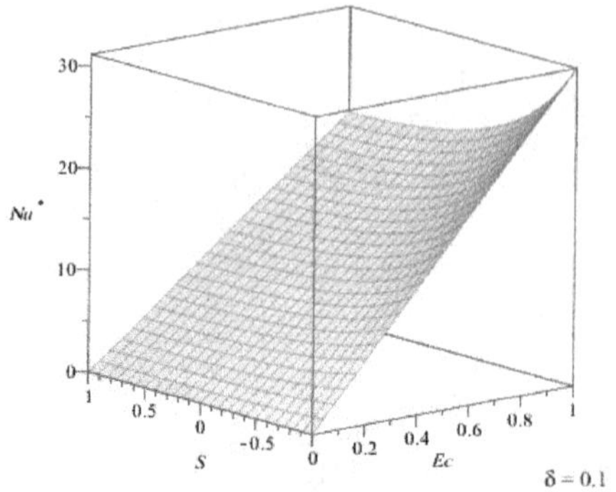

(a) $\phi = 0.06$, $\delta = 0.1$, and $Pr = 6.2$

(b)

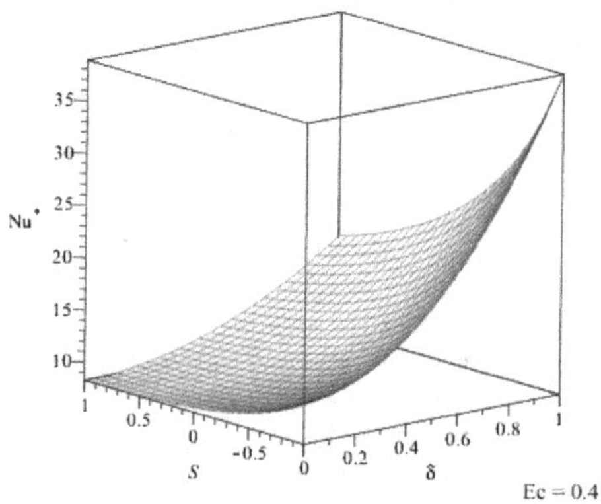

(c) $\phi = 0.06$, $Ec = 0.4$, and $Pr = 6.2$

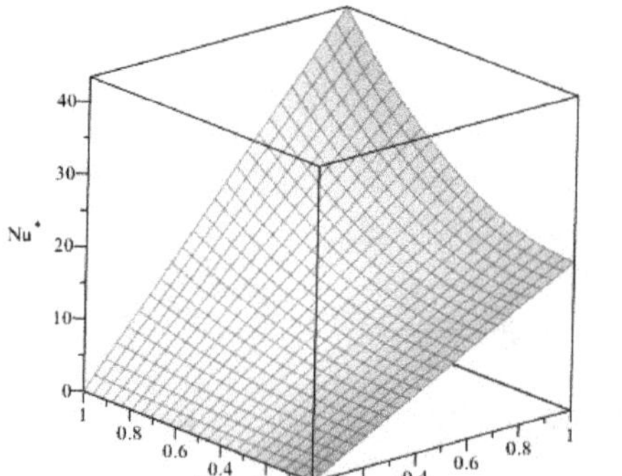

(d) $\phi = 0.06$, $Pr = 6.2$, and $S = 1$

Figure 2.4 Effect over Nusselt number of δ, Eckert number, and squeeze number.

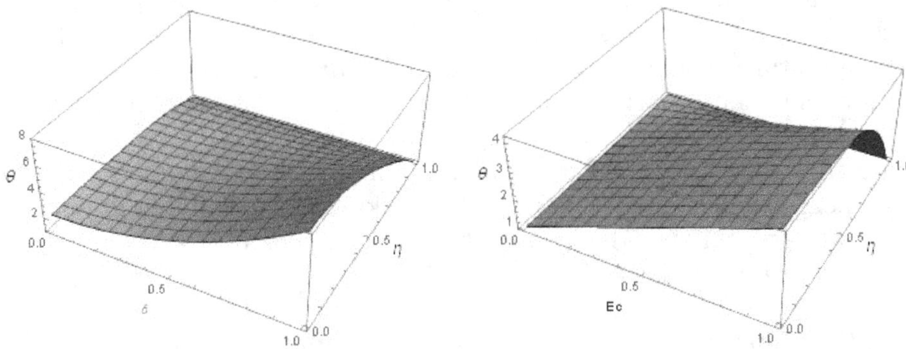

(a) S = 1, Ec = 0.1 (b) S = 1, δ = 0.1

Figure 2.5 Effect over temperature profile of δ and Eckert number with $Pr = 6.2$ and $\phi = 0.06$.

In this study HAM is implemented to study the heat transfter between two parallel platers in unsteady squeezing nanofluid flow. Table 2.2 and 2.3 shows the comparison of obtained solution via HAM and already available by Runge Kutta method of fourth order. Fig. 2.2 represents the residual error decreases as the order of approximation is increased, throughout paper we considered order of approximation as 16 which gives the error in terms of 10^{-30} and values of h_f and h_θ as -0.9087 and -0.9101, respectively.

The investigation's findings are represented visually, demonstrating how the heat transfer characteristics affected by the Eckert number, squeeze number, and nanofluid volume percentage. According to the squeeze number, Eckert number, volume percentage of nanofluid, and squeeze number, respectively, Figs. 2.3 and 2.5 show the temperature curve. The impacts of Squeeze number and nanoparticle volume fraction on Nusselt number are depicted in Fig. 2.2. The thickness of the nanofluid's thermal boundary layer is dependent on the volume fraction of nanoparticles due to increased thermal conductivity. In reality, as the thermal conductivity increases, the thermal diffusivity also increases. As the temperature gradients decrease due to strong thermal diffusivity, the boundary thickness grows as a result. While the Nusselt number is inversely proportional to the thermal boundary layer's thickness, however it is determined by product of the temperature gradient and the ratio of conductivities to the base fluid of the nanofluid i.e. the ratio of thermal conductivity. As the volume fraction of nanoparticles climbs so does the Nusselt number rise due to the fact that as the temperature gradient drops induced by nanoparticles is significantly smaller than the ratio of thermal conductivity. An increase in plate distance, a decrease in kinematic viscosity, and a faster speed are all related to a rise in the squeeze number. The absolute value of squeeze number, or $|S|$, falls as the thickness of thermal boundary layer rises when $S > 0$, but it increases when $S < 0$ as the squeeze number rises.

As the plates approach each other, it is clear that the thickness of the temperature boundary layer is relatively high.It is evident that when the Eckert number and δ are both increased, there is rise in the temperature. Fig. 2.4 depicts the relationship between the Nusselt number and the squeeze number, Eckert number, and δ. Viscous dissipation effects lead to a substantial rise in temperature which follows as the Eckert number climbs, the Nusselt number also increase. The rising functions of the Nusselt number are the Eckert number and δ.

2.5 Conclusion

The heat transmission in a nanofluid flow between parallel plates that is unsteadily compressing was examined in this chapter utilising HAM. The effect on Nusselt number has been studied with due respect to the nanofluid volume concentration, squeeze number, and Eckert number. An excellent agreement can be found between obtained HAM results and those already obtained from numerical solutions using Runge-Kutta method with iterations of the fourth order. When the squeeze number is below zero, the Nusselt number rises with the nanoparticle volume fraction, or δ, and it falls with the squeeze number when S is greater than 0.

References

[1] Choi, S. US, and Jeffrey A. Eastman. Enhancing thermal conductivity of fluids with nanoparticles. No. ANL/MSD/CP-84938; CONF-951135-29. Argonne National Lab. (ANL), Argonne, IL (United States), 1995.

[2] Khanafer, Khalil, Kambiz Vafai, and Marilyn Lightstone. "Buoyancy-driven heat transfer enhancement in a two-dimensional enclosure utilizing nanofluids." International journal of heat and mass transfer 46.19 (2003): 3639-3653.

[3] Sheikholeslami, M., M. Gorji-Bandpay, and D. D. Ganji. "Magnetic field effects on natural convection around a horizontal circular cylinder inside a square enclosure filled with nanofluid." International Communications in Heat and Mass Transfer 39.7 (2012): 978-986.

[4] Soleimani, Soheil, et al. "Natural convection heat transfer in a nanofluid filled semi-annulus enclosure." International Communications in Heat and Mass Transfer 39.4 (2012): 565-574.

[5] Domairry, Davood, et al. "Natural convection flow of a non-Newtonian nanofluid between two vertical flat plates." Proceedings of the Institution of Mechanical Engineers, Part N: Journal of Nanoengineering and Nanosystems 225.3 (2011): 115-122.

[6] M. Sheikholeslami, Soheil Soleimani, M. Gorji-Bandpy, D.D Ganji, S.M. Seyyedi, Natural convection of nanofluids in an enclosure between a circular and a sinusoidal cylinder in the presence of magnetic field, International Communications in Heat and Mass Transfer 39 (2012) 1435–1443.

[7] Stefan, Josef. "Versuche über die scheinbare Adhäsion." Annalen der Physik 230.2 (1875): 316-318.

[8] Mahmood, M., S. Asghar, and M. A. Hossain. "Squeezed flow and heat transfer over a porous surface for viscous fluid." Heat and mass Transfer 44.2 (2007): 165-173.

[9] Mustafa, M., T. Hayat, and S29663281293 Obaidat. "On heat and mass transfer in the unsteady squeezing flow between parallel plates." Meccanica 47.7 (2012): 1581-1589.

[10] Domairry, G., and A. Aziz. "Approximate analysis of MHD squeeze flow between two parallel disks with suction or injection by homotopy perturbation method." Mathematical Problems in Engineering 2009 (2009).

[11] Sheikholeslami, M., and D. D. Ganji. "Heat transfer of Cu-water nanofluid flow between parallel plates." Powder Technology 235 (2013): 873-879.

[12] Patel, Hardik S., and Ramakanta Meher. "Analytical investigation of Jeffery–Hamel flow by modified Adomian decomposition method." Ain Shams Engineering Journal 9.4 (2018): 599-606.

[13] Patel, H. S., and R. Meher."Simulation of imbibition phenomena in fluid flow through fractured heterogeneous porous media with different porous materials." Journal of Applied Fluid Mechanics 10.5

(2017): 1451-1460.
[14] Patel, Trushit, and Ramakanta Meher. "A Study on Convective-Radial Fins with Temperature-dependent Thermal Conductivity and Internal Heat Generation." Nonlinear Engineering 8.1 (2019): 145-156
[15] Patel, Trushit, and Ramakanta Meher. "Adomian decomposition sumudu transform method for solving fully nonlinear fractional order power-law fin-type problems." Int J Math Comput 27.2 (2015): 7-16.
[16] He, Ji-Huan. "Homotopy perturbation method: a new nonlinear analytical technique." Applied Mathematics and computation 135.1 (2003): 73-79.
[17] El-dabe, Nabil Tawfik, Mohamed Yahya Abou-zeid, and Ola S. Ahmed."Motion of a thin film of a fourth grade nanofluid with heat transfer down a vertical cylinder: Homotopy perturbation method application." Journal of Advanced Research in Fluid Mechanics and Thermal Sciences 66.2 (2020): 101-113.
[18] Meher, Ramakanta, and N. D. Patel."Numerical study of magnetohydrodynamics Jeffery–Hamel flow with cu-water nanofluid between two rectangular smooth walls with transverse magnetic field." International Journal of Computational Materials Science and Engineering 9.02 (2020): 2050010.
[19] Patel, N., and Ramakanta Meher. "Analytical investigation of Jeffery-Hemal flow with magnetic field by differential transform method." Int. J. Adv. Appl. Math. Mech 6 (2018): 1-9.
[20] Liao SJ. The proposed homotopy analysis technique for the solution of nonlinear problems (Doctoral dissertation, Ph. D. Thesis, Shanghai Jiao Tong University).1992.
[21] Liao S. A kind of linearity-invariance under homotopy and some simple applications of it in mechanics. 1992.
[22] Liao S. Notes on the homotopy analysis method: some definitions and theorems. Communications in Nonlinear Science and Numerical Simulation. 2009 Apr 1;14(4):983-97.
[23] Kumar S, Kumar A, Odibat ZM. A nonlinear fractional model to describe the population dynamics of two interacting species. Mathematical Methods in the Applied Sciences. 2017 Jul 30;40(11):4134-48.
[24] Kumar S, Singh J, Kumar D, Kapoor S. New homotopy analysis transform algorithm to solve Volterra integral equation. Ain Shams Engineering Journal. 2014 Mar 1;5(1):243-6.
[25] Martin O. On the homotopy analysis method for solving a par-

ticle transport equation. Applied Mathematical Modelling. 2013 Mar 15;37(6):3959-67.

[26] Massa F, Lallemand B, Tison T. Multi-level homotopy perturbation and projection techniques for the reanalysis of quadratic eigenvalue problems: The application of stability analysis. Mechanical Systems and Signal Processing. 2015 Feb 1;52:88-104.

[27] Molabahrami A, Khani F. The homotopy analysis method to solve the Burgers–Huxley equation. Nonlinear Analysis: Real World Applications. 2009 Apr 1;10(2):589-600.

[28] Nave O, Gol'dshtein V, Ajadi S. Singularly perturbed homotopy analysis method applied to the pressure driven flame in porous media. Combustion and Flame. 2015 Mar 1;162(3):864-73.

[29] Sardanyés J, Rodrigues C, Januário C, Martins N, Gil-Gómez G, Duarte J. Activation of effector immune cells promotes tumor stochastic extinction: A homotopy analysis approach. Applied Mathematics and Computation. 2015 Feb 1;252:484-95.

[30] Shivanian E, Abbasbandy S. Predictor homotopy analysis method: Two points second order boundary value problems. Nonlinear Analysis: Real World Applications. 2014 Jan 1;15:89-99.

[31] Yang Z, Liao S. A HAM-based wavelet approach for nonlinear partial differential equations: Two dimensional Bratu problem as an application. Communications in Nonlinear Science and Numerical Simulation. 2017 Dec 1;53:249-62.

[32] Liao S. Beyond perturbation: introduction to the homotopy analysis method. Chapman and Hall/CRC; 2003 Oct 27.

[33] Kumar, Ajay, and Ramakanta Meher."Solving Non-linear Partial Differential Equations Using Homotopy Analysis Method (HAM)." Nonlinear Dynamics and Applications. Springer, Cham, 2022. 955-964.

[34] Sartanpara, Parth, and Ramakanta Meher."Computational study of Klein-Gordan equation using homotopy analysis method." AIP Conference Proceedings. Vol. 2451. No. 1. AIP Publishing LLC, 2022.

[35] Verma, Lalchand, and Ramakanta Meher. "Solution of fuzzy differential equation using homotopy analysis method." AIP Conference Proceedings. Vol. 2451. No. 1. AIP Publishing LLC, 2022.

Chapter 3

Study of the Couple Stress on Micro-Polar Fluid Flow Saturating a Porous Medium with Hall Effect

Devilal Kumawat
Department of Mathematics and Statistics, Jai Narain Vyas University, Jodhpur-342001, Rajasthan, India

Abstract

Study of the couple-stress on micro polar fluid layer heated from below in the presence of uniform vertical magnetic field in a porous medium has been carried out. The dispersion relation has been find out using normal mode and the problem has been analyzed numerically by MATLAB. We observed that medium permeability and Hall parameter have destabilizing effect and the couple-stress parameter, coupling parameter, heat conduction parameter and micro-polar coefficient have stabilizing effect. The sufficient condition of over- stability has been obtained.

Keywords. Micro Polar Fluid, Porous Medium, Hall Effect, Couple Stress.

3.1 Introduction

In order to study of micro-polar fluids, the general theory of micro polar fluid was developed by Eringen [1, 2]. Chamkha et al. and Kumar Prathap [4, 10] discussed the theory of micro polar fluid and viscous fluid in a vertical channel. Liquid crystals, colloidal fluids, animal blood, polymeric suspension etc. are few examples of micro polar fluids [2, 8]. Oahimire and Olajuwon [9] discussed the effect of thermal radiation and MHD on micro polar fluid in porous medium. Sharma and Gupta [13] study the thermal convection on micro polar fluid in porous medium. Chamkha [3] investigated the problem of hydro magnetic two-phase flow in a channel. Rovin and Mehta [14] analyzed the medium permeability and other parameters on the visco elastic fluid flow with hall effect. Bhupander [6] study of the MHD on the micro polar fluid flow with hall effect.

Stokes [15] discussed the general theory of couple-stress fluid. The rotation on thermal instability in couple-stress viscous elastic fluid was introduced Kumar Pardeep et al. [11]. Banyal and Singh [5] study the rotation on the couple-stress fluid in a porous medium. Devilal Kumawat et al. [7] analyzed the effect of the couple-stress on micro polar rotating fluid flow. Xiong Pei-Ying et al. [12] investigated the couple stress on couple stress fluid flow between parallel plates with thermal convection.

Applications of this work in geophysics, engineering science, chemical science and industry include liquid crystal, blood flows, colloidal suspensions and clean engine lubricants. In this chapter, I attempt to study of the couple-stress on micro-polar fluid flow with Hall Effect.

3.2 Mathematical Formulation

An infinite, horizontal, incompressible micro-polar fluid layer of thickness d is assumed and has porosity $\in$ and medium permeability k_1. The upper limit $z = d$ and lower limit $z = 0$ are maintained at constant but varying temperatures T_0 and T_1 such that a study adverse temperature gradient $\beta = \left|\frac{dT}{dz}\right|$ has been continued. The rotation and gravity and magnetic field are applied along z-axis to the system.

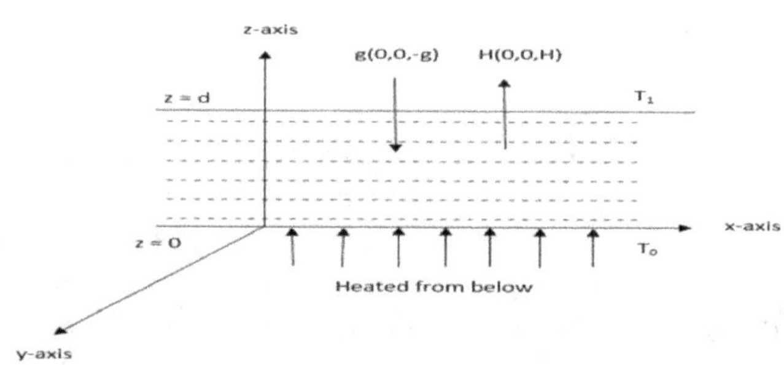

The equations of continuity, motion, internal angular momentum, temperature and basic's state are

$$\nabla . \vec{q} = 0, \tag{3.2.1}$$

$$\frac{\rho_0}{\epsilon}\left[\frac{\partial\vec{q}}{\partial t}+\frac{1}{\epsilon}(\vec{q}.\nabla)\vec{q}\right]=-\nabla P-\rho g\hat{e}_z+\left(\mu-\frac{\mu'}{\rho_0}\nabla^2\right)\nabla^2\vec{q}-\frac{1}{k_1}(\mu+\varsigma)\vec{q}$$
$$+\varsigma(\nabla\times\vec{v})+\frac{\mu_e}{4\pi}(\nabla\times\vec{H})\times\vec{H}, \quad (3.2.2)$$

$$\rho_0 J\left[\frac{\partial\vec{v}}{\partial t}+\frac{1}{\epsilon}(\vec{v}.\nabla)\vec{v}\right]=(\alpha'+\beta')\nabla(\nabla.\vec{v})+\gamma'\nabla^2\vec{v}+\frac{\varsigma}{\epsilon}(\nabla\times\vec{q})-2\varsigma\vec{v}, \quad (3.2.3)$$

$$[\epsilon\,\rho_0 C_v+(1-\epsilon)\rho_s C_s]\frac{\partial T}{\partial t}+\rho_0 C_v(\vec{q}.\nabla)T=\chi\nabla^2 T+\delta(\nabla\times\vec{v}).\nabla T, \quad (3.2.4)$$

$$\rho=\rho_0[1-\alpha(T-T_a)]. \quad (3.2.5)$$

where $\vec{q}$-Filter velocity, ρ -Fluid density, ρ_0-Reference density, $\vec{v}$-Spin (micro rotation), coefficient μ -Shear kinematic viscosity coefficient, ς -Coupling viscosity, P-Pressure, μ'-Couple stress viscosity, $\hat{e}_z$-Unit vector in z-direction, t-time, C_v-Specific heat at constant volume, C_s-Specific heat of solid (Porous Material Matrix), α'-Bulk spin viscosity coefficient, β'-Shear spin viscosity coefficient, γ'-Micro-polar viscosity coefficient, J-Micro inertia constant, ρ_s-Density of solid matrix, χ -Thermal conductivity, T-Temperature, δ -Micro-polar heat conduction coefficient, α -Coefficient of thermal expansion.

The Maxwell's equations are

$$\epsilon\frac{\partial\vec{H}}{\partial t}=\nabla\times(\vec{q}\times\vec{H})+\epsilon\,\gamma_m\nabla^2\vec{H}-\frac{\epsilon}{4\pi e n_e}\nabla\times[(\nabla\times\vec{H})\times\vec{H}], \quad (3.2.6)$$

$$\nabla.\vec{H}=0. \quad (3.2.7)$$

where $\vec{H}=(0,0,H_z), H_z$-Constant, n_e-Electron density, e-Charge on electron and γ_m-Magnetic viscosity .

3.3 Basic State of Problem

The basic states are

$$\vec{q}=\vec{q}_b(0,0,0), \vec{v}=\vec{v}_b(0,0,0), \rho=\rho_b(z) and P=P_b(z)$$

From equation (3.2.1) to (3.2.7)

$$\frac{dP_b}{dz} + \rho_b g = 0, \tag{3.3.1}$$

$$T = T_b(z) = -\beta z + T_a, \tag{3.3.2}$$

$$\rho_b = \rho_0(1 + \alpha\beta z). \tag{3.3.3}$$

3.4 Linearize Perturbation Equations

$$\nabla.\vec{q}' = 0, \tag{3.4.1}$$

$$\frac{\rho_0}{\epsilon}\frac{\partial \vec{q}'}{\partial t} = -\nabla P' - \rho' g\hat{e}_z + \left(\mu - \frac{\mu'}{\rho_0}\nabla^2\right)\nabla^2\vec{q}' - \frac{1}{k_1}(\mu + \varsigma)\vec{q}' + \varsigma(\nabla \times \vec{v}') + \frac{\mu_e}{4\pi}\left(\nabla \times \vec{h}\right) \times \vec{H}_z, \tag{3.4.2}$$

$$\rho_0 J\frac{\partial \vec{v}'}{\partial t} = (\alpha' + \beta')\nabla(\nabla.\vec{v}') + \gamma'\nabla^2\vec{v}' + \frac{\varsigma}{\epsilon}(\nabla \times \vec{q}') - 2\varsigma\vec{v}', \tag{3.4.3}$$

$$E\frac{\partial\theta}{\partial t} + (\vec{q}.\nabla)T_b = k_T\nabla^2\theta - \frac{\delta}{\rho_0 C_v}(\nabla \times \vec{v}')_z\beta + \beta(\vec{q}')_z, \tag{3.4.4}$$

$$\epsilon\frac{\partial\vec{h}}{\partial t} = H_z \times (\vec{q}' \times \hat{e}_z) + \epsilon\, \gamma_m\nabla^2\vec{h} - \frac{\epsilon H_z}{4\pi e n_e}\nabla \times \left[\left(\nabla \times \vec{h}\right) \times \hat{e}_z\right], \tag{3.4.5}$$

$$\nabla.\vec{h} = 0, \tag{3.4.6}$$

$$\rho' = -\rho_0\alpha\theta. \tag{3.4.7}$$

Converting equation (3.4.1) to (3.4.5) by the following transform

$x = dx*, y = dy*, z = dz*, \vec{q}' = \frac{k_T}{d}\vec{q}*, P' = \frac{\mu k_T}{d^2}P*, \vec{v}' = \frac{k_T}{d^2}\vec{v}*$
$, t = \frac{\rho_0 d^2}{\mu}t*, \nabla = \frac{\nabla*}{d}, \theta = \beta d\theta*,$ we have

$$\nabla.\vec{q} = 0, \tag{3.4.8}$$

$$\frac{1}{\epsilon}\frac{\partial\vec{q}}{\partial t} = -\nabla P + R\theta\hat{e}_z + (1 - F\nabla^2)\nabla^2\vec{q} - \frac{1}{K_1}(1 + K)\vec{q} + K(\nabla \times \vec{v}) + Q\left(\nabla \times \vec{h}\right) \times \hat{e}_z, \tag{3.4.9}$$

$$\bar{J}\frac{\partial \vec{v}}{\partial t} = C_1\nabla(\nabla.\vec{v}) - C_0\nabla(\nabla\times\vec{v}) + K\left\{\frac{1}{\in}(\nabla\times\vec{q}) - 2\vec{v}\right\}, \tag{3.4.10}$$

$$EP_r\frac{\partial\theta}{\partial t} = \nabla^2\theta - \bar{\delta}(\nabla\times\vec{v})_z + (\vec{q})_z, \tag{3.4.11}$$

$$\in P_r\frac{\partial\vec{h}}{\partial t} = \frac{\partial\vec{q}}{\partial z} + \frac{\in P_r}{P_m}\nabla^2\vec{h} - \in\beta_e^{1/2}\frac{\partial}{\partial z}\left(\nabla\times\vec{h}\right) \tag{3.4.12}$$

where $R = \frac{\rho_0 g\alpha\beta d^4}{\mu k_T}$ -Thermal Rayleigh number, $Q = \frac{\mu_e H_z^2 d^2}{4\pi\mu k_T}$ -Chandrasekhar number, $P_r = \frac{\mu}{\rho_0 k_T}$-Prandtl number, $P_m = \frac{\mu}{\rho_0\gamma_m}$ -Magnetic Prandtl number, $\beta_e = \left(\frac{H_z}{4\pi k_T e n_e}\right)^2$-Hall parameter $F = \frac{\mu'}{\rho_0 d^2}$, $E = \in + \frac{(1-\in)\rho_s C_s}{\rho_0 C_v}$, $\bar{J} = \frac{J}{d^2}, K_1 = \frac{k_1}{d^2}, \bar{\delta} = \frac{\delta}{\rho_0 C_v d^2}, C_0 = \frac{\gamma'}{\mu d^2}$, $C_1 = \frac{\alpha'+\beta'+\gamma'}{\mu d^2}$, and $W = \vec{q}.\hat{e}_z$.

3.5 Boundary conditions

$$W = \frac{d^2W}{dz^2} = 0, \theta = 0 at z = 0 and z = d. \tag{3.5.1}$$

3.6 Dispersion Relation

Taking curl on both sides of equation (3.4.9), we have

$$\left[\frac{1}{\in}\frac{\partial}{\partial t} + \left(\frac{1+K}{K_1}\right) - (1 - F\nabla^2)\nabla^2\right](\nabla\times\vec{q}) = R\left(\frac{\partial\theta}{\partial y}\hat{e}_x + \frac{\partial\theta}{\partial x}\hat{e}_y\right)$$
$$+K\nabla\times(\nabla\times\vec{v}) + \frac{2}{\in}\nabla\times(\vec{q}\times\Omega) + Q\frac{\partial}{\partial z}\left(\nabla\times\vec{h}\right). \tag{3.6.1}$$

Let $\nabla^2 = \frac{\partial^2}{\partial x^2} + \frac{\partial^2}{\partial y^2} + \frac{\partial^2}{\partial z^2}, \nabla_1^2 = \frac{\partial^2}{\partial x^2} + \frac{\partial^2}{\partial y^2}, \vec{h}_z = \vec{h}.\hat{e}_z, D = \frac{\partial}{\partial z}, \zeta_z = (\nabla\times\vec{q})_z, \Omega_z' = (\nabla\times\vec{v})_z$.

Taking curl and z-component of equations (3.6.1), (3.4.10) and (3.4.12), we have

$$\left[\frac{1}{\in}\frac{\partial}{\partial t} + \left(\frac{1+K}{K_1}\right) - (1 - F\nabla^2)\nabla^2\right]\nabla^2 W = R\nabla_1^2\theta + K\nabla^2\Omega_z'\hat{e}_z + QD\left(\nabla^2\vec{h}_z\right), \tag{3.6.2}$$

$$\bar{J}\frac{\partial \Omega_z'}{\partial t} = C_0 \nabla^2 \Omega_z' - K\left[\frac{1}{\epsilon}\nabla^2 W + 2\Omega_z'\right], \tag{3.6.3}$$

$$\epsilon P_r \frac{\partial m_z}{\partial t} = D\zeta_z + \frac{\epsilon P_r}{P_m}\nabla^2 m_z + \epsilon\, \beta_e^{1/2} D\left(\nabla^2 \vec{h}_z\right). \tag{3.6.4}$$

Taking z-component of equations (3.6.1), (3.4.11) and (3.4.12), we have

$$\left[\frac{1}{\epsilon}\frac{\partial}{\partial t} + \left(\frac{1+K}{K_1}\right) - (1 - F\nabla^2)\nabla^2\right]\zeta_z = QDm_z, \tag{3.6.5}$$

$$EP_r \frac{\partial \theta}{\partial t} = \nabla^2 \theta - \bar{\delta}\Omega_z' + W, \tag{3.6.6}$$

$$\epsilon P_r \frac{\partial \vec{h}_z}{\partial t} = DW + \frac{\epsilon P_r}{P_m}\nabla^2 \vec{h}_z - \epsilon\, \beta_e^{1/2} Dm_z. \tag{3.6.7}$$

3.6.1 Normal Mode Analysis

Let

$$[W, \zeta_z, \theta, \Omega_z', h_z, m_z] = [W(z), X(z), \Theta(z), G(z), B(z), M(z)]\ exp.\left[ik_x x + ik_y y + \sigma t\right]$$

Applying normal mode of equations (3.6.2) to (3.6.7), we have

$$\left[\frac{\sigma}{\epsilon} + \left(\frac{1+K}{K_1}\right) + F(D^2 - a^2)^2 - (D^2 - a^2)\right](D^2 - a^2)W$$
$$= -Ra^2\Theta + K(D^2 - a^2)G + QD(D^2 - a^2)B, \tag{3.6.8}$$

$$\left[\frac{\sigma}{\epsilon} + \left(\frac{1+K}{K_1}\right) + F(D^2 - a^2)^2 - (D^2 - a^2)\right]X = QDM, \tag{3.6.9}$$

$$[m\sigma + 2A - (D^2 - a^2)]G = -\frac{A}{\epsilon}(D^2 - a^2)W, \tag{3.6.10}$$

$$[EP_r\sigma - (D^2 - a^2)]\Theta = -\bar{\delta}G + W, \tag{3.6.11}$$

$$\left[\epsilon P_r\sigma - \frac{\epsilon P_r}{P_m}(D^2 - a^2)\right]M = DX + \epsilon\, \beta_e^{1/2} D(D^2 - a^2)B, \tag{3.6.12}$$

$$\left[\in P_r\sigma - \frac{\in P_r}{P_m}(D^2 - a^2)\right] B = DW - \in \beta_e^{1/2} DM. \tag{3.6.13}$$

where $a^2 = k_x^2 + k_y^2$ -wave number, $\sigma = \sigma_r + i\sigma_r$ -stability parameter and $m = \frac{\bar{J}A}{K}$, $A = \frac{K}{C_0}$, A-ratio between the micro-polar viscous effect and micro-polar diffusion effects.

Now the boundary condition becomes

$$W = D^2W = 0 = X = DX = G, \Theta = 0 atz = 0 toz = 1, \tag{3.6.14}$$

$$D^{2n}W = 0 atz = 0 toz = 1, \quad \text{where } n > 0.$$

Thus, the proper solution satisfying (3.6.14) can be taken as

$$W = W_0 \sin \pi z.$$

Eliminating Θ, G, Φ $andX$ from (3.6.8) to (3.6.13) and putting the value of W $andb = \pi^2 + a^2$, we have

$$b\left[\frac{\sigma}{\in} + \left(\frac{1+K}{K_1}\right) + Fb^2 + b\right][m\sigma + 2A + b][EP_r\sigma + b]\left[\left\{\frac{\sigma}{\in} + \left(\frac{1+K}{K_1}\right)\right.\right.$$
$$\left.+Fb^2 + b\right\}\left\{\left(\in P_r\sigma + \frac{\in P_r}{P_m}b\right)^2 + \in^2 \pi^2 b\beta_e\right\} + Q\pi^2\left(\in P_r\sigma + \frac{\in P_r}{P_m}b\right)\Bigg] =$$
$$Ra^2\left[(m\sigma + 2A + b) - \frac{\delta Ab}{\in}\right]\left[\left\{\frac{\sigma}{\in} + \left(\frac{1+K}{K_1}\right) + Fb^2 + b\right\}\left\{\left(\in P_r\sigma + \right.\right.\right.$$
$$\left.\left.\frac{\in P_r}{P_m}b\right)^2 + \in^2 \pi^2 b\beta_e\right\} + Q\pi^2\left(\in P_r\sigma + \frac{\in P_r}{P_m}b\right)\Bigg] + \frac{KAb^2}{\in}[EP_r\sigma +$$
$$b]\left[\left\{\frac{\sigma}{\in} + \left(\frac{1+K}{K_1}\right) + Fb^2 + b\right\}\left\{\left(\in P_r\sigma + \frac{\in P_r}{P_m}b\right)^2 + \in^2 \pi^2 b\beta_e\right\} +\right.$$
$$\left.Q\pi^2\left(\in P_r\sigma + \frac{\in P_r}{P_m}b\right)\right] - Q\pi^2 b[m\sigma + 2A + b][EP_r\sigma + b]\left[\left\{\frac{\sigma}{\in} + \left(\frac{1+K}{K_1}\right)\right.\right.$$
$$\left.\left.+Fb^2 + b\right\}\left(\in P_r\sigma + \frac{\in P_r}{P_m}b\right) + Q\pi^2\right] \tag{3.6.15}$$

3.7 Stationary Convection

Putting $\rho = 0$ in equation (3.6.15), we have

$$R = \frac{1}{a^2\left[2A + b - \frac{\bar{\delta}Ab}{\epsilon}\right]}\left[b^2(2A+b)\left(\frac{1+K}{K_1} + Fb^2 + b\right) - \frac{KAb^3}{\epsilon} + \frac{Q\pi^2 b^2(2A+b)\left[\left\{\left(\frac{1+K}{K_1}\right)+Fb^2+b\right\}\left(\frac{\epsilon P_r}{P_m}b\right)+Q\pi^2\right]}{\left[\left\{\left(\frac{1+K}{K_1}\right)+Fb^2+b\right\}\left\{\left(\frac{\epsilon P_r}{P_m}b\right)^2+\epsilon^2\pi^2 b\beta_e\right\}+Q\pi^2\frac{\epsilon P_r}{P_m}b\right]}\right]$$

To investigate the behaviors of medium permeability, rotation, couple-stress parameter, magnetic field, coupling parameter, micro-polar coefficient and micro-polar heat conduction parameter, we find the nature of $\frac{dR}{dK_1}, \frac{dR}{d\beta_e}, \frac{dR}{dF}, \frac{dR}{dQ}, \frac{dR}{dK}, \frac{dR}{dA}$ and $\frac{dR}{d\bar{\delta}}$ respectively, then

$$\frac{dR}{dK_1} = \frac{-b^2(2A+b)(1+K)}{a^2{K_1}^2\left[2A+b-\frac{\bar{\delta}Ab}{\epsilon}\right]}\left[1 - \frac{Q^2\pi^6\epsilon^2 b\beta_e}{\left[\left\{\left(\frac{1+K}{K_1}\right)+Fb^2+b\right\}\left\{\left(\frac{\epsilon P_r}{P_m}b\right)^2+\epsilon^2\pi^2 b\beta_e\right\}+Q\pi^2\frac{\epsilon P_r}{P_m}b\right]^2}\right], \tag{3.7.1}$$

$\frac{dR}{dK_1} < 0$ if $\sqrt{\beta_e} < \frac{\sqrt{b}}{\pi}\frac{P_r}{P_m}$ and $\bar{\delta} < \frac{\epsilon}{A}$.

From equation (3.7.1), we can say that the permeability has destabilizing effect.

$$\frac{dR}{d\beta_e} = \frac{-Q\pi^4 b^3\epsilon^2(2A+b)}{a^2\left[2A+b-\frac{\bar{\delta}Ab}{\epsilon}\right]}\left[\frac{\left\{\left(\frac{1+K}{K_1}\right)+Fb^2+b\right\}\left[\left\{\left(\frac{1+K}{K_1}\right)+Fb^2+b\right\}\left(\frac{\epsilon P_r}{P_m}b\right)+Q\pi^2\right]}{\left[\left\{\left(\frac{1+K}{K_1}\right)+Fb^2+b\right\}\left\{\left(\frac{\epsilon P_r}{P_m}b\right)^2+\epsilon^2\pi^2 b\beta_e\right\}+Q\pi^2\frac{\epsilon P_r}{P_m}b\right]^2}\right], \tag{3.7.2}$$

$\frac{dR}{d\beta_e} < 0$ if $\bar{\delta} < \frac{\epsilon}{A}$.

From equation (3.7.2) shows that the hall current parameter has destabilizing effect.

$$\frac{dR}{dF} = \frac{b^4(2A+b)}{a^2\left[2A+b-\frac{\bar{\delta}Ab}{\epsilon}\right]}\left[1 - \frac{Q^2\pi^6\epsilon^2 b\beta_e}{\left[\left\{\left(\frac{1+K}{K_1}\right)+Fb^2+b\right\}\left\{\left(\frac{\epsilon P_r}{P_m}b\right)^2+\epsilon^2\pi^2 b\beta_e\right\}+Q\pi^2\frac{\epsilon P_r}{P_m}b\right]^2}\right], \tag{3.7.3}$$

$\frac{dR}{dF} > 0 \ if \ \sqrt{\beta_e} < \frac{\sqrt{b}}{\pi}\frac{P_r}{P_m} \ and \ \bar{\delta} < \frac{\epsilon}{A}.$

It is clear that the couple-stress parameter has stabilizing effect.

$$\frac{dR}{dQ} = \frac{1}{a^2[2A+b-\frac{\bar{\delta}Ab}{\epsilon}]}$$

$$\left[\frac{\pi^2 b^2(2A+b)}{[\{(\frac{1+K}{K_1})+Fb^2+b\}\{(\frac{\epsilon P_r}{P_m}b)^2+\epsilon^2\pi^2 b\beta_e\}+Q\pi^2\frac{\epsilon P_r}{P_m}b]^2}\right]$$

$$[[\{(\frac{1+K}{K_1})+Fb^2+b\}\frac{\epsilon P_r}{P_m}b+2Q\pi^2][\{(\frac{1+K}{K_1})+Fb^2+b\}$$

$$\{(\frac{\epsilon P_r}{P_m}b)^2+\epsilon^2\pi^2 b\beta_e\}]+Q\pi^2\frac{\epsilon P_r}{P_m}b], \tag{3.7.4}$$

$$\frac{dR}{dQ} > 0 if \bar{\delta} < \frac{\epsilon}{A}$$

It is clear that the magnetic field has stabilizing effect.

$$\frac{dR}{dK} = \frac{1}{a^2\left[2A+b-\frac{\bar{\delta}Ab}{\epsilon}\right]}[Ab^2\left(\frac{2}{K_1}-\frac{b}{\epsilon}\right)+\frac{b^3}{K_1}$$

$$\left\{1 - \frac{Q^2\pi^2\epsilon^2(2A+b)\beta_e}{\left[\left\{\left(\frac{1+K}{K_1}\right)+Fb^2+b\right\}\left\{\left(\frac{\epsilon P_r}{P_m}b\right)^2+\epsilon^2\pi^2 b\beta_e\right\}+Q\pi^2\frac{\epsilon P_r}{P_m}b\right]^2}\right\}], \tag{3.7.5}$$

$\frac{dR}{dK} > 0 if \frac{2}{K} > \frac{b}{\epsilon}, \sqrt{\beta_e} < \frac{\sqrt{b}}{\pi}\frac{P_r}{P_m} and \bar{\delta} < \frac{\epsilon}{A}.$

Hence the coupling parameter has stabilizing effect.

$$\frac{dR}{dA}=\frac{1}{a^2\left[2A+b-\frac{\bar{\delta}Ab}{\epsilon}\right]^2}\left[\frac{Kb^4}{\epsilon}\left(\frac{\bar{\delta}}{K_1}-1\right)+\frac{\bar{\delta}b^4}{\epsilon}\left(\frac{1}{K_1}+Fb^2+b\right)\right.$$

$$\left.+\frac{\frac{\bar{\delta}b^4}{\epsilon^2}Q\pi^2\left[\left\{\left(\frac{1+K}{K_1}\right)+Fb^2+b\right\}\left(\frac{\epsilon Pr}{Pm}b\right)+Q\pi^2\right]}{\left[\left\{\left(\frac{1+K}{K_1}\right)+Fb^2+b\right\}\left\{\left(\frac{\epsilon Pr}{Pm}b\right)^2+\epsilon^2\pi^2 b\beta_e\right\}+Q\pi^2\frac{\epsilon Pr}{Pm}b\right]}\right], \tag{3.7.6}$$

$\frac{dR}{dA}>0 if \bar{\delta}>K_1.$

From equation (3.7.6), we can say that the micro-polar coefficient has stabilizing effect.

$$\frac{dR}{d\bar{\delta}}=\frac{\frac{Ab}{\epsilon}}{a^2\left[2A+b-\frac{\bar{\delta}Ab}{\epsilon}\right]^2}$$

$$\left[Kb^3\left(\frac{1}{K_1}-\frac{A}{\epsilon}\right)+\frac{b^3}{K_1}+2Ab^2\left(\frac{1+K}{K_1}\right)+(2A+b)b^2(Fb^2+b)\right.$$

$$\left.+\frac{Q\pi^2b^2(2A+b)\left[\left\{\left(\frac{1+K}{K_1}\right)+Fb^2+b\right\}\left(\frac{\epsilon Pr}{Pm}b\right)+Q\pi^2\right]}{\left[\left\{\left(\frac{1+K}{K_1}\right)+Fb^2+b\right\}\left\{\left(\frac{\epsilon Pr}{Pm}b\right)^2+\epsilon^2\pi^2 b\beta_e\right\}+Q\pi^2\frac{\epsilon Pr}{Pm}b\right]}\right], \tag{3.7.7}$$

$\frac{dR}{d\bar{\delta}}>0\ if\ \frac{1}{K_1}>\frac{A}{\epsilon}.$

From equation (3.7.7), shows that the micro-polar heat transfer parameter has stabilizing effect.

3.8 Oscillatory Convection

Putting $\sigma=i\sigma_i$ in equation (3.6.15), we get real and imaginary part, eliminating R between them, we have

$$f_0\sigma_i^8+f_1\sigma_i^6+f_2\sigma_i^4+f_3\sigma_i^2+f_4=0.$$

Putting $s=\sigma_i^2$, we have

$$f_0s^4+f_1s^3+f_2s^2+f_3s+f_4=0, \tag{3.8.1}$$

where

$$f_0 = a_1 q_1 - p_1 b_1,$$

$$f_1 = a_2 q_1 + a_1 q_2 - p_2 b_1 - p_1 b_2,$$

$$f_2 = a_3 q_1 + a_2 q_2 - p_3 b_1 - p_2 b_2 - p_1 b_3,$$

$$f_3 = a_4 q_1 + a_3 q_2 - p_3 b_2 - p_2 b_3,$$

$$f_4 = a_4 q_2 - p_3 b_3,$$

$$a_1 = -\frac{EP_r mb}{\in}, b_1 = m \in {P_r}^2 a^2,$$

$$b_2 = -a^2[m\left\{\in\left(\frac{P_r b}{P_m}\right)^2 + \in \pi^2 b\beta_e + 2\in^2 \frac{{P_r}^2 b}{P_m} + Q\pi^2 \in P_r\right\} + \left(2A + b - \frac{\bar{\delta}Ab}{\in}\right)\left\{2\in\frac{{P_r}^2 b}{P_m} + \in^2 P_r^2\left(\frac{1+K}{K_1} + Fb^2 + b\right)\right\}],$$

$$b_3 = a^2[\left(2A + b - \frac{\bar{\delta}Ab}{\in}\right)\left\{\{\left(\frac{\in P_r b}{P_m}\right)^2 + \in^2 \pi^2 b\beta_e\right\} \left(\frac{1+K}{K_1} + Fb^2 + b\right) + Q\pi^2 \frac{\in P_r b}{P_m}\}],$$

$$a_2 = \in {P_r}^2 b[\{\frac{(2A+b)b}{\in} + \{(2A+b)EP_r + mb\}\left(\frac{1+K}{K_1} + Fb^2 + b\right)\}] + b[mEP_r\left(\frac{1+K}{K_1} + Fb^2 + b\right) + \frac{(2A+b)EP_r + mb}{\in}][\{2\in\frac{P_r^2 b}{P_m} + \in^2 P_r^2 (\frac{1+K}{K_1} + Fb^2 + b)\}] + \frac{mEP_r b}{\in}[\{\in\left(\frac{P_r b}{P_m}\right)^2 + \in \pi^2 b\beta_e + 2\in^2 \frac{{P_r}^2 b}{P_m} + Q\pi^2 \in P_r\}] - \frac{KAb^2}{\in} EP_r^2 \in + Q\pi^2 bmEP_r^2,$$

$$q_1 = -a^2[m\left\{\frac{2\in P_r^2 b}{P_m} + \in^2 P_r^2\left(\frac{1+K}{K_1} + Fb^2 + b\right)\right\} + \in P_r^2\left(2A + b - \frac{\bar{\delta}Ab}{\in}\right)],$$

$$q_2 = a^2[m\left\{\left\{\left(\frac{\in P_r b}{P_m}\right)^2 + \in^2 \pi^2 b\beta_e\right\}\left(\frac{1+K}{K_1} + Fb^2 + b\right) + Q\pi^2 \frac{\in P_r b}{P_m}\right\}$$

$$+\left(2A+b-\frac{\bar{\delta}Ab}{\epsilon}\right)\left\{\epsilon\left(\frac{P_r b}{P_m}\right)^2+\epsilon\,\pi^2 b\beta_e+2\,\epsilon^2\frac{{P_r}^2 b}{P_m}+Q\pi^2\,\epsilon\,P_r\right\}],$$

$$P_1=b[\epsilon\,P_r^2\left\{mEP_r\left(\frac{1+K}{K_1}+Fb^2+b\right)+\frac{(2A+b)EP_r+mb}{\epsilon}\right\}$$
$$+\frac{mEP_r}{\epsilon}\left\{\frac{2\epsilon P_r^2 b}{P_m}+\epsilon^2\,P_r^2\left(\frac{1+K}{K_1}+Fb^2+b\right)\right\}],$$

$p_2=-\epsilon\,P_r^2(2A+b)b^2(\frac{1+K}{K_1}+Fb^2+b)-b\{mEP_r(\frac{1+K}{K_1}+Fb^2+$
$b)+\frac{(2A+b)EP_r+mb}{\epsilon}\}\quad\{\epsilon\,(\frac{P_r b}{P_m})^2+\epsilon\,\pi^2 b\beta_e+2\,\epsilon^2\frac{{P_r}^2 b}{P_m}+Q\pi^2\,\epsilon\,P_r\}-$
$b\{\frac{(2A+b)b}{\epsilon}+\{(2A+b)EP_r+mb\}(\frac{1+K}{K_1}+Fb^2+b)\}\{2\,\epsilon\,\frac{P_r^2 b}{P_m}+$
$\epsilon^2\,P_r^2(\frac{1+K}{K_1}+Fb^2+b)\}-b\,\frac{mEP_r}{\epsilon}\{\{(\frac{\epsilon P_r b}{P_m})^2+\epsilon^2\,\pi^2 b\beta_e\}(\frac{1+K}{K_1}+Fb^2+$
$b)+Q\pi^2\frac{\epsilon P_r b}{P_m}\}+\frac{KAb^2}{\epsilon}[EP_r\{2\,\epsilon\,\frac{{P_r}^2 b}{P_m}+\epsilon^2\,P_r^2(\frac{1+K}{K_1}+Fb^2+b)\}+b\,\epsilon$
$P_r^2]-Q\pi^2 b[mEP_r\{\epsilon\,P_r(\frac{1+K}{K_1}+Fb^2+b)+\frac{P_r b}{P_m}\}+P_r\{EP_r(2A+b)+$
$mb\}].$

From (3.8.1), we observed that $s=\sigma_i^2$ which is always positive, therefore the sum of roots equation of (8.1) is positive but this is impossible if $f_0>0$ and $f_1>0$, the sum of roots of equation (3.8.1) is $-{f_1}/{f_0}$ Thus, $f_0>0$ and $f_1>0$ are the sufficient condition for the non-existence of over stability.

Now $f_0>0$ and $f_1>0$ when $\bar{\delta}<\frac{\epsilon}{A},Q<\frac{Kb^2P_r}{2\pi^2}and\beta_e<\frac{KA{P_r}^2}{m\pi^2}.$

3.9 Numerical Computation

In this section, we have discussed the plotted variation of Rayleigh number R with K_1,Q,β_e in figures 3.10.1-3.10.8 and micro polar parameters $K,A,\bar{\delta}$ in figure 3.10.9-3.10.14.

Figure 3.10.1 and 3.10.2, shows the plots of Rayleigh number R versus medium permeability K_1 i.e. medium permeability increase then Rayleigh number decrease with the magnetic field $Q=10,20,30$ and hall parameter $\beta_e=0.05,0.10,0.15$ when $A=0.5,P_r=2,P_m=4,K=0.2,\bar{\delta}=0.05$. Figure 3.10.3 and 3.10.4, represent the plots of Rayleigh number R versus magnetic field Q i.e. magnetic field increase then Rayleigh number increase with the wave number $a = 1, 3, 5$ and $\beta_e = 0.05, 0.10, 0.15$ when $A = 0.5, P_r = 2, P_m = 4, K = 0.2, \bar{\delta} = 0.05$.

Similarly, Figure 3.10.5 and 3.10.6 shows the variation of Rayleigh number R with respect to hall parameterβ_e i.e. hall parameter increases then Rayleigh number decreases. Figure 3.10.7 and 3.10.8, represent the plot of Rayleigh number R versus couple stress parameter F i.e. couple stress parameter increase then Rayleigh number increase.

Figure 3.10.9 and 3.10.10, plot between Rayleigh number R and coupling parameter K i.e. coupling parameter increase then Rayleigh number increase with the magnetic field $Q = 10, 20, 30$ and wave number $a = 1, 3, 5$ when $A = 0.5, P_r = 2, P_m = 4, K = 0.2$, $\bar{\delta} = 0.05$. Figure 3.10.11 and 3.10.12, shows the variation of Rayleigh number R with respect to micro-polar coefficient A i.e. micro-polar coefficient increase then Rayleigh number increase with the magnetic field $Q = 10, 20, 30$ and hall parameter $\beta_e = 0.05, 0.10, 0.15$ when $a = 1, P_r = 2, P_m = 4, K = 0.2, \bar{\delta} = 0.05$. Figure 3.10.13 and 3.10.14, represent the plot of Rayleigh number R versus micro-polar heat conduction parameter $\bar{\delta}$ i.e. micro-polar heat conduction parameter increase then Rayleigh number increase.

3.10 Figures

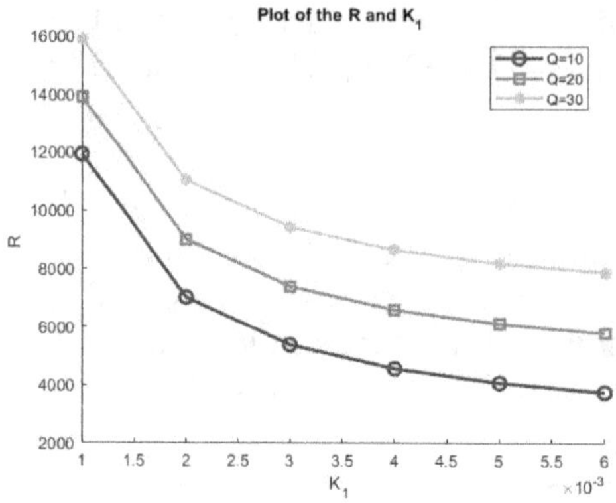

Figure 3.10.1

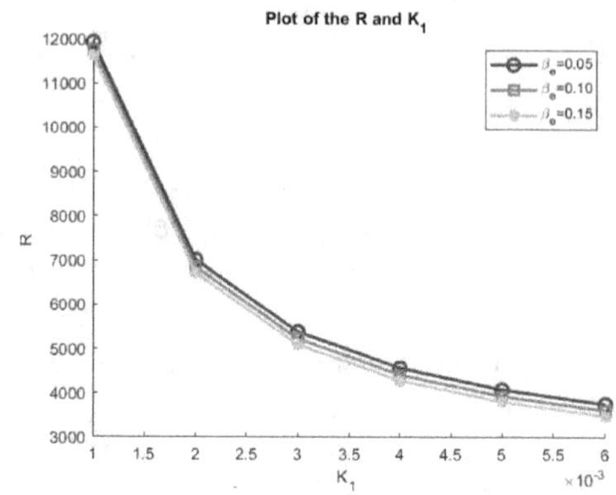

Figure 3.10.2

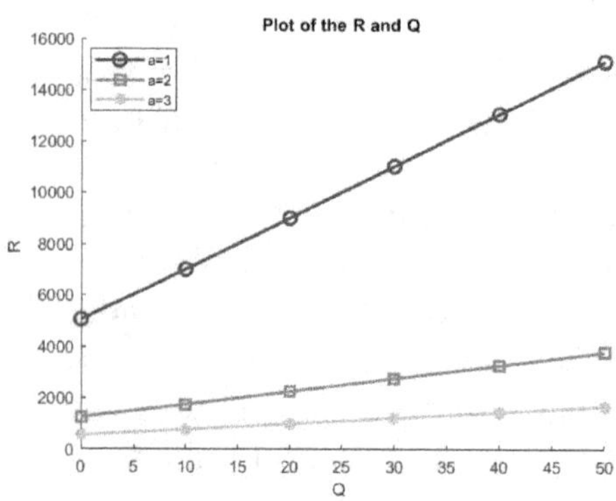

Figure 3.10.3

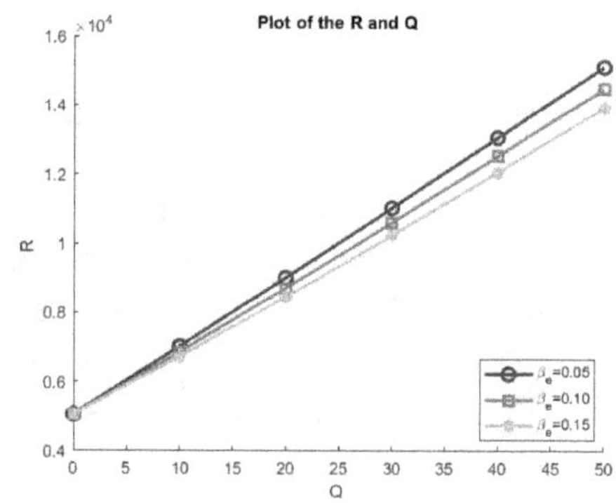

Figure 3.10.4

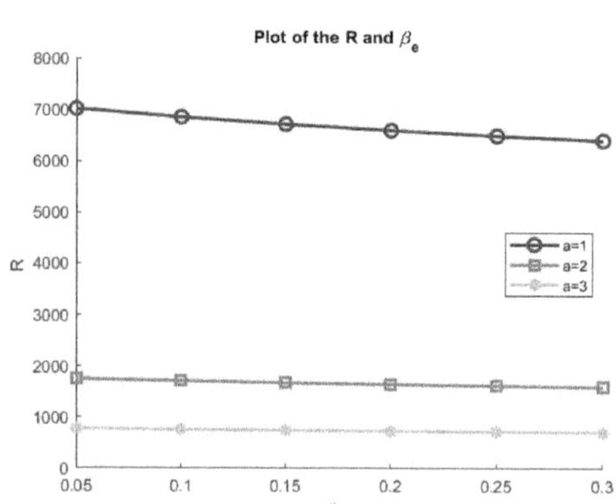

Figure 3.10.5

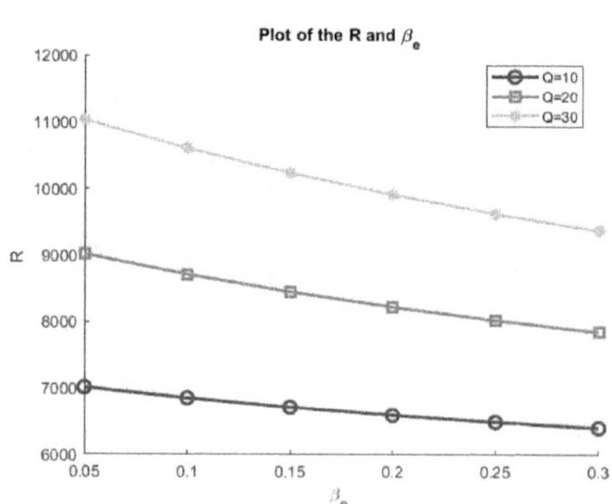

Figure 3.10.6

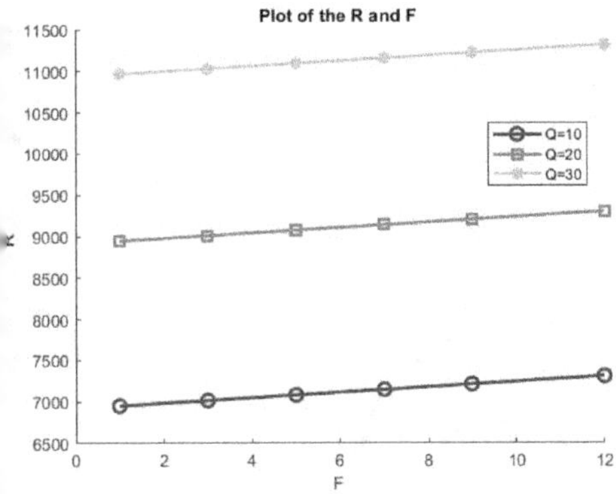

Figure 3.10.7

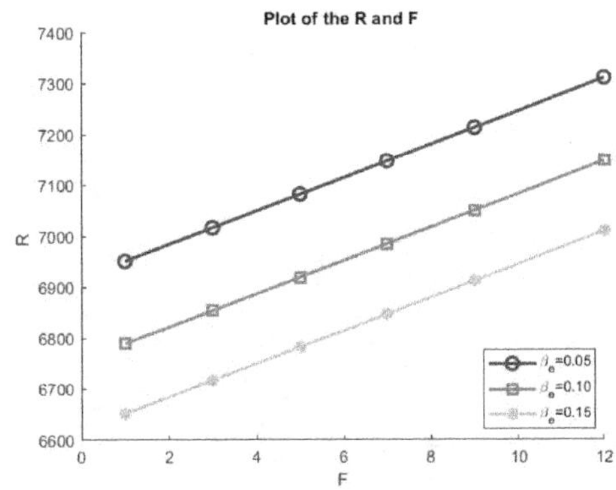

Figure 3.10.8

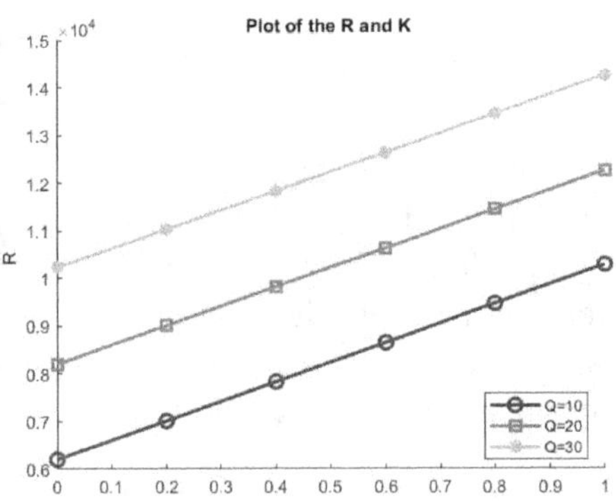

Figure 3.10.9

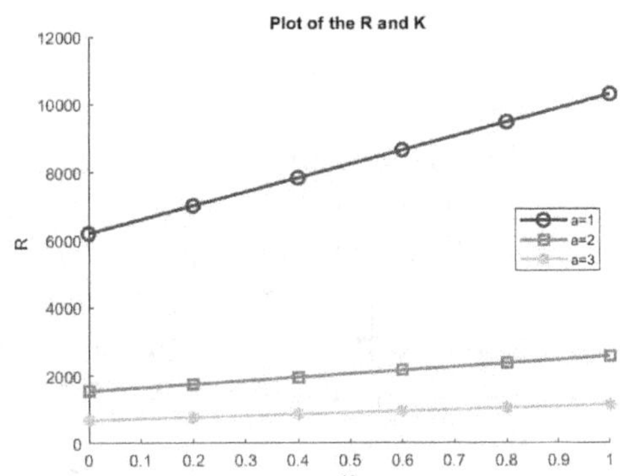

Figure 3.10.10

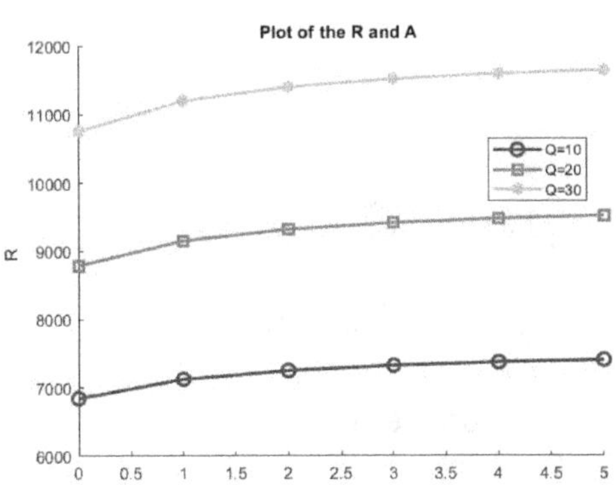

Figure 3.10.11

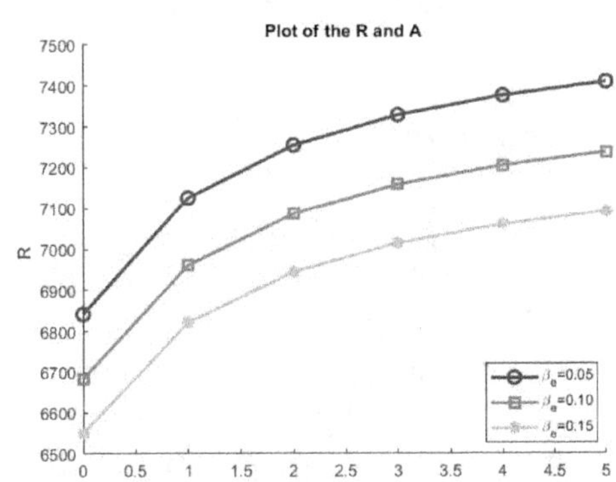

Figure 3.10.12

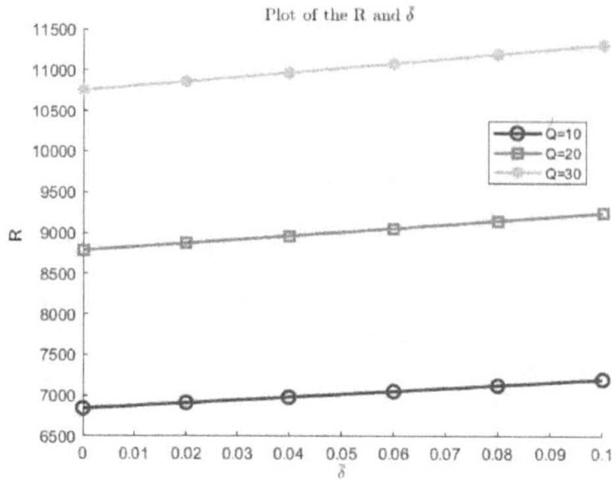

Figure 3.10.13

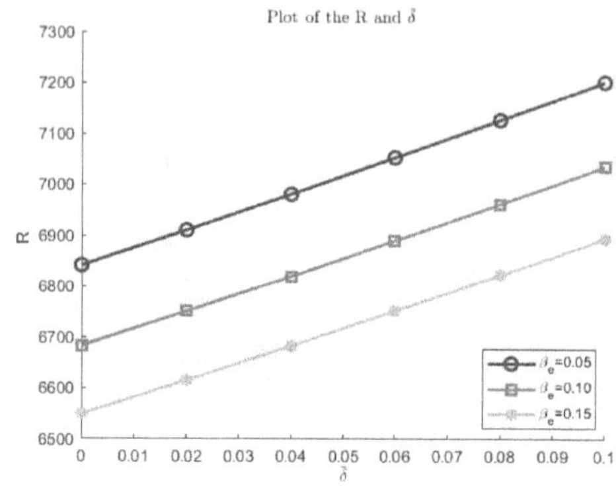

Figure 3.10.14

3.11 Conclusion

In this paper, according to sign of the derivatives and graphical representation we found that the effect of medium permeability and hall parameter are destabilizing. The effect of magnetic field, couple stress, coupling parameter, micropolar coefficient and micropolar heat conduction coefficient are stabilizing. Among them the most important result is that the hall parameter destabilizes the system.

The sufficient condition for the non-existence of over stability is given by $\bar{\delta} < \frac{\epsilon}{A}, Q < \frac{Kb^2P_r}{2\pi^2} and \beta_e < \frac{KAP_r{}^2}{m\pi^2}$.

References

[1] A.C. Eringen, 'Simple microfluids', International journal of engineering science, 205-217, May, 1964.

[2] A.C. Eringen, 'Theory of micropolar fluid', International journal of engineering science, 1-18, July 1966.

[3] A.J. Chamkha, 'Hydromagnetic two-phase flow in a channel', Int. J. engg. sci., 437-446, Feb 1995.

[4] A.J. Chamkha et al., 'Fully developed free convection of a micropolar fluid in a vertical channel', Int. comm. heat mass transfer, 1119-1127, Nov 2002.

[5] A.S. Banyal 'A characterization of couple-stress fluid heated from below in a porous medium in the presence of a rotation', International journal of advanced computer and mathematical science, 310-320, June 2012.

[6] B. Singh, 'Hall effect on MHD flow of visco-elastic micro-polar fluid layer heated from below saturating a porous medium', International journal of statistics and applied mathematics, 48-66, 2017.

[7] D. Kumawat et al., 'Effect of the couple-stress on micro polar rotating fluid flow saturating a porous medium', Journal of computational analysis and applications, 270-280, April, 2023.

[8] G. Lebon et al., 'Convective instability of a micropolar fluid layer by the method of energy', Int. J. Engg. Sci., 1321-1329, 1981.

[9] J.I. Oahimire and B.I. Olajuwon, 'Effect of hall current and thermal radiation on heat and mass transfer of a chemically reacting MHD flow of a micropolar fluid through a porous medium', Journal of king saud university – Engineering science, 112-121, July, 2013.

[10] J.P, Kumar et al., 'Fully developed free convection of micropolar and viscous fluid in a vertical channel', Applied mathematical modelling, 1175-1186, Aug 2009.

[11] P. Kumar et al., 'Effect of rotation on thermal instability in couple-stress elastio-viscous fluid', Zeitschrift fur Naturforschung, 407-411, June 2004.

[12] P.Y. Xiong, 'Two-phase flow of couple stress fluid thermally effected slip boundary conditions: Numerical analysis with variable liquids properties', Alexandria Engineering journal, 3821-3830, 2022.

[13] R.C. Sharma and U. Gupta, 'Thermal convection in micropolar fluids in porous medium', International journal of engineering science, 1887-1892, Oct 1995.

[14] R. Kumar and V. Mehta, 'Hall effect on MHD flow of visco-elastic fluid layer heated for below saturating a porous medium', International journal of statistics and applied mathematics, 250-260, Oct 2017.

[15] V.K. Stokes, 'Couple-stress in fluid', Physics of fluid, 1709-1715, May 1966.

Chapter 4

A Cryptographic Algorithm using Scalar Potential Technique with Matrix Formation

Kala Raja Mohan[1], R. Narmada Devi[2], Nagadevi Bala Nagaram[3], T. Bharathi[4] and Suresh Rasappan[5]

[1,2,3] Department of Mathematics, Vel Tech Rangarajan Dr. Sagunthala R&D Institute of Science and Technology, Avadi, Chennai – 600 062, Tamil Nadu, India
[4] Loyala College, Nungambakkam, Chennai-600 034, Tamil Nadu, India
[5] Mathematics Section, University of Technology and Applied Sciences, Ibri, Ibri 516, AI Dhahirah Governorate, Sultanate of Oman

Abstract

In this era of internet, people come across many user-friendly applications. These applications are readily available for usage in mobile itself, which is handy and makes each individual's job easy. To make more ease of the applications, these user-friendly applications comes with an option to save the password, This helps the user to make use of the application instantly without requirement of any recognition. Still, it asks for a password when it comes to important transaction. To do this, individual password is set up by the people themselves and maintained secretly to make their data safe.

Though there are benefits on these user-friendly applications, there are few drawbacks also. There is a possibility to be used by third person through mobile without the knowledge of the owner of the mobile. With a view to overcome such difficulty, various methods such as thumb recognition, face recognition and such methods are introduced. The option to enter OTP also occurs in most of the applications to maintain safety transactions.

Cryptography plays a major role in bringing out various methods to maintain a secrecy between the sender and the receiver. Mathematics remains the backbone in bringing out various cryptographic methods. This chapter aims at bringing out a cryptographic technique which involves mathematical concepts of gradient and scalar potential. The results obtained are thereafter converted to a matrix and the resulting matrix is represented as a diagram to make safety data transmission.

Keywords: Cryptography, Scalar Potential, Canonical Form, Matrix form, Graphical representation.

4.1 Introduction

Nowadays, an inevitable part of our life is mobile communications, computer networks which uses internet for its base. Still, the internet provides option for many hackers to steel the data stored in the system such as mobile or computer. It is very much essential to provide safety measure while sharing the information. Cryptography plays a major role in providing such information security. Cryptography ensure communication between any two individuals in a secured way.

S. Sujatha in 2013 made use of the application of Laplace Transform in the field of cryptography [1]. A. P. Hiwarekar in 2014 proposed two cryptographic techniques applying Laplace Transform and Hyperbolic functions [2,3]. A. K. H. Sedeeg et al in 2016 formulated a new cryptographic algorithm applying Aboodh Transform [4]. S. Dhingra et al proposed a network security method involving Laplace Transform [5]. M. Tuncay Gencoglu in 2017 introduced a crytographic process involving Laplace Transform with Hyperbolic functions [6]. M. Saha in 2017 utilized Laplace Mellin Transform in forming a cryptographic method for secured information sharing [7].

Dr. K. Hemant K. Undegaonkar introduced a secured communication method involving Laplace Transform [8]. C. H. Jayanthi and V. Srinivas in 2019 framed a new mathematical modelling involving Laplace Transform [9]. G. Nagalakshmi et al in 2020 involved Laplace Transform Laplace Transform using Asymmetric key for secured communication [10]. Kala Raja Mohan et al in 2022 applied Bilinear Transform with Probability in identifying a secured information sharing algorithm in one of their work [11]. The same author also introduced a cryptographic technique in involving Laplace transform and hyperbolic tangent function in cryptography and conformal mapping [12, 14]. Nagadevi Bala Nagaram et al made use of first order ODE together with variable separable method in their cryptographic analysis. This paper aims at developing a cryptographic algorithm using the concepts of scalar potential and matrix representation.

The standard definitions used in this crypto analysis are described in section 2. The representation describes the algorithm applied for encryption is given in section 3. The encryption algorithm applied is explained through an example in section 4. Section 5 depicts the algorithm applied for decryption. With the help of the cipher text obtained in Section 4, decryption process is explained in section 6. Section 7 is about the conclusion followed by references.

4.2 Standard definitions

The cryptographic analysis proposed in this chapter make use of the following standard definitions.

- Plain text refers to the information which is to be shared secretly to the other person.
- The message in encrypted form making use of the key specified for the process is known as the cipher text.
- Cipher refers to the process in which the plain text gets transformed into the cipher text.
- The process involved in converting plain text into secret message is encryption.
- The reverse process of encryption is decryption, which converts secret message to plain text.
- Objects consisting of both magnitude and direction is represented by a vector.
- For a 3-dimensional plane, $\vec{i}$, $\vec{j}$, $\vec{k}$ are considered to be the unit vectors along x-axis, y-axis and z-axis.
- The rate of change of something with respect to either distance or time is referred as gradient.
- A vector field in which the curl is always zero is termed as conservative vector field.

4.3 Algorithm for encryption

The various steps involved in the encryption process are as listed below.

- ➢ Form coding table.
- ➢ Assign numerals to the plain text using coding table.
- ➢ With each of the numerals obtained variables like x, y, z are multiplied to obtained polynomials
- ➢ The gradient vector is formed including $\vec{i}$, $\vec{j}$, $\vec{k}$ to each of the polynomials
- ➢ The method of scalar potential is applied to form the function involved
- ➢ The matrix representation of the function obtained above is shared as cipher text message

4.4 Encryption illustration

Though the concept can be extended for n-dimensional space, for ease understanding of the methodology, a three letter word has been chosen for illustration. The usual coding of alphabets is chosen for encryption starting with 1 for A, 2 for B and so on upto 26 for Z, which is shown in Table 4.4.1.

Table 4.1 Coding table

A	B	...	Z
1	2	...	26

The word 'CAT' is chosen to illustrate the methodology applied for this algorithm. The selected word is converted into numerical values as shown in Table 4.2.

Table 4.2 Assigned numerical values

C	A	T
3	1	20

Now, the assigned numerical values are converted into polynomials with coefficients as x, y, z in order.

Table 4.3 Formation of polynomials

3	1	20
3x	y	20z

Making use of the polynomials, the gradient vector is formed including $\vec{i}$ for the first polynomial, $\vec{j}$ for the second polynomial and $\vec{k}$ for the third polynomial. The gradient vector is represented as below.

$$\nabla \emptyset = 3x\vec{i} + y\vec{j} + 20\,z\vec{k}$$

At this stage the method of scalar potential is applied. The coefficient of each direction vector are assigned to the partial derivative of $\emptyset$ with respect to x, y and z in order as given below.

$$\frac{\partial \emptyset}{\partial x} = 3x; \frac{\partial \emptyset}{\partial y} = y; \frac{\partial \emptyset}{\partial z} = 20z$$

Applying integration on both sides with respect x, y and z, the following terms are obtained for $\emptyset$.

$$\emptyset = \frac{3}{2}x^2; \ \emptyset = \frac{1}{2}y^2; \ \emptyset = 10\,z^2$$

Combining the terms, $\emptyset$ is obtained as $\emptyset = \frac{3}{2}x^2 + \frac{1}{2}y^2 + 10\,z^2$
The diagonal matrix representation of $\emptyset$ is as mentioned below.

$$\begin{bmatrix} \frac{3}{2} & 0 & 0 \\ 0 & \frac{1}{2} & 0 \\ 0 & 0 & 10 \end{bmatrix}$$

The obtained diagonal matrix is shared as cipher text to the sender, which remains the encrypted message.

4.5 Algorithm for decryption

The various steps involved in the process of decryption are as listed below.

- Form the polynomial expression from the matrix received as cipher text.
- Find derivative of the polynomial with respect to the variables assigned.
- Using the derivative for the gradient vector.

- ➢ Isolate the coefficients of the gradient vector.
- ➢ Remove the variables from the coefficient obtained in previous step.
- ➢ Decode the numerals to obtain plain text using coding table.

4.6 Decryption illustration

The encrypted message sent to the receiver in matrix form is as given below.

$$\begin{bmatrix} \frac{3}{2} & 0 & 0 \\ 0 & \frac{1}{2} & 0 \\ 0 & 0 & 10 \end{bmatrix}$$

The polynomial expression relative to the matrix given above is formed and is assigned to $\emptyset$ which is given below and is assigned to.

$$\emptyset = \frac{3}{2}x^2 + \frac{1}{2}y^2 + 10\,z^2$$

Differentiating $\emptyset$ with respect to x, y, z, the derivatives obtained are as given below.

$$\frac{\partial \emptyset}{\partial x} = 3x;\ \frac{\partial \emptyset}{\partial y} = y;\ \frac{\partial \emptyset}{\partial z} = 20z$$

The derivative obtained are converted as a gradient vector, which is mentioned below.

$$\nabla\emptyset = 3x\vec{i} + y\vec{j} + 20\,z\vec{k}$$

The coefficients of the gradient vector are chosen and the variables are removed as given underneath.

Table 4.4 Numerals obtained removing variables

3x	y	20z
3	1	20

The alphabets corresponding to the numerals obtained in the above table are obtained using the coding table given in Table 4. 1.

Table 4.5 Numerals converted to alphabets using coding table

3	1	20
C	A	T

The plain text 'CAT' is obtained using the decryption algorithm.

4.7 Conclusion

In the process of secured data communication, cyptography plays a predominant role. Mathematics also plays a major background in the development of cryptographic algorithm. This chapter proposes a novel cryptographic algorithm using the concepts of scalar potential and matrix representation. Illustration of the proposed methodology has been performed using an example word.

References

[1] S. Sujatha, 'Application of Laplace Transforms in Cryptography', International Journal of Mathematical Archive, 4, 67-71, 2013.

[2] A. P. Hiwarekar, 'A new method of Cryptography using Laplace transform of Hyperbolic functions', International Journal of Mathematical Archive, 4, 208-213, 2013.

[3] A. P. Hiwarekar, 'New mathematical modeling for cryptography', Journal of Information Assurance and Security, 9, 027-033, 2014.

[4] A. K. H. Sedeeg, M. M. Abdelrahim Mahgoub, and M. A. Saif Saeed, 'An Application of the New Integral "Aboodh Transform" in Cryptography', Pure and Applied Mathematics Journal, 5, 151-154, 2016.

[5] S. Dhingra, A. A. Savalgi and S. Jain, 'Laplace Transformation based Cryptographic Technique in Network Security', International Journal of Computer Applications, 136, 6-10. 2016.

[6] M. Tuncay Gencoglu, 'Cryptanalaysis of a New Method of Cryptography using Laplace Transform Hyperbolic Functions'

Communications in Mathematics and Applications, 8, 183-189, 2017.

[7] M. Saha, 'Application of Laplace – Mellin Transform for Cryptography', Raj Journal of Technology Research & Innovation, 5, 12-17, 2017.

[8] Dr. K. Hemant K. Undegaonkar, 'Security in Communication By Using Laplace Transform and Cryptography', International Journal of Scientific & Technology Research, 8, 3207-3209, 2019.

[9] C. H. Jayanthi and V. Srinivas, 'Mathematical Modelling for Cryptography using Laplace Transform', International Journal of Mathematics Trends and Technology, 65, 10-15, 2019.

[10] G. Nagalakshmi, A. Chandra Sekhar and D. Ravi Sankar, 'Asymmetric key Cryptography using Laplace Transform', International Journal of Innovative Technology and Exploring Engineering, 9, 3083-3087, 2020.

[11] Kala Raja Mohan, Suresh Rasappan, Regan Murugesan, Sathish Kumar Kumaravel and Ahamed A. Elngar, "Secret Information Sharing Using Probability and Bilinear Transformation", Proceedings of 2nd International Conference on Mathematical Modeling and Computational Science, 115-122, 2022.

[12] Kala Raja Mohan, Suresh Rasappan and Sathish Kumar Kumaravel, 'Secret Information sharing Using Laplace Transform and Hyperbolic Tangent Function', AIP Conference Proceedings 2516, 12003, 1-6, 2022.

[13] Nagadevi Bala Nagaram, Kala Raja Mohan, Sathish Kumar Kumaravel, and Suresh Rasappan. "Secret Information Sharing Using First Order ODE and Variable Separable Method", Recent Trends in Computational Intelligence and its Applications, CRC Press, pp.515-518.

[14] Kala Raja Mohan, Nagadevi Bala Nagaram, Regan Murugesan and Suresh Rasappan. "A Cryptographic Technique Using Conformal Mapping, Recent Trends in Computational Intelligence and its Applications, CRC Press, pp.519-523

Chapter 5

A Study on the Application of the Homotopy Perturbation Method to Solve General Linear and Non-linear Fractional Partial Differential Equations

Kiran Dhirawat, Ramakanta Meher
Department of Mathematics, Sardar Vallabhbhai National Institute of Technology, Surat - 395007, Gujarat, India

Abstract

This paper presents a novel approximate analytical approach to solve general linear and non-linear time-fractional partial differential equations, specifically employing the Caputo fractional derivative. The core concept underlying the Homotopy Perturbation Method (HPM) is to formulate a solution as an infinite series, which exhibits rapid convergence towards the exact solution of the given problem. Notably, the distinguishing strength of the proposed HPM lies in its capability to address non-linear problems without resorting to linearization, discretization, perturbation, or any restrictive measures. The effectiveness and precision of the HPM are validated through various numerical examples. Convergence analyses conclusively demonstrate that this method is highly efficient and straightforward in resolving non-linear time-fractional partial differential equation problems.

Keywords: Fractional model; Riemann-Liouville integral; Caputo derivative; Approximate analytical solution.

5.1 Introduction

Differential Equations serve as fundamental mathematical tools that emerge when we try to create mathematical representations of a wide array of physical systems in different areas of science and engineering. These fields encompass physics, chemistry, biology, mechanics, economics, control theory, and circuit analysis, among others. Traditionally, we use ordinary or partial differential equations to capture the behavior of these physical systems. However, as we tackle more complex real-world scenarios, we often find that Fractional Differential Equations (FDEs) provide more efficient and accurate models. These FDEs involve fractional derivatives, which are a bit more sophisticated than the regular ones, but they help us describe these intricate systems more accurately. To truly grasp how these systems behave, we need to dive into the solutions of these differential equations. There are established classical methods that are excellent at finding exact solutions to differential equations for simpler cases. However, when we encounter non-linear and fractional-order differential equations, which are common in real-world physical systems, these classical methods fall short. This is where we turn to semi-analytical and numerical methods, which might not give us exact answers but provide close approximations, helping us understand these complex systems better. These non-linear fractional partial differential equations have widespread applications across various fields of science and engineering. This includes fields like mathematical physics, statistical mechanics, fluid dynamics, theoretical neuroscience, electro chemistry, viscoelasticity, population dynamics, porous media problems and mathematical problems in finance.

This paper focuses on introducing and extending a semi-analytical method known as the Homotopy Perturbation Method (HPM) for the solution of Differential Equations, particularly Linear and Non-linear Fractional-order Differential Equations. The Homotopy Perturbation Method, originally formulated by Ji-Huan He in 1999 [1][2], serves as a powerful tool for tackling both linear and non-linear ordinary or partial differential equations.

Unlike many traditional perturbation methods, which rely on the assumption of having a small parameter, HPM takes a different approach. In numerous real-world scenarios, non-linear situations don't possess easily identifiable small parameters. Identifying such parameters often requires specialized techniques. HPM, on the other hand, operates based on the artificial parameter method and the Homotopy analysis method, and the beauty of this approach lies in the fact that it doesn't require the presence of a small parameter. The primary goal of this paper is to obtain approximate analytical solutions for non-linear fractional partial differential equations, a task of significant importance in mathematics, science, and technology. Over time, various analytical and numerical methods have been developed for this purpose. Some of these methods include the variational iteration method (VIM) [3], homotopy analysis method (HAM) [4][17][12][7][18], homotopy perturbation transform method (HPTM) [6], new iterative method (NIM) [8], fractional Elzaki projected differential transform method (FEPDTM) [9], generalized differential transform method (GDTM)[10], modification of the reduced differential transform method (MRDTM)[11], fractional Taylor operational matrix method (FTOMM) [13], and fractional residual power series method (FRPSM)[14].

The methods known as ADM (Adomian Decomposition Method) [15][16] and HPM (Homotopy Perturbation Method) were initially introduced to provide numerical solutions for a specific category of Nonlinear Fractional Partial Differential Equations (NFPDEs). Both ADM and HPM employ an iterative algorithm, enabling them to efficiently yield power series solutions for both linear and nonlinear fractional differential equations, all without the need for linearization, perturbation, or discretization techniques. Notably, in contrast to alternative series-based approaches, ADM and HPM do not require the matching of coefficients among similar conditions, nor do they necessitate the use of repeated connections.

5.2 The Basic Idea of Homotopy Perturbation Method

Consider the equation

$$D(w) - g(r) = 0, \qquad r \in \Omega, \tag{5.2.1}$$

with boundary condition

$$A\left(w, \frac{\partial w}{\partial n}\right) = 0, \qquad r \in \Gamma. \tag{5.2.2}$$

Where D is a general differential operator, A is a boundary operator, $g(r)$ is a known analytical function, and Γ is the boundary of the domain Ω.

The operator D can be divided into two parts, which are λ and η, where λ is a linear and η is a non-linear operator. (5.2.1) can be, therefore, rewritten as follows:

$$\lambda[w] + \eta[w] - g(r) = 0. \tag{5.2.3}$$

An artificial parameter p can be embedded in Eq. (3) as follows:

$$\lambda[w]+p\big(\eta[w]-g(r)\big)=0. \tag{5.2.4}$$

Where $p\in[0,1]$ is an embedded parameter.

Using the homotopy approach, we construct the following homotopy $u(r,p):\Omega\times[0,1]\to\mathcal{R}$, which satisfies

$$\mathbb{H}(u,p)=(1-p)(\lambda[u]-\lambda[w_0])+p\big(\lambda[u]+\eta[u]-g(r)\big)=0,$$
$$\mathbb{H}(u,p)=\lambda[u]-\lambda[w_0]+p(\lambda[w_0])+p\big(\eta[u]-g(r)\big)=0. \tag{5.2.5}$$

Where w_0 is the initial approximation of (5.2.3), which satisfies the given condition.

By substitution $p=0$ and $p=1$ in (5.2.5), we have the following set of equation

$$\mathbb{H}(u,0)=\lambda[u]-\lambda[w_0]=0, \tag{5.2.6}$$

and

$$\mathbb{H}(u,1)=\lambda[u]+\eta[u]-g(r)=0. \tag{5.2.7}$$

The basic assumption is that the solution of (5.2.5) can be written as a power series in p:

$$u=u_0+pu_1+p^2u_2+\ldots.. \tag{5.2.8}$$

Finally, the approximate solution can be obtained as:

$$w=\lim_{p\to1}u=u_0+u_1+u_2+\ldots.. \tag{5.2.9}$$

Provided that series is convergent.

The fusion of perturbation and homotopy methods gives rise to the Homotopy Perturbation Method (HPM), which effectively overcomes the constraints commonly associated with traditional perturbation techniques. Conversely, HPM retains the advantageous attributes of conventional perturbation methods. In many instances, the series represented by equation (5.2.10) exhibits convergence. Nevertheless, He [5] has proposed certain recommendations for assessing the convergence rate in the context of non-linear operators:

• The second derivative of $\eta[u]$ concerning u must be small because the parameter may be relatively large, i.e., $p \to 1$.

• The norm of $\lambda^{-1}\frac{\partial \eta}{\partial u}$ must be smaller than one so that the series converges.

Theorem 5.2.1 Consider Banach spaces $\mathbb{A}$ and $\mathbb{B}$, and let $\mathbb{C}: \mathbb{X} \to \mathbb{Y}$ represent a non-linear contraction mapping, defined such that

$$\forall v, \tilde{v} \in \mathbb{X}; \|\mathbb{C}(v) - \mathbb{C}(\tilde{v})\| \leq \gamma\|v - \tilde{v}\|, 0 < \gamma < 1. \tag{5.2.10}$$

In accordance with Banach's fixed-point theorem, there exists a point denoted as u for which the condition $\mathbb{C}(u) = u$ holds true. The outcome derived from employing the homotopy perturbation method will be recognized as a sequence where the relationship between the terms can be expressed as $\mathbb{W}_n = \mathbb{C}(\mathbb{W}_{n-1}), \mathbb{W}_{n-1} = \sum_{i=0}^{n-1} u_i, n = 1,2,3,\ldots$ and suppose that $\mathbb{W}_0 = v_0 = u_0 \in \mathbb{B}_r(u)$ where $\mathbb{B}_r(u) = \{u^* \in \mathbb{X}: \|u^* - u\| < r\}$then we have the following statements

(1) $\|\mathbb{W}_n - u\| \leq \gamma^n\|v_0 - u\|$,

(2) $\mathbb{W}_n \in \mathbb{B}_r(u)$,

(3) $\lim_{n\to\infty} \mathrm{W}_n = u$.

Proof

(1) By the induction method on n, for $n = 1$ we have

$$\|\mathbb{W}_1 - u\| = \|\mathbb{C}(\mathbb{W}_0) - \mathbb{C}(u)\| \leq \gamma\|v_0 - u\|. \tag{5.2.11}$$

Assume that $\|\mathbb{W}_{n-1} - u\| \leq \gamma^{n-1}\|v_0 - u\|$ as an induction hypothesis, then

$$\begin{aligned}\|\mathbb{W}_n - u\| = \|\mathbb{C}(\mathbb{W}_{n-1}) - \mathbb{C}(u)\| \leq \gamma\|\mathbb{W}_{n-1} - u\| \leq \gamma\gamma^{n-1}\|v_0 - u\| \\ = \gamma^n\|v_0 - u\|.\end{aligned} \tag{5.2.12}$$

(2) Using (1) , we have

$$\|\mathbb{W}_n - u\| \leq \gamma^n\|v_0 - u\| \leq \gamma^n r < r \Rightarrow \mathbb{W}_n \in \mathbb{B}_r(u). \tag{5.2.13}$$

(3) Because of $\|\mathbb{W}_n - u\| \le \gamma^n \|v_0 - u\|$, and $\lim_{n\to\infty} \gamma^n = 0$, we can derive

$$\lim_{n\to\infty} \|W_n - u\| = 0, \text{ that is, } \lim_{n\to\infty} W_n = u.$$

5.3 Basic Definitions

Fractional calculus serves as a unifying framework that extends and generalizes the concepts of both integer-order differentiation and the process of integrating multiple times. In this paper, we present fundamental definitions and properties from the theory of fractional calculus that will find application in our study. There are many ways to define differ-integrals such as the Riemann-Liouville definition, Grunwald-Letnikov definition, Caputo definition. Grunwald definition has technical difficulties of computation, but Riemann and Caputo's definition is more elegant, and here we mainly focus on these two.

5.3.1 The Riemann-Liouville and Caputo Differintegral

Riemann's idea of fractional derivatives is based on Cauchy's formula for nth Integral

$$I_a^n f(t) = \frac{1}{\Gamma(n-1)} \int_a^t (t-x)^{n-1} f(x)dx. \tag{5.3.1}$$

Now, if we substitute a positive real number α instead of n in equation (5.3.1), then it becomes

$$I_a^n f(t) = \frac{1}{\Gamma(\alpha)} \int_a^t (t-x)^{\alpha-1} f(x)dx, \alpha > 0. \tag{5.3.2}$$

For $\alpha = 0$, $I_a^n f(t) = f(t)$.

Now the Riemann fractional derivative is defined as follows:

$$\boldsymbol{D}^\alpha f(t) = \begin{cases} \frac{1}{\Gamma(n-\alpha)} \frac{d^n}{dt^n} \int_a^t (t-x)^{n-\alpha-1} f(x)dx, & \alpha > 0,\ n-1 < \alpha \le n, \\ \frac{\partial^n f(t)}{\partial t^n}, & \alpha = n, n \in \mathcal{N} \end{cases} \tag{5.3.3}$$

and the Caputo fractional derivative is defined as:

$$D_*^\alpha f(x,t) = \begin{cases} \dfrac{1}{\Gamma(m-\alpha)}\displaystyle\int_a^t (t-x)^{m-\alpha-1}\frac{\partial^m}{\partial x^m}f(x)dx, & ,m-1<\alpha\le m \\ \dfrac{\partial^m f(t)}{\partial t^m}, & \alpha = m, m \in \mathcal{N} \end{cases}$$

Here we use the symbol $\boldsymbol{D}$ for Riemann-Liouville fractional derivatives and D_* use for Caputo fractional derivatives.
The main difference between Riemann and Caputo definition of fractional derivatives is that Caputo derivatives of a constant function is zero while Riemann fractional derivatives of a constant function is non-zero.

5.3.2 Basic Properties of Differintegrals

Linearity

Let A and B are real constant and $f(t)$, $g(t)$ are two arbitrary functions.

Then the linearity property can be defined as:

$$D_a^\alpha\big(Af(t)+Bg(t)\big) = AD_a^\alpha f(t) + BD_a^\alpha g(t) \tag{5.3.4}$$

Where D indicate one of the differ-integral mention above.
The differ-integral can be systematically applied to each term within an infinite series.

Composition

First, we look at the composition of integral. We use the symbol D to point out the independence of the process. Let $f(t)$ is an integrable function $\alpha \in \mathcal{R}$, α, $\beta > 0$.

$$D_a^{-\alpha}\left(D_a^{-\beta}f(t)\right) = \tfrac{1}{\Gamma(\alpha)}\textstyle\int_a^t (t-\mu)^{\alpha-1}\tfrac{1}{\Gamma(\beta)}\int_a^t (\mu-\tau)^{\beta-1}f(\tau)\,d\tau d\mu \tag{5.3.5}$$

$$= \frac{1}{\Gamma(\alpha)\Gamma(\beta)} \int_a^t \int_a^{\mu} (t-\mu)^{\alpha-1} (\mu-\tau)^{\beta-1} f(\tau) d\tau d\mu \tag{5.3.6}$$

Now upon changing the order of integration, we find

$$= \frac{1}{\Gamma(\alpha)\Gamma(\beta)} \int_a^t f(\tau) \int_{\tau}^t (t-\mu)^{\alpha-1} (\mu-\tau)^{\beta-1}\, d\tau d\mu \tag{5.3.7}$$

$$= \frac{1}{\Gamma(\alpha)\Gamma(\beta)} \int_a^t f(\tau)\, (t-\tau)^{\alpha+\beta-1} \boldsymbol{B}(\alpha,\beta) d\tau \tag{5.3.8}$$

Upon using the property $\boldsymbol{B}(\alpha,\beta) = \frac{\Gamma(\alpha)\Gamma(\beta)}{\Gamma(\alpha+\beta)}$, it becomes

$$= \frac{1}{\Gamma(\alpha+\beta)} \int_a^t (t-\tau)^{\alpha+\beta-1} f(\tau) d\tau = D^{-(\alpha+\beta)} f(t) \tag{5.3.9}$$

Hence, the fractional integrals are commutative. This raises a question, does fractional differential is commutative ?. The answer is certainly no.

Existence and Uniqueness Theorem

Let us consider the following linear IVP of the form

$$D_0^{\alpha_i} y(t) + \sum_{i=1}^{k-1} P_i(t)\, D_0^{\alpha_i} y(t) + P_0(t) y(t) = g(t) \tag{5.3.10}$$

$$D_0^{\alpha_{i-1}} y(t)\big|_{t=0} = b_i \tag{5.3.11}$$

$$i = 1,2,\ldots\ldots,k \quad \text{and} \quad 0 < t < T < \infty. \tag{5.3.12}$$

Theorem 5.3.1 If $g(t)$ is bounded on $(0,T)$ and $P_i(t)$ for $i \in \{0,1,2,\ldots\ldots,k\}$ are continuous functions in the $[0,T]$. Then the above IVP has the unique solution $y(t)$.

5.4 Homotopy Perturbation Method for Fractional differential equation

Consider the following equation

$$D_{*t}^{\alpha} w = \lambda[w] + \eta[w] + f(x,t)$$

$w^j(x,0) = g_j(x) j = 0,1,\ldots,k-1$ (5.4.1)

In the context of Caputo fractional derivatives, we consider the operator λ as linear, and the operator η as non-linear. The non-linear operator η can encompass fractional-order derivatives of degrees lower than α, where α falls within the range of $k-1 < \alpha \leq k$.

Now we construct homotopy:

$$(1-p)[\lambda(v) - \lambda(w_0)] - p[\lambda(v) + \eta(v) - D_{*t}^{\alpha} v + f(x,t)] = 0 \tag{5.4.2}$$

and

$$\frac{\partial^k w}{\partial t^k} - f(x,t) = p\left[\frac{\partial^k w}{\partial t^k} + \lambda(w, w_x, u_{xx}) + \eta(w, w_x, w_{xx}) - D_{*t}^{\alpha} w\right]. \tag{5.4.3}$$

The parameter p belongs to the interval [0, 1]. The homotopy parameter, denoted as p, continuously varies from 0 to 1. When p equals 0, equation (5.4.2) transforms into a linear equation.

$$\frac{\partial^k w}{\partial t^k} = f(x,t) + \lambda(w, w_x, w_{xx}) \tag{5.4.4}$$

and the eq.(5.4.3) become

$$\frac{\partial^k w}{\partial t^k} = f(x,t) \tag{5.4.5}$$

and when $p = 1$ then the eq. (5.4.2) or eq. (5.4.3) turn out to be the original equation (5.4.1). The basic assumption is that the solution of Eq. (5.4.2) or eq. (5.4.3) can be written as a power series in p:

$$w = w_0 + p w_1 + p^2 w_2 + \ldots.. \tag{5.4.6}$$

Substituting Eq.(5.4.6) in Eq.(5.4.2) or Eq.(5.4.3) and comparing the power of embedded parameter p, we obtain the set of the linear equation of the form:

$$p^0: \left(\frac{\partial^k w_0}{\partial t^k}\right) - \lambda_0(w_0) = f(x,t), w_0^K(0) = g_k \tag{5.4.7}$$

$$p^1: \left(\frac{\partial^k w_1}{\partial t^k}\right) - \lambda_1(w_0, w_1) = \frac{\partial^k w_0}{\partial t^k} + \eta_0(w_0) - D_{*t}^{\alpha} w_0, w_1^K(0) = 0 \tag{5.4.8}$$

$$p^2: \left(\frac{\partial^k w_2}{\partial t^k}\right) - \lambda_2(w_0, w_1, w_2) = \frac{\partial^k w_1}{\partial t^k} + \eta_1(w_0, w_1) - D_{*t}^{\alpha} w_1, w_2^K(0) = 0 \tag{5.4.9}$$

$$p^3: \left(\frac{\partial^k w_3}{\partial t^k}\right) - \lambda_3(w_0, w_1, w_2, w_3) = \frac{\partial^k w_2}{\partial t^k} + \eta_2(w_0, w_1, w_2) - D_{*t}^{\alpha} w_2, \quad w_3^K(0) = 0$$

$$\vdots$$

or of the form

$$p^0: \left(\frac{\partial^k w_0}{\partial t^k}\right) = f(x,t), w_0^K(0) = g_k \tag{5.4.10}$$

$$p^1: \left(\frac{\partial^k w_1}{\partial t^k}\right) = \frac{\partial^k w_0}{\partial t^k} + \lambda_0(w_0) + \eta_0(w_0) - D_{*t}^{\alpha} w_0, u_1^K(0) = 0 \tag{5.4.11}$$

$$p^2: \left(\frac{\partial^k w_2}{\partial t^k}\right) = \frac{\partial^k w_1}{\partial t^k} + \lambda_1(w_0, u_1) + \eta_1(w_0, w_1) - D_{*t}^{\alpha} w_1, w_2^K(0) = 0 \tag{5.4.12}$$

$$p^3: \left(\frac{\partial^k w_3}{\partial t^k}\right) = \frac{\partial^k w_2}{\partial t^k} + \lambda_2(w_0, w_1, w_2) + \eta_2(w_0, w_1, w_2) - D_{*t}^{\alpha} w_2, \quad w_3^K(0) = 0$$

$$\vdots$$

where the terms $\eta_0, \eta_1, N_2, \lambda_1, \lambda_2$ are satisfied the following relation

$$\lambda(w_0 + pw_1 + p^2 w_2 + \ldots\ldots) = \lambda_0(w_0) + p\lambda_1(w_0, w_1) + p^2\lambda_2(w_0, w_1, w_2) + .. \tag{5.4.14}$$

$$N(w_0 + pw_1 + p^2 w_2 + \ldots\ldots) = N_0(w_0) + pN_1(w_0, w_1) + p^2 N_2(w_0, w_1, w_2) + .. \tag{5.4.15}$$

Now setting $p = 1$ in Eq. (5.4.6) gives the solution of original equation (5.4.1).The components $w_0, w_1, w_2 \ldots.$ can easily obtained from the set of linear Eq. (5.4.14) or Eq. (5.4.15). Finally, we can approximate the solution, provided that the series (5.4.6) is convergent as p tends to unity,

$$w(x,t) = \sum_{i=0}^{\infty} w_i(x,t). \tag{5.4.16}$$

5.5 Numerical experiments

In this section, we will demonstrate the novel HPM algorithm through several examples. These examples may appear somewhat contrived since the exact solutions are known in advance, particularly in the special cases of $\alpha = 1$ or 2. Furthermore, the boundary conditions and initial conditions are directly derived from these known solutions. However, this approach is essential for assessing the precision of the analytical method and for investigating how altering the order of the time fractional derivative impacts the solution's behavior.

Additionally, we will compare the outcomes obtained via the Adomian decomposition method. All computations will be carried out using the symbolic calculus software Mathematica.

Fractional Riccati differential equation

This section considers a numerical and analytical solution for Riccati fractional differential equation of the form

$$\frac{d^\alpha u}{dt^\alpha} = f(t) + g(t)u + h(t)u^2, t > 0, m-1 < \alpha \le m \tag{5.5.1}$$

with initial condition

$$u_k(0) = c_k, k = 0,1,2,3,\dots, m-1, \tag{5.5.2}$$

using the Homotopy perturbation method.

Here $f(t)$, $g(t)$, $h(t)$ are given function and c_k ; $k = 0,1,2,3,\dots, m-1$ are the arbitrary constant and α is a parameter which describes the order of the differential equation.

Example 5.5.1

$$\frac{d^\alpha u}{dt^\alpha} = 1 - u^2, t > 0, 0 < \alpha \le 1, \tag{5.5.3}$$

with initial condition

$$u(0) = 0. \tag{5.5.4}$$

Solution using HPM

Because of the homotopy perturbation method, we construct the following homotopy for Eq. (5.5.3)

$$\frac{du}{dt} - 1 = p\left[\frac{du}{dt} - u^2 - \frac{d^\alpha u}{dt^\alpha}\right] \tag{5.5.5}$$

Substituting $u = u_0 + pu_1 + p^2u_2 + \ldots..$ and the initial condition (5.5.4) in Eq. (5.5.5), we obtain the set of linear differential equations:

$$\frac{du_0}{dt} = 1, u_0(0) = 0 \tag{5.5.6}$$

$$\frac{du_1}{dt} = \frac{du_0}{dt} - {u_0}^2 - \frac{d^\alpha u_0}{dt^\alpha}, u_1(0) = 0 \tag{5.5.7}$$

$$\frac{du_2}{dt} = \frac{du_1}{dt} - 2u_0u_1 - \frac{d^\alpha u_1}{dt^\alpha}, u_2(0) = 0 \tag{5.5.8}$$

$$\frac{du_3}{dt} = \frac{du_2}{dt} - 2u_0u_2 - {u_1}^2 - \frac{d^\alpha u_2}{dt^\alpha}, u_3(0) = 0$$

$$\vdots$$

For $\alpha = 1$, the first few components of homotopy perturbation solution of (5.5.3) can be derived as follows:

$$u_0 = t, \tag{5.5.9}$$

$$u_1 = -\frac{t^3}{3}, \tag{5.5.10}$$

$$u_2 = \frac{2t^5}{15}, \tag{5.5.11}$$

$$u_3 = -\frac{17t^7}{315},$$

$$\vdots$$

In the same manner, the remaining components can be computed. The fourth term approximation of Eq. (5.5.3)

$$u_4 = t - \frac{t^3}{3} + \frac{2t^5}{15} - \frac{17t^7}{315}$$

Only for $\alpha = 1$ the exact solution of Eq. (5.5.3) is known. That is

$$u(t) = \frac{e^{2t}-1}{e^{2t}+1} \tag{5.5.12}$$

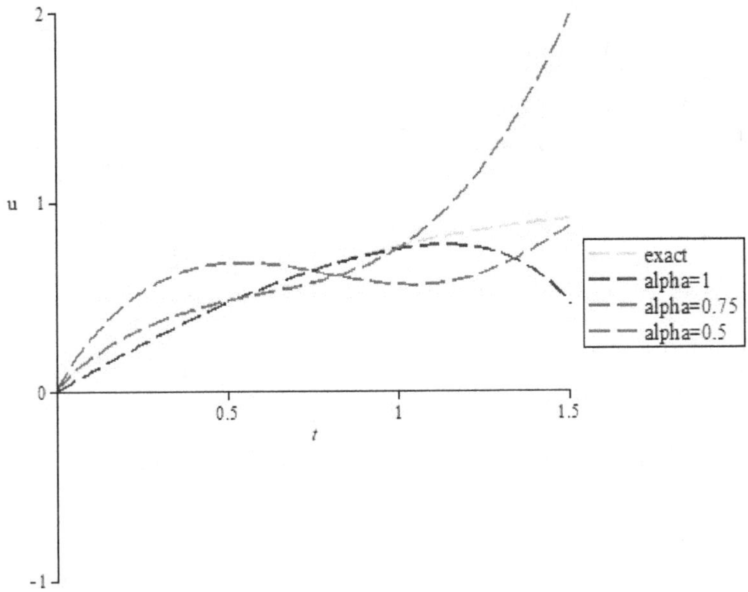

Figure 5.1 Exact solution and homotopy perturbation solution for different values of α.

Figure 5.1 shows the exact solution and perturbation solution for different values of α. The figure shows that the homotopy perturbation solution is in high agreement with the exact solution for a small value of $t > 0$ or in the neighbourhood of initial point zero.

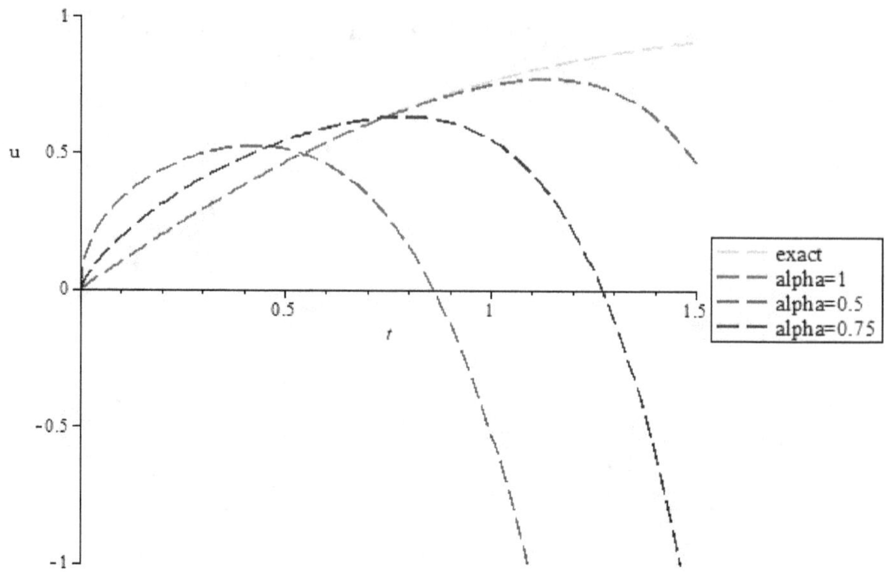

Figure 5.2 Exact solution and decompose solution for different values of α.

Table 5.1 Numerical values when α = 0.5, 0.75, 1

t	$\alpha = 1/2$		$\alpha = 3/4$		$\alpha = 1$		
	u_{hpm}	u_{adm}	u_{hpm}	u_{adm}	u_{hpm}	u_{adm}	u_{exact}
0	0	0	0	0	0	0	0
0.1	2.7387×10^{-1}	3.3012×10^{-1}	1.7196×10^{-1}	1.9010×10^{-1}	0.9966×10^{-1}	0.9966×10^{-1}	0.9966×10^{-1}
0.2	4.5412×10^{-1}	4.3740×10^{-1}	2.8313×10^{-1}	3.0997×10^{-1}	1.9737×10^{-1}	1.9737×10^{-1}	1.9737×10^{-1}
.0.3	5.7393×10^{-1}	5.0958×10^{-1}	3.6533×10^{-1}	4.0462×10^{-1}	2.9131×10^{-1}	2.9131×10^{-1}	2.9131×10^{-1}
.0.4	6.4442×10^{-1}	5.7432×10^{-1}	4.2642×10^{-1}	4.8176×10^{-1}	3.7994×10^{-1}	3.7994×10^{-1}	3.7994×10^{-1}
.0.5	6.7413×10^{-1}	6.5491×10^{-1}	4.7235×10^{-1}	5.4585×10^{-1}	4.6207×10^{-1}	4.6212×10^{-1}	4.6211×10^{-1}
.0.6	6.7198×10^{-1}	7.8025×10^{-1}	5.0973×10^{-1}	6.0098×10^{-1}	5.3685×10^{-1}	5.3707×10^{-1}	5.3704×10^{-1}
.0.7	6.4800×10^{-1}	9.8869×10^{-1}	5.4647×10^{-1}	6.5244×10^{-1}	6.0363×10^{-1}	6.0451×10^{-1}	6.0436×10^{-1}
.0.8	6.1330×10^{-1}	13.303×10^{-1}	5.9184×10^{-1}	7.0857×10^{-1}	6.6170×10^{-1}	6.6464×10^{-1}	6.6403×10^{-1}
.0.9	5.7964×10^{-1}	18.689×10^{-1}	6.5617×10^{-1}	7.8329×10^{-1}	7.0991×10^{-1}	7.1839×10^{-1}	7.1629×10^{-1}
01	5.5855×10^{-1}	26.830×10^{-1}	7.5033×10^{-1}	8.9935×10^{-1}	7.4603×10^{-1}	7.6790×10^{-1}	7.6159×10^{-1}

Example 5.5.2

$$\frac{d^\alpha u}{dt^\alpha} = 1 + 2u - u^2, t > 0, 0 < \alpha \leq 1, \tag{5.5.13}$$

with initial condition

$$u(0) = 0. \tag{5.5.14}$$

Solution using HPM

Because of the homotopy perturbation method, we construct the following homotopy for Eq. (5.5.13)

$$\frac{du}{dt} - 1 = p\left[\frac{du}{dt} + 2u - u^2 - \frac{d^\alpha u}{dt^\alpha}\right], \tag{5.5.15}$$

substituting $u = u_0 + pu_1 + p^2u_2 + \ldots..$ and the initial condition (5.5.14) in Eq. (5.5.13), we obtain the set of linear differential equations:

$$\frac{du_0}{dt} = 1, u_0(0) = 0 \tag{5.5.16}$$

$$\frac{du_1}{dt} = \frac{du_0}{dt} + 2u_0 - {u_0}^2 - \frac{d^\alpha u_0}{dt^\alpha}, u_1(0) = 0 \tag{5.5.17}$$

$$\frac{du_2}{dt} = \frac{du_1}{dt} + 2u_1 - 2u_0u_1 - \frac{d^\alpha u_1}{dt^\alpha}, u_2(0) = 0 \tag{5.5.18}$$

$$\frac{du_3}{dt} = \frac{du_2}{dt} + 2u_2 - 2u_0u_2 - {u_1}^2 - \frac{d^\alpha u_2}{dt^\alpha}, u_3(0) = 0 \tag{5.5.19}$$

$\vdots$

Few components for the solution of Eq. (5.5.13) are derived for α = 1:

$$u_0 = t, \tag{5.5.20}$$

$$u_1 = -\frac{t^3}{3} + t^2, \tag{5.5.21}$$

$$u_2 = \frac{2t^5}{15} - \frac{2t^4}{3} + \frac{2t^3}{3}, \tag{5.5.22}$$

$$u_3 = -\frac{17}{315}t^7 + \frac{17}{45}t^6 - \frac{11}{15}t^5 + \frac{1}{3}t^4 \tag{5.5.23}$$

$\vdots$

The rest of the components of the perturbation solution of Eq.(5.5.13) can be derived in the same manner. Fourth term approximation is given by

$$u(t) = t + \frac{t^3}{3} + t^2 - \frac{3}{5}t^5 - \frac{1}{3}t^4 - \frac{17}{315}t^7 + \frac{17}{45}t^6. \tag{5.5.24}$$

The exact solution of Eq. (5.5.13) for α = 1

$$u(t) = 1 + \sqrt{2}\, tanh\left(\sqrt{2}t + \frac{1}{2}log\left(\frac{\sqrt{2}-1}{\sqrt{2}+1}\right)\right) \tag{5.5.25}$$

From the Figure (5.3), it is clear that the homotopy perturbation solution for Eq. (5.5.13) is consistent with the exact solution at α = 1 for small values t.

From the Table (5.2), the solution obtained by HPM and ADM greatly deal with the exact solution at $\alpha = 1$. However, ADM is more accurate and consistent in comparison with HPM.

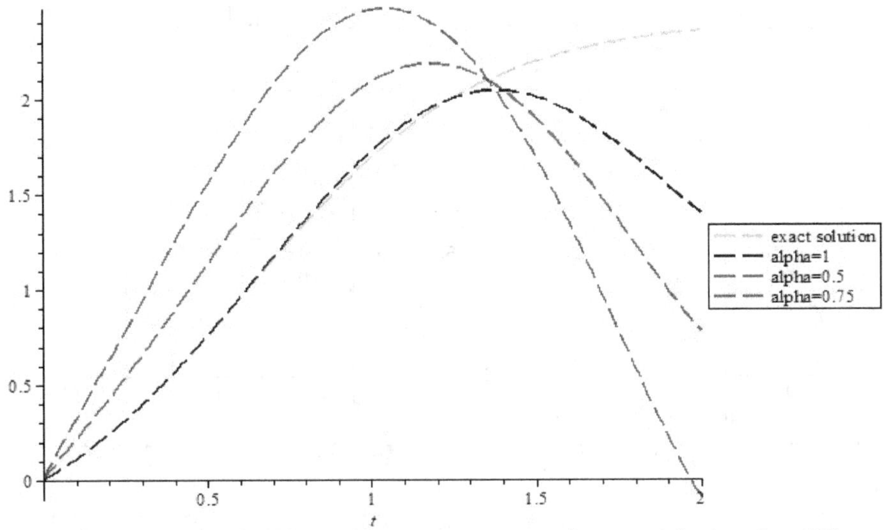

Figure 5.3 Exact solution and homotopy perturbation solution for different values of α.

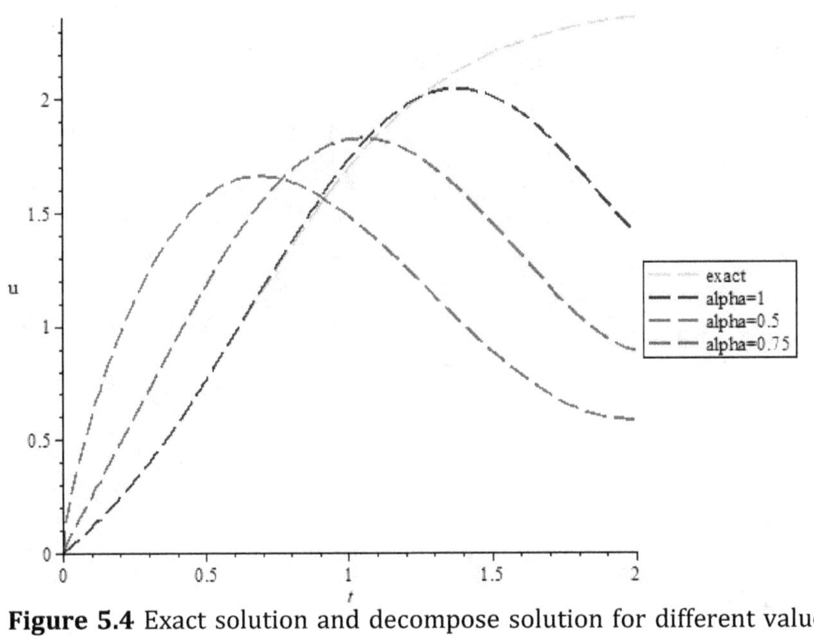

Figure 5.4 Exact solution and decompose solution for different values of α.

Table 5.2 Numerical values when $\alpha = 0.5,\ 0.75,\ 1$ for Eq. (5.5.13)

$\alpha = 1/2$			$\alpha = 3/4$		$\alpha = 1$		
t	u_{hpm}	u_{adm}	u_{hpm}	u_{adm}	u_{hpm}	u_{adm}	u_{exact}
0	0	0	0	0	0	0	0
0.1	3.2172×10^{-1}	5.9454×10^{-1}	2.1686×10^{-1}	2.4537×10^{-1}	1.1029×10^{-1}	1.1029×10^{-1}	1.1029×10^{-1}
0.2	6.2966×10^{-1}	9.5638×10^{-1}	4.2889×10^{-1}	4.7520×10^{-1}	2.4196×10^{-1}	2.4196×10^{-1}	2.4197×10^{-1}
0.3	9.4094×10^{-1}	12.320×10^{-1}	6.5461×10^{-1}	7.1271×10^{-1}	3.9510×10^{-1}	3.9510×10^{-1}	3.9510×10^{-1}
0.4	12.507×10^{-1}	14.323×10^{-1}	8.9140×10^{-1}	9.4921×10^{-1}	5.6811×10^{-1}	5.6811×10^{-1}	5.6781×10^{-1}
0.5	15.494×10^{-1}	15.649×10^{-1}	11.327×10^{-1}	11.739×10^{-1}	7.5756×10^{-1}	7.5756×10^{-1}	7.5601×10^{-1}
0.6	18.254×10^{-1}	16.375×10^{-1}	13.702×10^{-1}	13.765×10^{-1}	9.5825×10^{-1}	9.5825×10^{-1}	9.5356×10^{-1}
0.7	20.665×10^{-1}	16.579×10^{-1}	15.942×10^{-1}	15.480×10^{-1}	11.634×10^{-1}	11.634×10^{-1}	11.529×10^{-1}
0.8	22.606×10^{-1}	16.342×10^{-1}	17.948×10^{-1}	16.816×10^{-1}	13.652×10^{-1}	13.652×10^{-1}	13.463×10^{-1}
0.9	23.968×10^{-1}	15.746×10^{-1}	19.622×10^{-1}	17.724×10^{-1}	15.549×10^{-1}	15.549×10^{-1}	15.269×10^{-1}
1	24.660×10^{-1}	14.871×10^{-1}	20.873×10^{-1}	18.182×10^{-1}	17.238×10^{-1}	17.238×10^{-1}	16.894×10^{-1}

Example 5.5.3

$$\frac{d^{\frac{1}{2}}u}{dt^{\frac{1}{2}}} = \Gamma\left(\frac{3}{2}\right)(u^2 - t + 1), t > 0, 0 < \alpha \leq 1 \qquad (5.5.26)$$

with initial condition

$$u(0) = 0. \qquad (5.5.27)$$

Solution using HPM

Because of standard homotopy perturbation method, we construct the following homotopy:

$$\frac{d^{\frac{1}{2}}u}{dt^{\frac{1}{2}}} = p\left[\Gamma\left(\frac{3}{2}\right)u^2 + \Gamma\left(\frac{3}{2}\right)(1-t)\right], \tag{5.5.28}$$

substituting $u = u_0 + pu_1 + p^2u_2 + \ldots..$ and the initial condition in (5.5.27), we obtain the set of linear differential equations:

$$p^0: \frac{d^{\frac{1}{2}}u_0}{dt^{\frac{1}{2}}} = 0, u_0(0) = 0 \tag{5.5.29}$$

$$p^1: \frac{d^{\frac{1}{2}}u_1}{dt^{\frac{1}{2}}} = \Gamma\left(\frac{3}{2}\right)u_0{}^2 + \Gamma\left(\frac{3}{2}\right)(1-t), u_1 = 0 \tag{5.5.30}$$

$$p^2: \frac{d^{\frac{1}{2}}u_2}{dt^{\frac{1}{2}}} = \Gamma\left(\frac{3}{2}\right)2u_0u_1, u_2 = 0 \tag{5.5.31}$$

$$\vdots$$

On solving the above equation for the first three-term of perturbation solutions, we have

$$u_0 = 0, \tag{5.5.32}$$

$$u_1 = \sqrt{t} - \frac{2}{3}t^{\frac{3}{2}}, \tag{5.5.33}$$

$$u_2 = 0 \tag{5.5.34}$$

$$\vdots$$

The remaining components can be computed using maple. So the general solution can be written as

$$u(t) = \sqrt{t} - \frac{2}{3}t^{\frac{3}{2}} + \ldots.. \tag{5.5.35}$$

It is also depicted that decompose solution is more consistent as compared to perturbation solution.

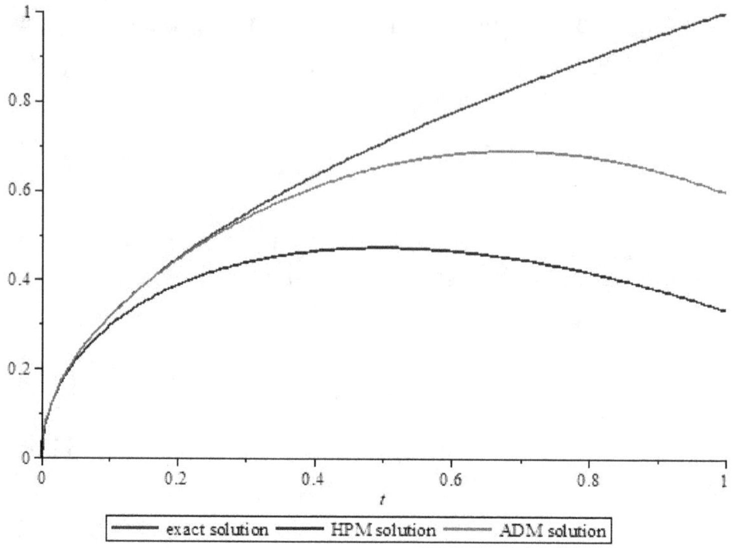

Figure 5.5 Exact solution and versus approximate solution.

Linear Fractional Differential equation

Let us consider a linear fractional differential equation of the form

$$\frac{\partial^{\alpha} u}{\partial t^{\alpha}} + L[u] = R(x,t), t > 0, x \in R \tag{5.5.36}$$

Example 5.5.4

$$\frac{\partial^{\alpha} u}{\partial t^{\alpha}} = \frac{\partial^2 u}{\partial x^2}, t > 0, 0 < \alpha \leq 1 \tag{5.5.37}$$

with initial condition

$$u(x,0) = sin\, x. \tag{5.5.38}$$

First, we construct the following homotopy

$$\frac{\partial u}{\partial t} = p\left[\frac{\partial u}{\partial t} + \frac{\partial^2 u}{\partial x^2} - \frac{\partial^{\alpha} u}{\partial t^{\alpha}}\right] \tag{5.5.39}$$

Now we substitute $u = u_0 + pu_1 + p^2u_2 + \ldots..$ in Eq. (5.5.37) and comparing the power of p.

We have,

$$p^0: \frac{\partial u_0}{\partial t} = 0, u(x,0) = \sin x \quad (5.5.40)$$

$$p^1: \frac{\partial u_1}{\partial t} = \frac{\partial u_0}{\partial t} + \frac{\partial^2 u_0}{\partial x^2} - \frac{\partial^\alpha u_0}{\partial t^\alpha}, u_1(x,0) = 0 \quad (5.5.41)$$

$$p^2: \frac{\partial u_2}{\partial t} = \frac{\partial u_1}{\partial t} + \frac{\partial^2 u_1}{\partial x^2} - \frac{\partial^\alpha u_1}{\partial t^\alpha}, u_2(x,0) = 0 \quad (5.5.42)$$

$\vdots$

On solving the above equation for u_0, u_1, u_2, we have

$$u_0(x,t) = \sin x, \quad (5.5.43)$$

$$u_1(x,t) = -t \sin x, \quad (5.5.44)$$

$$u_2(x,t) = \left(-t + \frac{t^2}{2} + \frac{t^{(2-\alpha)}}{\Gamma(3-\alpha)} \right) \sin x \quad (5.5.45)$$

$\vdots$

Table 5.3 Numerical values when $\alpha = 0.5,\ 0.75,\ 1$ for Eq. (5.5.37)

		$\alpha = 1/2$		$\alpha = 3/4$		$\alpha = 1$		
t	x	u_{adm}	u_{hpm}	u_{adm}	u_{hpm}	u_{adm}	u_{hpm}	u_{exact}
	0.25	1.5927×10^{-1}	1.5580×10^{-1}	1.8094×10^{-1}	1.7925×10^{-1}	2.0254×10^{-1}	2.0254×10^{-1}	2.02557×10^{-1}
	0.5	3.0864×10^{-1}	3.0191×10^{-1}	3.5064×10^{-1}	3.4736×10^{-1}	3.9248×10^{-1}	3.9248×10^{-1}	3.9252×10^{-1}
00.2	0.75	4.3883×10^{-1}	4.2926×10^{-1}	4.9853×10^{-1}	4.9387×10^{-1}	5.5803×10^{-1}	5.5803×10^{-1}	5.5807×10^{-1}
	1	5.4172×10^{-1}	5.2991×10^{-1}	6.1543×10^{-1}	6.0967×10^{-1}	6.8888×10^{-1}	6.8888×10^{-1}	6.8893×10^{-1}
	0.25	1.3696×10^{-1}	1.1364×10^{-1}	1.4674×10^{-1}	1.4386×10^{-1}	1.6559×10^{-1}	1.6559×10^{-1}	1.6583×10^{-1}
	0.5	2.6542×10^{-1}	2.2022×10^{-1}	2.8435×10^{-1}	2.7878×10^{-1}	3.2089×10^{-1}	3.2089×10^{-1}	3.2136×10^{-1}
00.4	0.75	3.7737×10^{-1}	3.1311×10^{-1}	4.0429×10^{-1}	3.9637×10^{-1}	4.5624×10^{-1}	4.5624×10^{-1}	4.5691×10^{-1}
	1	4.6585×10^{-1}	3.8653×10^{-1}	4.9909×10^{-1}	4.8931×10^{-1}	5.6322×10^{-1}	5.6322×10^{-1}	5.6405×10^{-1}
	0.25	1.2325×10^{-1}	1.0020×10^{-1}	1.1963×10^{-1}	1.2470×10^{-1}	1.3458×10^{-1}	1.3458×10^{-1}	1.3577×10^{-1}
	0.5	2.3884×10^{-1}	1.9418×10^{-1}	2.3183×10^{-1}	2.4165×10^{-1}	2.6080×10^{-1}	2.6080×10^{-1}	2.6311×10^{-1}
00.6	0.75	3.3958×10^{-1}	2.7608×10^{-1}	3.2961×10^{-1}	3.4357×10^{-1}	3.7081×10^{-1}	3.7081×10^{-1}	3.7409×10^{-1}
	1	4.1920×10^{-1}	3.4082×10^{-1}	4.0690×10^{-1}	4.2413×10^{-1}	4.5776×10^{-1}	4.5776×10^{-1}	4.6180×10^{-1}

The remaining component of the homotopy perturbation method can be computed by Maple software. The third term of approximation for Eq. (5.5.37) is given by

$$u(x,t) = sin\,x - t\,sin\,x + \left(-t + \frac{t^2}{2} + \frac{t^{(2-\alpha)}}{\Gamma(3-\alpha)}\right) sin\,x + \ldots.. \qquad (5.5.46)$$

Table 5.3 incorporates the approximation up to the third-order term. To enhance the efficiency of both methods, it is advisable to calculate additional terms beyond this point. $u(x,t)$.

Example 5.5.5

$$\frac{\partial^\alpha u}{\partial t^\alpha} = \frac{x^2}{2}\left(\frac{\partial^2 u}{\partial x^2}\right), t > 0,\ , x \in \mathcal{R},\ 1 < \alpha \le 2. \qquad (5.5.47)$$

Subjected to initial condition

$$u(x,0) = x,\ \frac{\partial u(x,0)}{\partial t} = x^2 \qquad (5.5.48)$$

Proceeding with HPM, we construct the homotopy as follows:

$$\frac{\partial^2 u}{\partial t^2} = p\left[\frac{\partial^2 u}{\partial t^2} + \frac{x^2}{2}\left(\frac{\partial^2 u}{\partial x^2}\right) - \frac{\partial^\alpha u}{\partial t^\alpha}\right] \tag{5.5.49}$$

Now we substitute $u(x,t) = u_0(x,t) + pu_1(x,t) + p^2u_2(x,t) + \ldots\ldots$ and compare the power of p. We have

$$p^0: \frac{\partial^2 u_0}{\partial t^2} = 0, u_0(x,0) = x, \frac{\partial u_0(x,0)}{\partial t} = x^2 \tag{5.5.50}$$

$$p^1: \frac{\partial^2 u_1}{\partial t^2} = \frac{\partial^2 u_0}{\partial t^2} + \frac{x^2}{2}\left(\frac{\partial^2 u_0}{\partial x^2}\right) - \frac{\partial^\alpha u_0}{\partial t^\alpha}, u_1(x,0) = 0 \tag{5.5.51}$$

$$p^2: \frac{\partial^2 u_2}{\partial t^2} = \frac{\partial^2 u_1}{\partial t^2} + \frac{x^2}{2}\left(\frac{\partial^2 u_1}{\partial x^2}\right) - \frac{\partial^\alpha u_1}{\partial t^\alpha}, u_2(x,0) = 0 \tag{5.5.52}$$

$\vdots$

On solving above Eq. (5.5.50)-(5.5.52) for $u_0, u_1, u_2, \ldots$ we have

$$u_0(x,t) = x + x^2 t, \tag{5.5.53}$$

$$u_1(x,t) = x^2\frac{t^3}{6}, \tag{5.5.54}$$

$$u_2(x,t) = x^2\left(\frac{t^3}{6} + \frac{t^5}{5!} - \frac{t^{(5-\alpha)}}{\Gamma(6-\alpha)}\right) \tag{5.5.55}$$

$\vdots$

The rest of the components can be computed similarly. The third term approximation for Eq. (5.5.47) is given by

$$u(x,t) = x + x^2 t + x^2\frac{t^3}{6} + x^2\left(\frac{t^3}{6} + \frac{t^5}{5!} - \frac{t^{(5-\alpha)}}{\Gamma(6-\alpha)}\right) + \ldots\ldots \tag{5.5.56}$$

Table 5.4 presents the estimated solutions for Equation (5.5.47) across various values of α, employing both the ADM and HPM methods.

Both methods are robust for finding an exact and approximate solution of the linear fractional partial differential equation. They provide a more realistic series solution.

Non-linear Fractional Partial Differential Equation

This section discusses the solution of the non-linear time-fractional differential equation using the Homotopy perturbation method.

Here we illustrate two examples for which the exact solution is known for $\alpha = 1$ or $\alpha = 2$.

Table 5.4 Numerical values when $\alpha = 1.5,\ 1.75,\ 2$ for Eq. (5.5.47)

		$\alpha = 3/2$		$\alpha = 7/4$		$\alpha = 2$	
t	x	u_{adm}	u_{hpm}	u_{adm}	u_{hpm}	u_{adm}	u_{hpm}
00.2	0.25	0.262840616	0.262696933	0.262669900	0.262648505	0.262583500	0.262583500
	0.5	0.551362462	0.550787732	0.550679598	0.550594020	0.550334001	0.550334001
	0.75	0.865565539	0.864272398	0.864029096	0.863836546	0.863251501	0.863251501
	1	1.205449847	1.203150930	1.202718392	1.202376082	1.201336003	1.201336003
00.4	0.25	0.276971138	0.276427395	0.276156689	0.276076148	0.275672020	0.275672020
	0.5	0.607884554	0.605709581	0.604626756	0.604304593	0.602688081	0.602688081
	0.75	0.992740246	0.987846558	0.985410200	0.984685333	0.981048183	0.981048183
	1	1.431538215	1.422838324	1.418507022	1.417218370	1.410752325	1.410752325
00.6	0.25	0.293094809	0.291986160	0.291090098	0.290922330	0.289790847	0.289790847
	0.5	0.672379236	0.667944641	0.664360392	0.663689320	0.659163389	0.659163389
	0.75	1.137853281	1.127875442	1.119810882	1.118300970	1.108117624	1.108117624
	1	1.689516943	1.671778564	1.657441567	1.654757279	1.636653554	1.636653554

Example 5.5.6

$$\frac{\partial^\alpha u}{\partial t^\alpha} + u\left(\frac{\partial u}{\partial x}\right) = x + xt^2, t > 0\ , x \in \mathcal{R},\ 0 < \alpha \le 1 \qquad (5.5.57)$$

with initial condition $u(x,0) = 0$ Proceeding with HPM, we have the following homotopy:

$$\frac{\partial u}{\partial t} - x - xt^2 = p\left[\frac{\partial u}{\partial t} - u\left(\frac{\partial u}{\partial x}\right) - \frac{\partial^\alpha u}{\partial t^\alpha}\right] \qquad (5.5.58)$$

Now substituting $u(x,t) = u_0(x,t) + pu_1(x,t) + p^2u_2(x,t) + \ldots$ in Eq.(5.5.57) with initial condition and equating the terms the power of p, we have the following set of linear partial differential equation

$$p^0: \frac{\partial u_0}{\partial t} = x + xt^2, u_0(x,0) = 0 \tag{5.5.59}$$

$$p^1: \frac{\partial u_1}{\partial t} = \left[\frac{\partial u_0}{\partial t} - u_0\left(\frac{\partial u_0}{\partial x}\right) - \frac{\partial^\alpha u_0}{\partial t^\alpha}\right], u_1(x,0) = 0 \tag{5.5.60}$$

$$p^2: \frac{\partial u_2}{\partial t} = \left[\frac{\partial u_1}{\partial t} - u_0\left(\frac{\partial u_1}{\partial x}\right) - u_1\left(\frac{\partial u_0}{\partial x}\right) - \frac{\partial^\alpha u_1}{\partial t^\alpha}\right], u_2(x,0) = 0 \tag{5.5.61}$$

$\vdots$

To initiate our analysis, we will derive the initial components of the homotopy perturbation solution in the following manner:

$$u_0(x,t) = x\left(\frac{t^3}{3} + t\right), \tag{5.5.62}$$

$$u_1(x,t) = x\left(-\frac{2t^2}{15} + t - \frac{t^7}{63} - \frac{t^{(2-\alpha)}}{\Gamma(3-\alpha)} - \frac{2t^{(4-\alpha)}}{\Gamma(5-\alpha)}\right), \tag{5.5.63}$$

$$\begin{aligned} u_2(x,t) = x\Bigg[& t - \frac{t^3}{3} - \frac{2t^5}{15} + \frac{t^7}{45} + \frac{2t^9}{567} - \frac{4t^{11}}{2475} - \frac{4t^{13}}{12285} - \frac{t^{15}}{59535} \\ & - \frac{2t^{(2-\alpha)}}{\Gamma(3-\alpha)} + \frac{t^{(3-\alpha)}}{\Gamma(4-\alpha)} + \left(\frac{2}{\Gamma(3-\alpha)} - \frac{2}{\Gamma(4-\alpha)}\right)\frac{t^{(4-\alpha)}}{(4-\alpha)} \\ & + \left(\frac{4}{\Gamma(5-\alpha)} + \frac{16}{\Gamma(6-\alpha)}\right)\frac{t^{(6-\alpha)}}{(6-\alpha)} + \left(\frac{80}{\Gamma(8-\alpha)} - \frac{4}{15\Gamma(3-\alpha)}\right) \\ & \frac{t^{(8-\alpha)}}{(8-\alpha)} - \left(\frac{8}{15\Gamma(5-\alpha)} + \frac{2}{63\Gamma(3-\alpha)}\right)\frac{t^{(10-\alpha)}}{(10-\alpha)} \\ & - \frac{4t^{(12-\alpha)}}{63(12-\alpha)\Gamma(5-\alpha)} + \left(\frac{2}{\Gamma(5-2\alpha)} - \frac{1}{\Gamma(3-\alpha)^2}\right)\frac{t^{(5-2\alpha)}}{(5-2\alpha)} \\ & - \frac{4t^{(7-2\alpha)}}{(7-2\alpha)\Gamma(3-\alpha)\Gamma(5-\alpha)} - \frac{4t^{(9-2\alpha)}}{(9-2\alpha)\Gamma(5-\alpha)^2}\Bigg] \end{aligned} \tag{5.5.64}$$

The remaining component can be computed by Maple software. The third-order approximation solution is given by

$$\begin{aligned} u(x,t) = x\Bigg[& 3t - \frac{4t^5}{15} + \frac{t^7}{315} + \frac{2t^9}{567} - \frac{4t^{11}}{2475} - \frac{4t^{13}}{12285} - \frac{t^{15}}{59535} \\ & - \frac{2t^{(2-\alpha)}}{\Gamma(3-\alpha)} + \frac{t^{(3-\alpha)}}{\Gamma(4-\alpha)} + \left(\frac{2}{\Gamma(3-\alpha)} - \frac{2}{\Gamma(4-\alpha)}\right)\frac{t^{(4-\alpha)}}{(4-\alpha)} \\ & + \left(\frac{4}{\Gamma(5-\alpha)} + \frac{16}{\Gamma(6-\alpha)}\right)\frac{t^{(6-\alpha)}}{(6-\alpha)} \end{aligned}$$

$$+\left(\frac{80}{\Gamma(8-\alpha)}-\frac{4}{15\Gamma(3-\alpha)}\right)\frac{t^{(8-\alpha)}}{(8-\alpha)}$$
$$-\left(\frac{8}{15\Gamma(5-\alpha)}+\frac{2}{63\Gamma(3-\alpha)}\right)\frac{t^{(10-\alpha)}}{(10-\alpha)}-\frac{4t^{(12-\alpha)}}{63(12-\alpha)\Gamma(5-\alpha)}$$
$$+\left(\frac{2}{\Gamma(5-2\alpha)}-\frac{1}{\Gamma(3-\alpha)^2}\right)\frac{t^{(5-2\alpha)}}{(5-2\alpha)}$$
$$\left.-\frac{4t^{(7-2\alpha)}}{(7-2\alpha)\Gamma(3-\alpha)\Gamma(5-\alpha)}-\frac{4t^{(9-2\alpha)}}{(9-2\alpha)\Gamma(5-\alpha)^2}\right] \tag{5.5.65}$$

Table (5.5) shows the solution for different values of α using ADM and HPM. Only for $\alpha = 1$, the exact solution is known $u(x,t) = xt$. From Table (5.5) can be concluded that ADM is more accurate.

Example 5.5.7

$$\frac{\partial^\alpha u}{\partial t^\alpha} = \frac{\partial}{\partial x}\left(u\frac{\partial u}{\partial x}\right), t > 0,\ x \in \mathcal{R},\ 0 < \alpha \le 1 \tag{5.5.66}$$

with initial conditions: $u(x,0) = x^2, u_t(x,0) = -2x^2$ The homotopy for Eq.(5.5.66) can be constructed as

$$\frac{\partial^2 u}{\partial t^2} = p\left[\frac{\partial^2 u}{\partial t^2} + u\left(\frac{\partial^2 u}{\partial x^2}\right) + \left(\frac{\partial u}{\partial x}\right)^2 - \frac{\partial^\alpha u}{\partial t^\alpha}\right] \tag{5.5.67}$$

$u(x,t) = u_0(x,t) + pu_1(x,t) + p^2u_2(x,t) + \ldots..$ in Eq. (107) and equating the power of p, we obtained the following set of linear partial differential equations

$$p^0: \frac{\partial^2 u_0}{\partial t^2} = 0, u_0(x,0) = x^2, u_{0t}(x,0) = -2x^2 \tag{5.5.68}$$

$$p^1: \frac{\partial^2 u_1}{\partial t^2} = \left[\frac{\partial^2 u_0}{\partial t^2} + u_0\left(\frac{\partial^2 u_0}{\partial x^2}\right) + \left(\frac{\partial u_0}{\partial x}\right)^2 - \frac{\partial^\alpha u_0}{\partial t^\alpha},\right]$$

$$u_1(x,0) = 0, u_{1t}(x,0) = 0 \tag{5.5.69}$$

$$p^2: \frac{\partial^2 u_2}{\partial t^2} = \left[\frac{\partial^2 u_1}{\partial t^2} + u_0\left(\frac{\partial^2 u_1}{\partial x^2}\right) + u_1\left(\frac{\partial^2 u_0}{\partial x^2}\right) + 2\left(\frac{\partial u_0}{\partial x}\right)\left(\frac{\partial u_1}{\partial x}\right) - \frac{\partial^\alpha u_1}{\partial t^\alpha}\right],$$

$$u_2(x,0) = 0, u_{2t}(x,0) = 0 \tag{5.5.70}$$

⋮

Table 5.5 Numerical values when $\alpha = 0.5,\ 0.75,\ 1$ for Eq. (5.5.57)

		$\alpha = 1/2$		$\alpha = 3/4$		$\alpha = 1$		
t	x	u_{hpm}	u_{adm}	u_{hpm}	u_{adm}	u_{hpm}	u_{adm}	u_{exact}
	0.25	0.115330	0.110979	0.079678	0.078787	0.050000	0.050000	0.050000
	0.5	0.230659	0.221958	0.159356	0.157574	0.100000	0.100000	0.100000
	0.75	0.345989	0.332937	0.239034	0.236361	0.150000	0.150000	0.150000
00.2	1	0.461319	0.443916	0.318712	0.315148	0.200000	0.200000	0.200000
	0.25	0.172130	0.139706	0.079678	0.128941	0.099998	0.100000	0.100000
	0.5	0.344261	0.279412	0.159356	0.257881	0.199997	0.200000	0.200000
	0.75	0.516391	0.419118	0.239034	0.386822	0.299995	0.300000	0.300000
00.4	1	0.688521	0.558824	0.318712	0.515763	0.399994	0.400000	0.400000
	0.25	0.195479	0.117882	0.079678	0.177253	0.149937	0.150000	0.150000
	0.5	0.390959	0.235764	0.159356	0.354505	0.299874	0.300000	0.300000
	0.75	0.586438	0.353646	0.239034	0.531758	0.449811	0.450000	0.450000
00.6	1	0.781917	0.471528	0.318712	0.709010	0.599748	0.600000	0.600000

The first few components of the solution can be derived as follows:

$$u_0(x,t) = x^2(1-2t), \tag{5.5.71}$$

$$u_1(x,t) = x^2(3t^2 - 4t^3 + 2t^4), \tag{5.5.72}$$

$$\begin{aligned} u_2(x,t) = x^2 \Big(3t^2 - 4t^3 + 5t^4 - 6t^5 + \frac{174}{30}t^6 - \frac{192}{42}t^7 + \frac{168}{56}t^8 \\ + \frac{96}{72}t^9 + \frac{24}{90}t^{10}\Big) \\ + x^2 \left(\frac{-(6)}{\Gamma(5-\alpha)}t^{(4-\alpha)} + \frac{24}{\Gamma(6-\alpha)}t^{(5-\alpha)} \\ - \frac{48}{\Gamma(7-\alpha)}t^{(6-\alpha)}\right) \end{aligned}$$

⋮

Table 5.6 Numerical values when $\alpha = 15,\ 1.75,\ 2$ for Eq. (5.5.66)

		$\alpha = 3/2$		$\alpha = 7/4$		$\alpha = 2$		
t	x	u_{adm}	u_{hpm}	u_{adm}	u_{hpm}	u_{adm}	u_{hpm}	u_{exact}
	0.25	0.60049×10^{-1}	0.50725×10^{-1}	0.48780×10^{-1}	0.47439×10^{-1}	0.43403×10^{-1}	0.43403×10^{-1}	0.43403×10^{-1}
	0.5	2.40195×10^{-1}	2.02899×10^{-1}	1.95119×10^{-1}	1.89755×10^{-1}	1.73610×10^{-1}	1.73610×10^{-1}	1.73611×10^{-1}
00.2	0.75	5.40439×10^{-1}	4.56523×10^{-1}	4.39018×10^{-1}	4.26949×10^{-1}	3.90623×10^{-1}	3.90623×10^{-1}	3.90625×10^{-1}
	1	9.60781×10^{-1}	8.11596×10^{-1}	7.80477×10^{-1}	7.59021×10^{-1}	6.94441×10^{-1}	6.94441×10^{-1}	6.94444×10^{-1}
	0.25	0.74536×10^{-1}	0.50138×10^{-1}	0.45434×10^{-1}	0.41336×10^{-1}	0.31853×10^{-1}	0.31853×10^{-1}	0.31888×10^{-1}
	0.5	2.98142×10^{-1}	2.00552×10^{-1}	1.81736×10^{-1}	1.65344×10^{-1}	1.27412×10^{-1}	1.27412×10^{-1}	1.27551×10^{-1}
00.4	0.75	6.70821×10^{-1}	4.51241×10^{-1}	4.08907×10^{-1}	3.72023×10^{-1}	2.86677×10^{-1}	2.86677×10^{-1}	2.86990×10^{-1}
	1	1.92570×10^{-1}	8.02207×10^{-1}	7.26946×10^{-1}	6.61375×10^{-1}	5.09649×10^{-1}	5.09649×10^{-1}	5.10204×10^{-1}
	0.25	0.91469×10^{-1}	0.49723×10^{-1}	0.45907×10^{-1}	0.36903×10^{-1}	0.23920×10^{-1}	0.23920×10^{-1}	0.24414×10^{-1}
	0.5	3.65878×10^{-1}	1.98892×10^{-1}	1.83626×10^{-1}	1.47611×10^{-1}	0.95680×10^{-1}	0.95680×10^{-1}	0.97656×10^{-1}
00.6	0.75	8.23225×10^{-1}	4.47507×10^{-1}	4.13159×10^{-1}	3.32124×10^{-1}	2.15280×10^{-1}	2.15280×10^{-1}	2.19727×10^{-1}
	1	4.63511×10^{-1}	7.95567×10^{-1}	7.34504×10^{-1}	5.90442×10^{-1}	3.82720×10^{-1}	3.82720×10^{-1}	3.90625×10^{-1}

Similarly, we can extend this process to compute the remaining components. Consequently, we can express the approximate solution for the third-order term in Equation (5.5.66) as follows:

$$u(x,t) = x^2\left(1 - 2t + 6t^2 - 8t^3 + 7t^4 - 6t^5 + \frac{174}{30}t^6 - \frac{192}{42}t^7 + \frac{168}{56}t^8 + \frac{96}{72}t^9 + \frac{24}{90}t^{10}\right) + x^2\left(\frac{-(6)}{\Gamma(5-\alpha)}t^{(4-\alpha)} + \frac{24}{\Gamma(6-\alpha)}t^{(5-\alpha)} - \frac{48}{\Gamma(7-\alpha)}t^{(6-\alpha)}\right) \quad (5.5.74)$$

Table (5.6) shows the solution obtained for different values of α. For $\alpha = 2$ the exact solution is known.

5.6 Conclusion

The modified homotopy perturbation method, as introduced in this paper, represents a particularly efficient and valuable approach for the computation of approximate solutions to a class of complex mathematical problems known as non-linear fractional-order partial differential equations. When applied, this method demonstrates a remarkable capability for achieving high levels of precision in comparison to alternative techniques like the Adomian decomposition method. Moreover, the results obtained through the suggested method exhibit a remarkable proximity to the exact solutions of these intricate equations, underlining its robustness and accuracy in solving such challenging mathematical problems.

References

[1]J.H. He, Homotopy perturbation technique, Comput. Methods. Appl. Mech. Engg. vol. 178, P-257-262, 1999.

[2]Ji-Huan He, A coupling method of a homotopy technique and a perturbation technique for non-linear problems, International Journal of Non-Linear Mechanics, Volume 35, Issue 1,2000, Pages 37-43, ISSN 0020-7462,

[3]He JH, Latifizadeh H. A general numerical algorithm for nonlinear differential equations by the variational iteration method. International Journal of Numerical Methods for Heat and Fluid Flow. 2020 Oct 15;30(11):4797-810.

[4]Sartanpara P, Meher R. Computational study of Klein-Gordan equation using homotopy analysis method. InAIP Conference Proceedings 2022 Oct 7 (Vol. 2451, No. 1). AIP Publishing.

[5]He JH. A simple perturbation approach to Blasius equation. Applied Mathematics and Computation. 2003 Aug 10;140(2-3):217-22.

[6]Jleli M, Kumar S, Kumar R, Samet B. Analytical approach for time fractional wave equations in the sense of Yang-Abdel-Aty-Cattani via the homotopy perturbation transform method. Alexandria Engineering Journal. 2020 Oct 1;59(5):2859-63.

[7]Prajapati VJ, Meher R. Solution of Time-Fractional Rosenau-Hyman Model Using a Robust Homotopy Approach via Formable Transform. Iranian Journal of Science and Technology, Transactions A: Science. 2022 Oct;46(5):1431-44.

[8]Bhalekar S, Daftardar-Gejji V. New iterative method: application

to partial differential equations. Applied Mathematics and Computation. 2008 Sep 15;203(2):778-83.

[9]Elzaki TM, Alamri BA. Projected differential transform method and Elzaki transform for solving system of nonlinear partial differential equations. World Applied Sciences Journal. 2014;32(9):1974-9.

[10] Odibat Z, Momani S, Erturk VS. Generalized differential transform method: application to differential equations of fractional order. Applied Mathematics and Computation. 2008 Apr 1;197(2):467-77.

[11] Wang KL, Wang KJ. A modification of the reduced differential transform method for fractional calculus. Thermal Science. 2018;22(4):1871-5.

[12] Kumar A, Meher R. Solving Non-linear Partial Differential Equations Using Homotopy Analysis Method (HAM). InNonlinear Dynamics and Applications: Proceedings of the ICNDA 2022 2022 Oct 6 (pp. 955-964). Cham: Springer International Publishing.

[13] Avcı I, Mahmudov NI. Numerical solutions for multi-term fractional order differential equations with fractional Taylor operational matrix of fractional integration. Mathematics. 2020 Jan 7;8(1):96.

[14] Abu Arqub O. Application of residual power series method for the solution of time-fractional Schrödinger equations in one-dimensional space. Fundamenta Informaticae. 2019 Jan 1;166(2):87-110.

[15] Wazwaz AM. A reliable modification of Adomian decomposition method. Applied mathematics and computation. 1999 Jul 1;102(1):77-86.

[16] Duan JS, Rach R, Baleanu D, Wazwaz AM. A review of the Adomian decomposition method and its applications to fractional differential equations. Communications in Fractional Calculus. 2012 Oct;3(2):73-99.

[17] Verma L, Meher R. Effect of heat transfer on Jeffery–Hamel

Cu/Ag–water nanofluid flow with uncertain volume fraction using the double parametric fuzzy homotopy analysis method. The European Physical Journal Plus. 2022 Mar 21;137(3):372.

[18] Dhirawat K, Meher R. Semi-analytical Approach to Nonlinear Partial Differential Equations Using Homotopy Analysis Technique (HAM). Contemporary Mathematics. 2023 Oct 24:721-32.

Chapter 6

One-Dimensional Mathematical Model of Pollutant Transport in Rivers with Different Input Sources

Manan A. Maisuria, Priti V. Tandel
Department of Mathematics, Veer Narmad South Gujarat University, Surat - 395007, Gujarat, India

Abstract

In this study, a mathematical model of pollutant movement in a river that considers sources of pollutant input is presented. The current model is investigated for two distinct scenarios: one without an input source and another with an input source. The present study examines the effects of three distinct input source functions, including constant, linear, and exponential forms. These functions are analysed separately. The reduced differential transform method is used to find an approximate analytical solution to the governing equation, which is a one-dimensional advection-diffusion equation. This equation governs the behaviour of the system. In addition, we have examined the convergence of the analytical solution for each case. In each of the situations, both numerical and graphical representations of the concentration levels of the pollutant at different distances and time levels has been addressed. The comparison of pollutant level for the models with and without sources have been examined in 2D and 3D. Also, the pollutant levels for three different input sources have been compared graphically.

Keywords: Input pollutant source, Pollutant transport, Reduced differential transform method, Mathematical modelling.

6.1 Introduction

Poor air quality, land pollution, water pollution, and groundwater contamination are the four types of pollution. Its origin might be natural or man-made. A point source is one form of pollution source [1].

Water contamination is one of the most serious issues facing the area's water resources. The cause of this issue is the growth of the town, the development of industrial units, and the extension of agricultural lands. Because rivers transport water and nutrients to places around them, polluted water impacts human life and is the world's largest source of illness. The discharge of untreated point and nonpoint source contaminants is a cause of water contamination. Pollutants released into a river from a single discrete place or point, such as a factory or domestic wastewater, are referred to as point-source pollution [2].

Water contamination monitoring and evaluation from both a technical and economic perspective is a critical topic in ecosystems [3]. Mass and heat transfer mechanisms are an important class of processes in the geophysics and environmental sciences. Pollution transfer in the environment is a significant aspect of environmental modelling. Advection and diffusion mechanisms are used to convey the particles. The advection-diffusion equation models have shown to be quite effective in understanding and forecasting these environments [4]. In the atmospheric sector, many modelling software linked to the mobility of accidentally released dangerous compounds has been created [5].

In Chemistry, advection refers to the movement of contaminants downstream in a river by bulk water flow, whereas diffusion refers to the distribution of gas through water. A pulse-type source's longitudinal solute transport and time-dependent solute dispersion were solved analytically in a two-dimensional superficial aquifer [6]. Several analytical solutions to the one-dimensional advection–diffusion equation (ADE) with variable dispersion coefficient and velocity could be determined by using Green's function technique [7]. The answer to the inverse issue of identifying the source was established mathematically. The assumption was made that the flow conditions would be unstable and non-uniform. The finite volume approach was used to solve the advection-dispersion equation for sources of unit intensity in varied locations.

This article discusses the 1-D mathematical model developed to study how contaminants flow via rivers. The governing equation is solved using the Reduced Differential Transport Method (RDTM). We have observed two separate scenarios within the framework of this model: one without an input source and one with an input source. The concentration of the pollutant was observed at various points and periods downstream from the point where it was introduced into the river as part of three separate ways that were each assessed. For the present study, we have considered the pollutant substance NO_3 of river Choga. The initial condition of the concentration function is generated from past collected data of this river. The mathematical formulation of this problem is covered in section [1]. The basic concepts of the reduced differential transform method have been mentioned in section [2]. An analytic series solution is obtained using the reduced differential transform approach. Therefore, in the next part (section [3]), we discussed how the convergence of the analytic series solution derived by RDTM could be obtained. The graphical depiction of the pollutant for fixed t (or fixed x) and the 3D graph of each concentration function for each case is discussed in section [4]. The conclusion is included in section [5].

6.2 Mathematical formulation of the problem

The advection-diffusion equation, a partial differential equation of the parabolic type, explains the concentration of pollutants as they are transported from a source through the flow field through an air or water medium. It is accomplished by using the theory of mass conservation as well as Fick's laws of diffusion to the problem [1]. The advection–diffusion equation, derived from the law of conservation of mass for fluid flow, is used to calculate the concentration field of a tracer [9]. It is important to solve the advection-diffusion equation in order to calculate the pollutant's concentration in the stream water [10,11], which is given by

$$A\frac{\partial c}{\partial t} = \frac{\partial}{\partial x}\left(AD\frac{\partial c}{\partial x}\right) - Av\frac{\partial c}{\partial x} - A\beta c + AS \qquad (6.2.1)$$

$c(x,t) =$ Pollutant Concentration (x-distance, t-time) $\left(\frac{mg}{l}\right)$

$A(x) =$ The cross-sectional area of the river at a distance $x(m^2)$

$D =$Diffusion Coefficient$\left(\frac{m^2}{sec}\right)$

$v =$Mean flow velocity$\left(\frac{m}{sec}\right)$

$\beta =$non-conservation coefficient$\left(\frac{1}{sec}\right)$

$S(x,t) =$ input source$\left(\frac{mg}{secm^3}\right)$

For the entire study, the cross-sectional area of the river is considered as a constant. Hence equation (6.2.1) becomes,

$$\frac{\partial c}{\partial t}=\frac{\partial}{\partial x}\left(D\frac{\partial c}{\partial x}\right)-v\frac{\partial c}{\partial x}-\beta c+S \tag{6.2.2}$$

For the study of this problem, we have used concentration of polluting substance of river Choga. Using past collected data, at time (any fixed time), the concentration is assumed as a polynomial of degree two and following values of parameters are used to obtain the solutions [3].

D (Mean diffusion factors) $=0.18350\frac{m^2}{sec}$

v (Mean flow velocity) $=0.68429\frac{m}{sec}$

α (non-conservation coefficient) $=0\ \frac{1}{sec}$

Hence, equation (6.2.2) becomes

$$\frac{\partial c}{\partial t}=0.1835\frac{\partial^2 c}{\partial x^2}-0.68429\frac{\partial c}{\partial x}+S(x,t) \tag{6.2.3}$$

In this article, the concentration level of contamination is discussed for both without and with various kinds of input sources. Various cases have been examined to determine the concentration function's value and visual depiction. We're looking at three distinct sorts of known sources to see the pollution level in rivers. Here, Sdenotes the rate of the pollutant that enters the beginning border of the portion of the river that is being examined per unit of time and per unit of volume. According to the value of $S(x,t)$, these three types of input sources include constant, linear, and exponential sources.

Case-1: Without an input source

For this case, $S(x,t)=0$

Hence from equation (6.2.3), we have

$$\frac{\partial c}{\partial t}=0.1835\frac{\partial^2 c}{\partial x^2}-0.68429\frac{\partial c}{\partial x} \tag{6.2.4}$$

Case-2: With an input source

In this case, we have considered the various forms of input sources as follow:

(a) Constant input source

For this case, the pollutant input is assumed in the form of $S(x,t) = \frac{\alpha}{x^3}$

Hence, from equation (6.2.3), we have

$$\frac{\partial c}{\partial t} = 0.1835\frac{\partial^2 c}{\partial x^2} - 0.68429\frac{\partial c}{\partial x} + \frac{\alpha}{x^3} \tag{6.2.5}$$

(b) Linear input source

This case is discussed when the input pollutant source is lineally entered. i.e. $S(x,t) = \frac{\alpha t}{x^3}$

Hence, from equation (6.2.3), we have

$$\frac{\partial c}{\partial t} = 0.1835\frac{\partial^2 c}{\partial x^2} - 0.68429\frac{\partial c}{\partial x} + \frac{\alpha t}{x^3} \tag{6.2.6}$$

(c) Exponential input source

This model is discussed when the pollutant is entered into the river exponentially. i.e. $S(x,t) = \frac{e^{\alpha t}}{x^3}$

Hence, from equation (6.2.3), we have

$$\frac{\partial c}{\partial t} = 0.1835\frac{\partial^2 c}{\partial x^2} - 0.68429\frac{\partial c}{\partial x} + \frac{e^{\alpha t}}{x^3} \tag{6.2.7}$$

For the present study, we have considered $\alpha = 0.0001, \alpha > 0$ and the initial Condition which is generated from past collected data as follows [3]:

$$c(x,0) = 6e - 09x^2 - 0.0002x + 3.7454 \tag{6.2.8}$$

6.3 Reduced Differential Transform Method

Let $b(\xi,\tau)$ be a two-variable function. Assume that $b(\xi,\tau)$ can be described as the combination of two single variable functions. i.e., $b(\xi,\tau) = m(\xi)l(\tau)$. By properties of differential transform, $b(\omega,\tau)$ can be represented as

$$b(\xi,\tau) = \sum_{i=0}^{\infty} M_{(i)}\,\xi^i \sum_{j=0}^{\infty} L_{(j)}\,\tau^j = \sum_{k=0}^{\infty} B_k\,(\xi)\tau^k \qquad (6.3.1)$$

where, $B_k(\xi)$ is called t-dimensional spectrum function of $b(\xi,\tau)$.

$$B_m(\xi) = \frac{1}{m!}\left[\frac{\partial^m}{\partial\tau^m} b(\xi,\tau)\right]_{\tau=0} \qquad (6.3.2)$$

In this paper, Uppercase letters $B(\xi)$ denote changed functions, whereas lowercase letters $b(\xi,\tau)$ denote original functions. The differential inverse transforms of $B_k(\xi)$ is defined by

$$b(\xi,\tau) = \sum_{m=0}^{\infty} B_m\,(\xi)\tau^m \qquad (6.3.3)$$

From Equations (6.3.2) and (6.3.3), we get

$$b(\xi,\tau) = \sum_{m=0}^{\infty} \frac{1}{m!}\left[\frac{\partial^m}{\partial\tau^m} b(\xi,\tau)\right]_{\tau=0} \tau^m \qquad (6.3.4)$$

Basic concept of RDTM, Consider the nonlinear differential partial differential equation:

$$Lb(\xi,\tau) + Rb(\xi,\tau) + Nb(\xi,\tau) = v(\xi,\tau) \qquad (6.3.5)$$

with initial condition $b(\xi,0) = y(\xi)$, where $L = \frac{\partial}{\partial\tau}, Rb(\xi,\tau)$ is a linear operator which has partial derivatives, while $Nb(\xi,\tau)$ is a non-linear term, and $v(\xi,\tau)$ is an inhomogeneous term. By applying the transform on equation (6.3.5), we get

$$(k+1)B_{k+1}(\xi) = \Psi_k(\xi) - RB_k(\xi) - NB_k(\xi) \qquad (6.3.6)$$

where $B_m(\xi)$, $\Psi_k(\xi), RB_k(\xi)$ and $NB_k(\xi)$ are transform of $b(\xi,\tau), v(\xi,\tau), Rb(\xi,\tau)$ and $Nb(\xi,\tau)$ respectively. From initial condition, we can write

$$B_0(\xi) = y(\xi) \tag{6.3.7}$$

From equations (6.3.6) and (6.3.7), we get values of $B_k(\xi)$. After that, an approximation solution is produced by carrying out an inverse transformation on the set of values $\{B_k(\xi)\}_{k=0}^{n}$. This transformation yields an approximation solution as

$$\tilde{b}_n(\xi,\tau) = \sum_{k=0}^{n} B_k\,(\xi)\tau^m \tag{6.3.8}$$

Where n is an approximation order solution. Hence, the exact solution to the problem is stated by

$$b(\xi,\tau) = \lim_{n\to\infty} \tilde{b}_n(\xi,\tau) \tag{6.3.9}$$

Table 6.1

Function	Transformation
$b(\xi,\tau)$	$B_m(\xi) = \frac{1}{m!}\left[\frac{\partial^m}{\partial\tau^m} b(\xi,\tau)\right]_{\tau=0}$
$\alpha f(\xi,\tau) \pm \beta g(\xi,\tau)$	$\alpha F_m(\xi) + \beta G_m(\xi)$
$\xi^k \tau^n$	$\xi^k \delta(m-n)$
$\xi^k \tau^n b(\xi,\tau)$	$\xi^k B_{m-n}(\xi)$
$l(\xi,\tau) = f(\xi,\tau)g(\xi,\tau)$	$L_m(\xi) = \sum_{r=0}^{m} F_r\,(\xi)G_{m-r}(\xi)$
$\frac{\partial^r}{\partial\tau^r} b(\xi,\tau)$	$\frac{(m+r)!}{m!} B_{m+r}(\xi)$
$\frac{\partial}{\partial\xi} b(\xi,\tau)$	$\frac{\partial}{\partial\xi} B_m(\xi)$

6.4 Convergence of RDTM

To understand the convergence, let us consider the solution of equation (6.3.5) in power series form as follows:

$$b(\xi,\tau) = \sum_{n=0}^{\infty} B_n\,(\xi)\tau^n = \sum_{n=0}^{\infty} B_n\,\tau^n \tag{6.4.1}$$

which is obtained by equation (6.3.9) [15,16,17].

Theorem 6.4.1: If $\sum_{n=0}^{\infty} B_n \tau^n$ is given series,
[1] $\exists 0 < \beta < 1 \ni \frac{\|B_{n+1}\|}{\|B_n\|} \leq \beta \Rightarrow$ series is convergent.
[2] $\exists \beta > 1 \ni \frac{\|B_{n+1}\|}{\|B_n\|} \geq \beta \Rightarrow$series is divergent.

Proof: Let $(C[l], \|, \|)$ represent the Banach space that contains all continuous functions on l that satisfy the norm $\|.\|$. Also, let's suppose that $\|B_0(\xi)\| \leq M$, where M is an integer in the positive range. Let $\{\delta_n\}_{n=0}^{\infty}$be a partial sum

$$\delta_n = B_0 + B_1 + B_2 + \ldots + B_n$$

If we can prove that $\{\delta_n\}_{n=0}^{\infty}$ is a Cauchy sequence in Banach Space, then we can conclude that the sequence of partial sum is convergent in Banach Space. As a result, at this point, we shall demonstrate that the series of partial sums follows the Cauchy sequence. We take

$$\|\delta_{n+1} - \delta_n\| = \|B_{n+1}\| \leq \beta\|B_n\| \leq \ldots \leq \beta^{n+1}\|B_0\| \leq \beta^{n+1}M$$

Therefore, $\forall n, m \in N, n \geq m$, We have

$$\begin{aligned} \|\delta_n - \delta_m\| &= \|(\delta_n - \delta_{n-1}) + (\delta_{n-1} - \delta_{n-2}) + \ldots + (\delta_{m+1} - \delta_m)\| \\ &\leq \|\delta_n - \delta_{n-1}\| + \|\delta_{n-1} - \delta_{n-2}\| + \ldots + \|\delta_{m+1} - \delta_m\| \\ &\leq \frac{1-\beta^{n-m}}{1-\beta} \beta^{m+1}\|B_0\| \end{aligned} \quad (6.4.2)$$

Now, here $0 < \beta < 1$, we obtained

$$\lim_{n,m \to \infty} \|\delta_n - \delta_m\| = 0$$

Hence, $\{\delta_n\}_{n=0}^{\infty}$ is Cauchy sequence in Banach Space. Therefore, given series is convergent.

6.5 Results and Discussion

Now applying RDTM on equation (6.2.3), we get

$$(n+1)C_{n+1} = 0.1835\frac{\partial^2 C_n}{\partial x^2} - 0.68429\frac{\partial C_n}{\partial x} + \chi_n; n \in N \cup \{0\} \quad (6.5.1)$$

Where χ_n is transform of $S(x,t)$. Now we discuss the solution obtained from RDTM for different form of pollutant (NO_3) sources added in river.

Case 1: Without an input source

Solving equation (6.2.4) by RDTM with initial condition (6.2.8), we get

$$c(x,t) = M_0 + M_1 t + M_2 t^2 + M_3 t^3 + M_4 t^4 + \ldots \tag{6.5.2}$$

Where

$$\begin{aligned} M_0 &= 6e-09x^2 - 0.0002x + 3.7454 \\ M_1 &= 1.3686e-04 - 8.2115e-09x \\ M_2 &= 2.8095e-09 \\ M_3 &= 0 \\ M_4 &= 0 \end{aligned}$$

Here,

$$\frac{||M_1||}{||M_0||} = 2.0487e-05 < 1, \frac{||M_2||}{||M_1||} = 6.0376e-05 < 1, \frac{||M_3||}{||M_2||} = 0 < 1$$

Therefore, the solution function $c(x,t)$ is convergent. Also, given solution is exact.

Table 6.2 $c(x,t)$ for Case-1

$x(m)$ \ $t(sec)$	0	1600	3200	4800	6400	8000
400	3.66636	3.887273	4.122571	4.372254	4.636322	4.914774
1400	3.47716	3.684935	3.907095	4.143639	4.394568	4.659882
2400	3.29996	3.494597	3.703618	3.927024	4.164815	4.41699
3400	3.13476	3.316258	3.512141	3.722409	3.947061	4.186098
4400	2.98156	3.14992	3.332664	3.529794	3.741308	3.967207
5400	2.84036	2.995581	3.165188	3.349179	3.547554	3.760315
6400	2.71116	2.853243	3.009711	3.180564	3.365801	3.565423
7400	2.59396	2.722905	2.866234	3.023948	3.196047	3.382531
8400	2.48876	2.604566	2.734758	2.879333	3.038294	3.211639
9400	2.39556	2.498228	2.615281	2.746718	2.89254	3.052747
10400	2.31436	2.40389	2.507804	2.626103	2.758787	2.905856

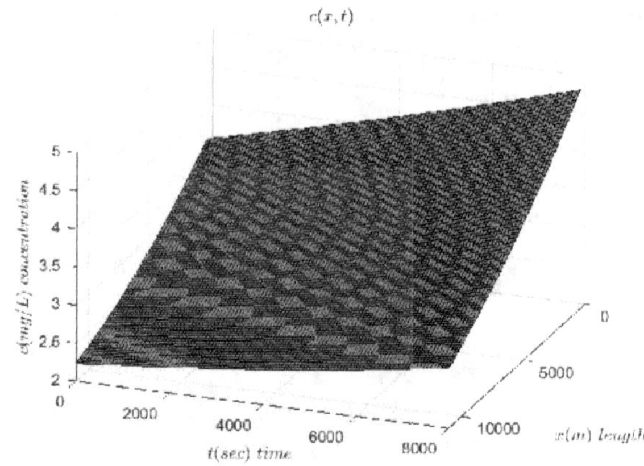

Figure 6.1 Graphical representation of $c(x,t)$ for Case-1.

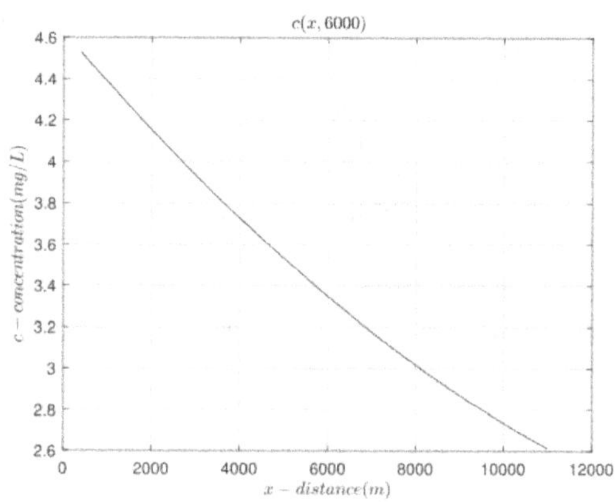

Figure 6.2 $c(x,t)$ for t=6000 second.

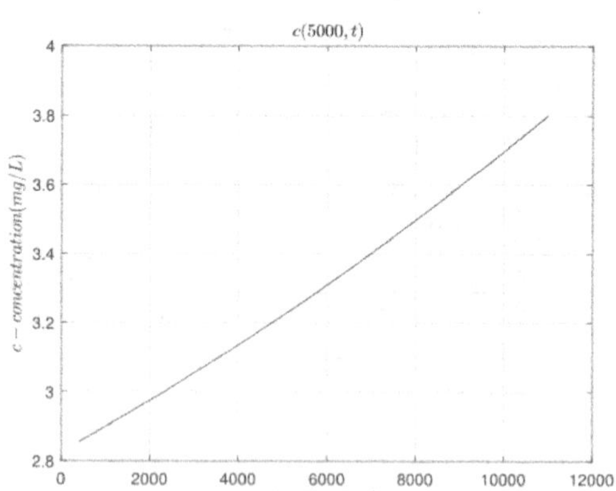

Figure 6.3 $c(x,t)$ for x=5000 m.

Case 2: With an input source

(a) Constant input source

Using RDTM, solving equation (6.2.5) with the initial condition (6.2.8), we get

$$c(x,t) = A_0 + A_1 t + A_2 t^2 + A_3 t^3 + A_4 t^4 + \ldots \tag{6.5.3}$$

Where

$$\begin{aligned}
A_0 &= 3.7454 - 2e - 04x + 6e - 09x^2 \\
A_1 &= 0.0001x^{-3} - 8.2115e - 09x + 1.3686e - 04 \\
A_2 &= 1.0264e - 04x^{-4} + 1.1010e - 04x^{-5} + 2.8095e - 09 \\
A_3 &= 9.3651e - 05x^{-5} + 2.5113e - 04x^{-6} + 2.0203e - 04x^{-7} \\
A_4 &= 8.0105e - 05x^{-6} + 3.8666e - 04x^{-7} + 7.2581e - 04x^{-8} + 5.1902e - 04
\end{aligned}$$

Here

$$\frac{||A_1||}{||A_0||} = 3.6433e - 05 < 1, \frac{||A_2||}{||A_1||} = 2.1033e - 05 < 1,$$
$$\frac{||A_3||}{||A_2||} = 3.2771e - 09 < 1, \frac{||A_4||}{||A_3||} = 0.0021 < 1$$

Therefore, the solution function $c(x,t)$ is convergent.

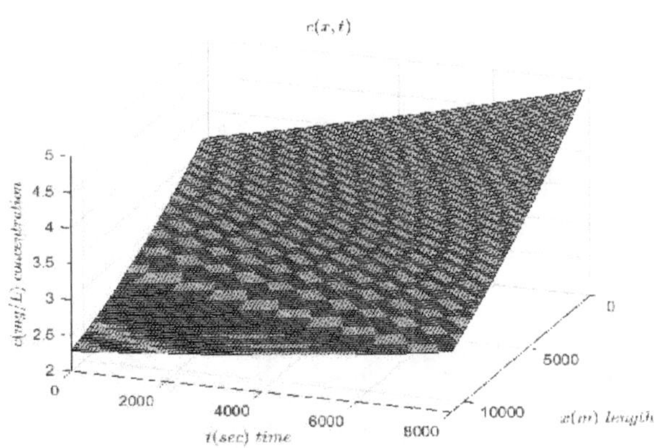

Figure 6.4 Graphical representation of $c(x,t)$ for Case-2(a).

Table 6.3 $c(x,t)$ for Case-2(a)

$x(m)$ \\$t(sec)$	0	1600	3200	4800	6400	8000
400	3.66636	3.887274	4.122574	4.372266	4.636358	4.91486
1400	3.47716	3.684935	3.907095	4.143639	4.394568	4.659882
2400	3.29996	3.494597	3.703618	3.927024	4.164815	4.41699
3400	3.13476	3.316258	3.512141	3.722409	3.947061	4.186098
4400	2.98156	3.14992	3.332664	3.529794	3.741308	3.967207
5400	2.84036	2.995581	3.165188	3.349179	3.547554	3.760315
6400	2.71116	2.853243	3.009711	3.180564	3.365801	3.565423
7400	2.59396	2.722905	2.866234	3.023948	3.196047	3.382531
8400	2.48876	2.604566	2.734758	2.879333	3.038294	3.211639
9400	2.39556	2.498228	2.615281	2.746718	2.89254	3.052747
10400	2.31436	2.40389	2.507804	2.626103	2.758787	2.905856

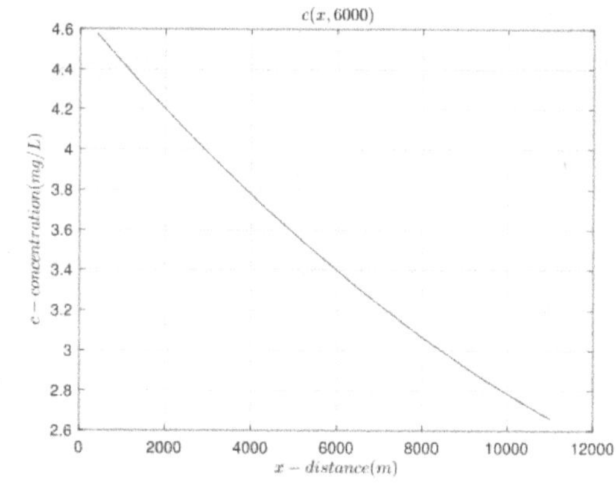

Figure 6.5 $c(x,t)$ for t=6000 second.

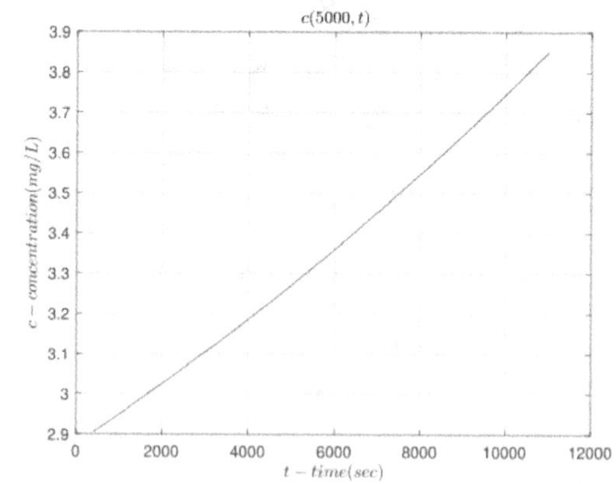

Figure 6.6 $c(x,t)$ for x=5000 m.

(b) Linear input source

Solving equation (6.2.6) using RDTM with the initial condition (6.2.8), we get

$$c(x,t) = K_0 + K_1 t + K_2 t^2 + K_3 t^3 + K_4 t^4 + \ldots \quad (6.5.4)$$

Where,

$$K_0 = 3.7454 - 2e - 4x + 6e - 9x^2$$

$$K_1 = 1.3686e - 04 - 8.2115e - 09x$$

$$K_2 = 5.0e - 05x^{-3} + 2.8095e - 09$$

$$K_3 = 3.4215e - 05x^{-4} + 3.6700e - 05$$

$$K_4 = 2.3413e - 05x^{-5} + 6.2784e - 05x^{-6} + 5.0508e - 05x^{-7}$$

Hence,

$$\frac{\|K_1\|}{\|K_0\|} = 3.6433e - 05 < 1, \frac{\|K_2\|}{\|K_1\|} = 2.1039e - 05 < 1, \frac{\|K_3\|}{\|K_2\|} = 4.7685e - 07 < 1, \frac{\|K_4\|}{\|K_3\|} = 0.0017 < 1$$

Therefore, the solution function $c(x,t)$ is convergent.

Table 6.4 $c(x,t)$ for Case-2(b)

$x(m)$ \ $t(sec)$	0	1600	3200	4800	6400	8000
400	3.66636	3.887296	4.122865	4.373642	4.640567	4.924938
1400	3.47716	3.684935	3.907096	4.143643	4.394579	4.659906
2400	3.29996	3.494597	3.703618	3.927024	4.164816	4.416992
3400	3.13476	3.316258	3.512141	3.722409	3.947062	4.186099
4400	2.98156	3.14992	3.332664	3.529794	3.741308	3.967207
5400	2.84036	2.995582	3.165188	3.349179	3.547554	3.760315
6400	2.71116	2.853243	3.009711	3.180564	3.365801	3.565423
7400	2.59396	2.722905	2.866234	3.023948	3.196047	3.382531
8400	2.48876	2.604566	2.734758	2.879333	3.038294	3.211639
9400	2.39556	2.498228	2.615281	2.746718	2.89254	3.052747
10400	2.31436	2.40389	2.507804	2.626103	2.758787	2.905856

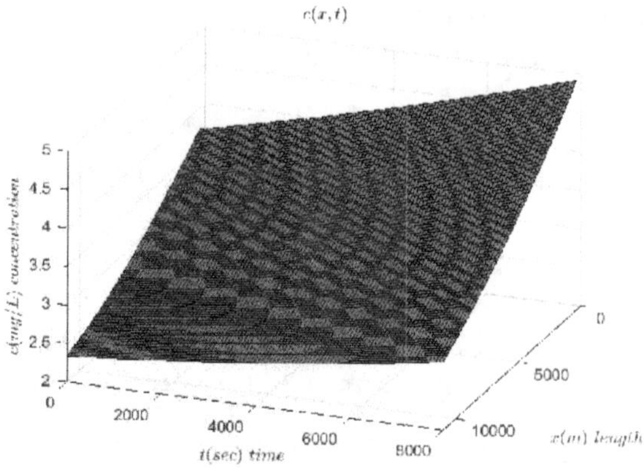

Figure 6.7 Graphical representation of $c(x,t)$ for Case-2(b).

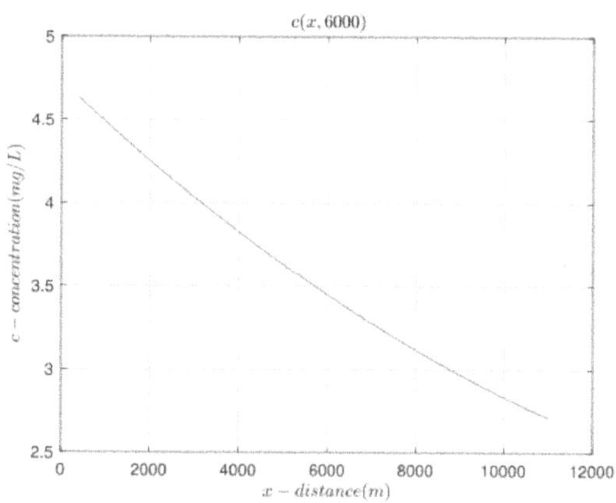

Figure 6.8 $c(x,t)$ for t=6000 second.

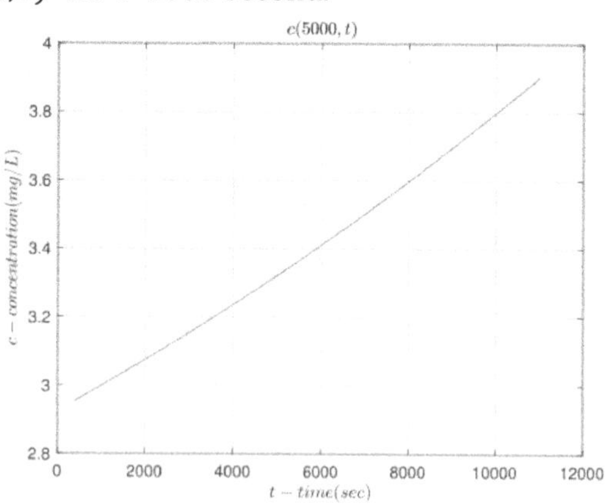

Figure 6.9 $c(x,t)$ for x=5000 m.

(c) Exponential input source

Using RDTM, solving equation (6.2.7) with the initial condition (6.2.8), we get

$$c(x,t) = L_0 + L_1 t + L_2 t^2 + L_3 t^3 + L_4 t^4 + \ldots \tag{6.5.5}$$

Where

$$\begin{aligned}
L_0 &= 3.7454 - 2e - 04x + 6e - 09x^2 \\
L_1 &= x^{-3} - 8.2115e - 09x + 1.3686e - 04 \\
L_2 &= 5e - 05x^{-3} + 1.0264x^{-4} + 1.101x^{-5} + 2.8095e - 09 \\
L_3 &= 1.6667e - 09x^{-3} + 3.4215e - 05x^{-4} + 0.9365x^{-5} + 2.5113x^{-6} + 2.0203x^{-7} \\
L_4 &= 4.1667e - 14x^{-3} + 8.5536e - 10x^{-4}2.3414e - 05x^{-5} + 0.8011x^{-6} + 3.8666x^{-7} \\
&\quad + 7.2581x^{-8} + 5.1902x^{-9}
\end{aligned}$$

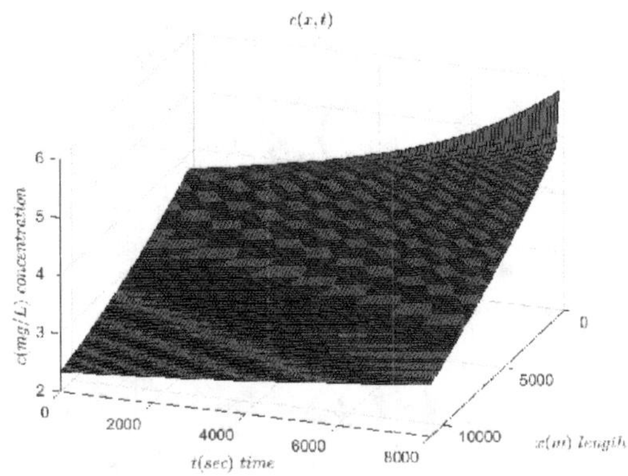

Figure 6.10 Graphical representation of $c(x,t)$ for Case-2(c).

Hence,

$$\frac{||L_1||}{||L_0||} = 3.6437e - 05 < 1, \frac{||L_2||}{||L_1||} = 2.1337e - 05 < 1,$$

$$\frac{||L_3||}{||L_2||} = 3.2779e - 05 < 1, \frac{||L_4||}{||L_3||} = 0.0021 < 1$$

Therefore, the solution function $c(x,t)$ is convergent.

Table 6.5 $c(x,t)$ for Case-2(c)

$x(m)$ $\backslash t(sec)$	0	1600	3200	4800	6400	8000
400	3.66636	3.889099	4.147103	4.489922	4.998602	5.785693
1400	3.47716	3.684938	3.907116	4.143727	4.394817	4.660454
2400	3.29996	3.494597	3.703619	3.927029	4.164828	4.417018
3400	3.13476	3.316258	3.512142	3.72241	3.947063	4.186103
4400	2.98156	3.14992	3.332665	3.529794	3.741308	3.967208
5400	2.84036	2.995582	3.165188	3.349179	3.547555	3.760315
6400	2.71116	2.853243	3.009711	3.180564	3.365801	3.565423
7400	2.59396	2.722905	2.866234	3.023948	3.196047	3.382531
8400	2.48876	2.604566	2.734758	2.879333	3.038294	3.211639
9400	2.39556	2.498228	2.615281	2.746718	2.89254	3.052747
10400	2.31436	2.40389	2.507804	2.626103	2.758787	2.905856

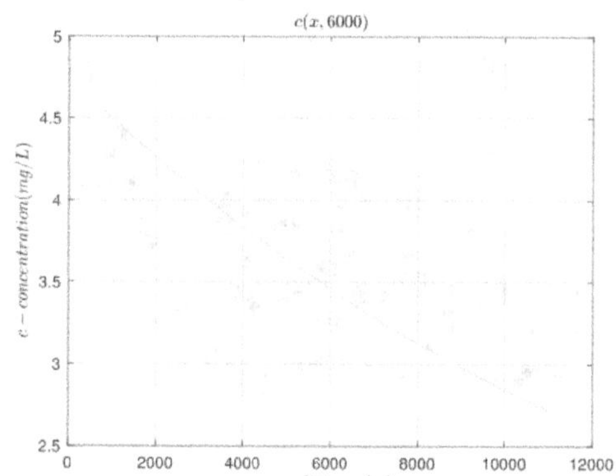

Figure 6.11 $c(x,t)$ for t=6000 second.

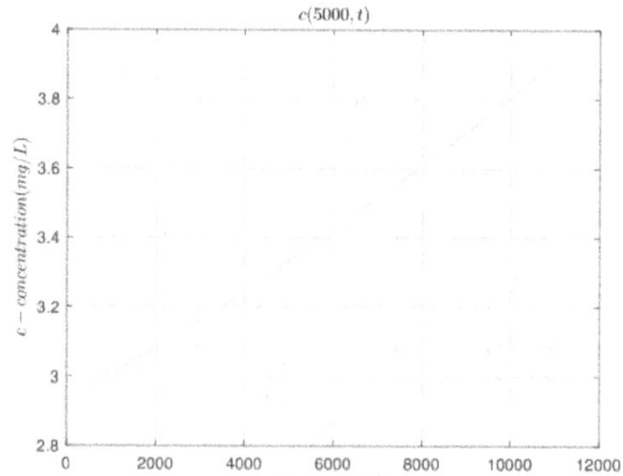

Figure 6.12 $c(x,t)$ for $x = 5000\ m$.

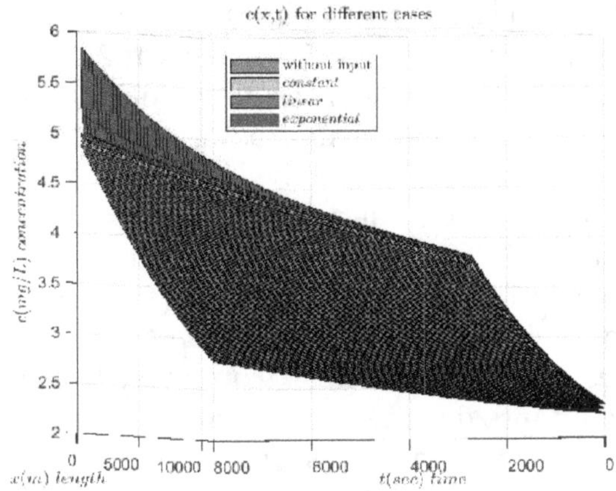

Figure 6.13 3D-Graphical comparison of $c(x, t)$.

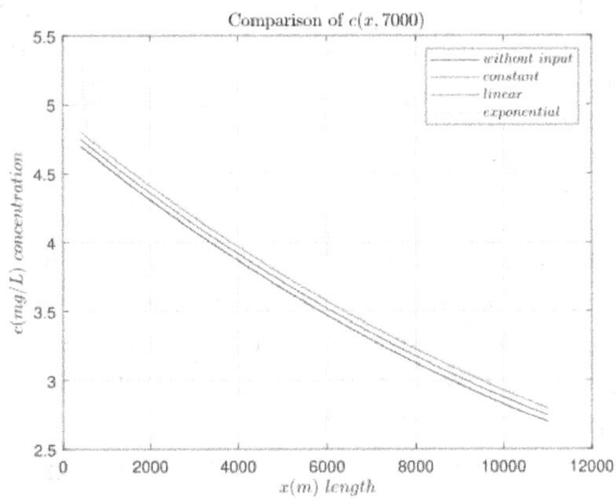

Figure 6.14 2D-Graphical comparison of $c(x, 7000)$.

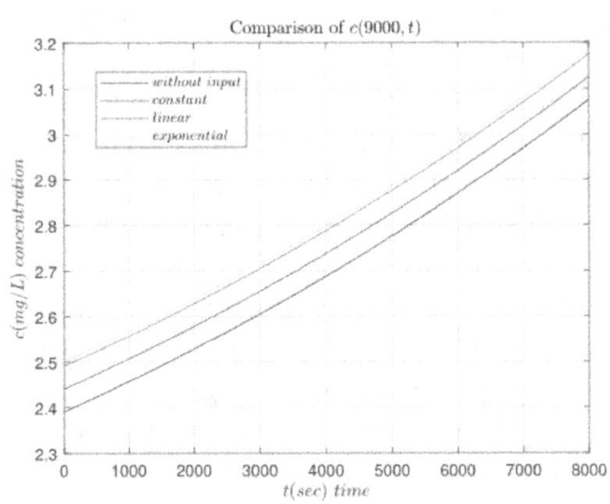

Figure 6.15 2D-Graphical comparison of $c(9000, t)$.

The values of the concentration at various times and distances are shown in tables 6.2, 6.3, 6.4 and 6.5 for cases 1, 2(a), 2(b) and 2(c) respectively. A three-dimensional graphical depiction of the concentration function is shown in figures 6.1, 6.4, 6.7 and 6.10 for cases 1, 2(a), 2(b) and 2(c) respectively. Graphs 6.2, 6.5, 6.8 and 6.11 depict the concentration at a fixed time (t = 6000 second) for cases 1, 2(a), 2(b) and 2(c) respectively. The graphical representation of the concentration at a given distance (x = 5000 m) is shown in figure 6.3, figure 6.6, figure 6.9 and figure 6.12 for cases 1, 2(a), 2(b) and 2(c) respectively. It is clear from these data that the concentration of pollutants declines with increasing distance and that the concentration of pollutants rises with increasing time.

Figure 6.13 gives a three-dimensional comparison of the solution concentration function of each case. Figures 6.14 and 6.15 give a 2D dimension comparison of the concentration function of each case for fixed t = 7000 sec and x = 9000 m, respectively.

6.6 Conclusion

We have analysed the present problem for two different cases: one without an input source and one with an input source. We come up with solutions for three distinct kinds of input pollution sources: linear, constant, and exponential. We have observed that the value of concentration acquired from an exponential input source is greater than the value of concentration obtained from a linear input source for a variety of combinations of distance and time. When accounting for differences in distance and time, the value of concentration that can be acquired from a linear input source is shown to be greater than the value of concentration that can be obtained from a constant input source. The value of concentration that can be acquired when the input source remains constant is greater than the value obtained when the input source is eliminated throughout a range of distances and times. Hence, we conclude that the pollutant level is higher in the model with an input source compared to the model without an input source. Also, we conclude that in the model with various input sources, the concentration level in the exponential form is higher compared to a linear and constant input source. Through the use of the Reduced Differential Transform Method, we can solve an advection-diffusion equation in one dimension in this study. The initial condition is obtained from the previously collected data in a second-degree polynomial. We have obtained the exact solution for the case without an input source, which shows the advantage of this method. This approach has the potential to be expanded to include advection-diffusion equations in both two and three-dimensions.

Acknowledgements

Manan A. Maisuria is thankful to the Council of Scientific and Industrial Research (CSIR), India, for his Junior Research Fellowship.

References

[1]Yadav, S. K., Kumar, A., Jaiswal, D. K., Kumar, N., One-dimensional unsteady solute transport along unsteady flow through inhomogeneous medium, Journal of earth system science, 120, 205-213(2011).

[2]Manitcharoen, N., Pimpunchat, B., Analytical and numerical solutions of pollution concentration with uniformly and exponentially increasing forms of sources, Journal of Applied Mathematics, 2020, 1-9(2020).

[3]Kachiashvili, K., Gordeziani, D., Lazarov, R., Melikdzhanian, D., Modeling and simulation of pollutants transport in rivers, Applied mathematical modelling, 31, 1371-1396(2007).

[4]Union, J. I. G., Advection diffusion equation models in near-surface geophysical and environmental sciences, J. Ind. Geophys. Union, 17, 117-127(2013).

[5]Lee, M. E., Seo, I. W., 2D finite element pollutant transport model for accidental mass release in rivers, KSCE Journal of Civil Engineering, 14, 77-86(2010).

[6]Yadav, S. K., Kumar, A., Kumar, N., Horizontal solute transport from a pulse type source along temporally and spatially dependent flow: Analytical solution, Journal of Hydrology, 412, 193-199(2012).

[7]Sanskrityayn, A., Kumar, N., Analytical solution of advection-diffusion equation in heterogeneous infinite medium using Green's function method, Journal of Earth System Science, 125, 1713-1723(2016).

[8]Mazaheri, M., Mohammad Vali Samani, J., Samani, H. M. V., Mathematical model for pollution source identification in rivers, Environmental Forensics, 16(4), 310-321(2015).

[9] Pereira, F. F., Fragoso Jr, C. R., Uvo, C. B., Collischonn, W., Motta Marques, D. M. L., Assessment of numerical schemes for solving the advection-diffusion equation on unstructured grids: case study of the Guaíba River, Brazil, Nonlinear Processes in Geophysics, 20, 1113-1125(2013).

[10] Kachiashvili, K., Gordeziani, D., Melikdzhanian, D., Mathematical models of Pollutants Transport with Allowance for Many Affecting Pollution Sources, In Urban Drainage Modeling, 692-700(2001).

[11] Marziali, L., Valsecchi, L., Schiavon, A., Mastroianni, D., Viganò, L., Vertical profiles of trace elements in a sediment core from the Lambro River (northern Italy): Historical trends and pollutant transport to the Adriatic Sea, Science of the Total Environment, 782, 146766(2021).

[12] Al-Amr, M. O., New applications of reduced differential transform method, Alexandria Eng. J, 53, 243-247(2014).

[13] Keskin, Y., Oturanc, G., Reduced differential transform method for generalized KdV equations, Mathematical and Computational applications, 15, 382-393(2010).

[14] Srivastava, V. K., Awasthi, M. K., Kumar, S., Analytical approximations of two and three dimensional time-fractional telegraphic equation by reduced differential transform method, Egyptian Journal of Basic and Applied Sciences, 1, 60-66(2014).

[15] Moosavi Noori, S. R., Taghizadeh, N., Study of convergence of reduced differential transform method for different classes of differential equations, International Journal of Differential Equations, 2021, 1-16(2021).

[16] Saeed, R. K., Mustafa, A. A., Numerical solution of Fisher-KPP equation by using reduced differential transform method, 020045, AIP Publishing LLC, In AIP Conference Proceedings (2017).

[17] Maisuria, M. A., Tandel, P. V., Mathematical modelling and application of reduced differential transform method for river pollution, Ratio Mathematica, 47,342-366(2023).

Chapter 7

A New Computational Technique for FitzHugh-Nagumo Model

Manikandan Mariappan
Department of Mathematics, School of Engineering Presidency University, Bengaluru – 560064, Karnataka, India

Abstract

In this chapter, a new computational technique is developed for FitzHugh-Nagumo model. The main objective of this chapter is to develop a computational algorithm which works well for all values of the parameter ϵ occurring in FitzHugh-Nagumo model. The technique is developed in such a way that it resolves the FitzHugh-Nagumo model when the parameter ϵ takes a large value and it traces the singular perturbation nature of a two-time scale system when the parameter ϵ takes a small value. Illustrations included well agree with the literature on both the FitzHugh-Nagumo model and a two-time scale system.

Keywords: FitzHugh-Nagumo model, a two-time scale system, finite difference scheme, the continuation method.

7.1 Introduction

FitzHugh-Nagumo model narrates a prototype of an excitable system. The nonlinear system of differential equations given below in (7.2.1)-(7.2.2) represents the most general form of FitzHugh-Nagumo model in which the component $w(t)$ denotes the recovery variable and the component $v(t)$ denotes the membrane potential. The parameter ϵ occurring in this model takes a large value [1], [2]. In this case the component $v(t)$ changes rapidly near $t = 0$ whereas the component $w(t)$ is smooth and linear throughout the domain.

On the other hand, when the parameter ϵ occurring in this model takes a small value then it represents a two-time scale system [3], [4]. Here the component $w(t)$ denotes the fast state and the component $v(t)$ denotes the slow state. In this case the component $w(t)$ changes rapidly near $t = 0$ whereas the component $v(t)$ is smooth throughout the domain.

Hence it is obvious that based on the magnitude of the parameter ϵ the components $w(t)$ and $v(t)$ behave totally opposite in FitzHugh-Nagumo model and in a two-time scale system.

Numerical methods for FitzHugh-Nagumo model are constructed in [5], [6] and for a two-time scale system in [7]. These methods are developed either for FitzHugh-Nagumo model or for a two-time scale system only.

Developing a new computational technique which works well for the nonlinear system represented in (7.2.1)-(7.2.2), irrespective of the magnitude of the parameter ϵ and helps to resolve both FitzHugh-Nagumo model and a two-time scale system is highly complicated. No technique is available in the literature which involves both FitzHugh-Nagumo model and a two-time scale system.

In the present work, a new computational technique is developed for the nonlinear system mentioned in (7.2.1)-(7.2.2) which works well for all values of the parameter ϵ. A new type of piecewise-uniform mesh based on a parameter η which involves the parameter ϵ is constructed in this work. This helps to resolve both FitzHugh-Nagumo model and a two-time scale system, irrespective of the magnitude of the parameter ϵ.

It is worth observing that all the characteristics of both FitzHugh-Nagumo model and a two-time scale system reported in the literature are preserved by the present computational technique.

From Tables 1-9, in this present work, it is evident that the maximum point-wise error (D^N) decreases monotonically and on the other hand the order of convergence (p^N) increases when the number of mesh points N is increased. This shows the consistency of the present computational technique.

7.2 The main problem

The following nonlinear system of differential equations is considered in this chapter.

$$\epsilon \frac{dw(t)}{dt} + b\,w(t) + v(t) - a = 0 \tag{7.2.1}$$

$$\frac{dv(t)}{dt} + \epsilon\left(\frac{(v(t))^3}{3} - v(t) - w(t) - I\right) = 0, \quad t \in (0,1]$$
with $w(0) = \alpha$ and $v(0) = \beta$. (7.2.2)

Here α, β, a, b, I and ϵ are known values. The nonlinear system (7.2.1)-(7.2.2) is written as follows

$$\epsilon \frac{dw(t)}{dt} + f\big(t, w(t), v(t)\big) = 0 \tag{7.2.3}$$

$$\frac{dv(t)}{dt} + g\big(t, w(t), v(t)\big) = 0, \quad t \in (0,1]$$
with $w(0) = \alpha$ and $v(0) = \beta$ (7.2.4)

where $f\big(t, w(t), v(t)\big) = b\,w(t) + v(t) - a$, and
$g\big(t, w(t), v(t)\big) = \epsilon\left(\frac{(v(t))^3}{3} - v(t) - w(t) - I\right)$.

7.3 Mesh and the discrete problem

To capture both the characteristics of FitzHugh-Nagumo model and the singular perturbation nature of a two-time scale system, on [0,1] a piecewise-uniform is constructed based on a parameter η such that the magnitude of the parameter η is determined in terms of the parameter ϵ associated with the nonlinear system (7.2.3)-(7.2.4). On [0,1], a piecewise-uniform mesh with N mesh-intervals is designed as follows. [0,1] is divided into 2 sub-intervals $[0,\eta]$ and $(\eta,1]$ such that $[0,1] = [0,\eta] \cup (\eta,1]$. The parameter η is defined to be

$$\eta = \min\left\{\frac{1}{2}, \epsilon \ln N\right\}.$$

From the total N mesh points, $\frac{N}{2}$ mesh points are placed uniformly on each of the sub-domains $[0,\eta]$ and $[\eta,1]$.

The discrete problem corresponding to (7.2.3)-(7.2.4) is defined to be

$$\epsilon\, D^- W(t_j) + f\left(t_j, W(t_j), V(t_j)\right) = 0 \tag{7.3.1}$$

$$\epsilon\, D^- V(t_j) + g\left(t_j, W(t_j), V(t_j)\right) = 0, \quad j = 1,2,\dots,N$$

with $W(t_0) = w$ and $V(t_0) = v(0)$. (7.3.2)

Here $D^- \psi(t_j) = \frac{\psi(t_j) - \psi(t_{j-1})}{h_j}$, $h_j = t_j - t_{j-1}$.

7.4 The continuation method

The nonlinear system of ordinary differential equations in (7.2.3)-(7.2.4) is modified to an artificial nonlinear system of partial differential equations as given below. For $(t,x) \in (0,1] \times (0,X]$,

$$-\frac{\partial w(t,x)}{\partial x} + \epsilon\,\frac{\partial w(t,x)}{\partial t} + f(t, w(t,x), v(t,x)) = 0, \tag{7.4.1}$$

$$-\frac{\partial v(t,x)}{\partial x} + \frac{\partial v(t,x)}{\partial t} + g(t, w(t,x), v(t,x)) = 0$$

with $\begin{bmatrix} w(0,x) \\ v(0,x) \end{bmatrix} = \begin{bmatrix} \alpha \\ \beta \end{bmatrix}$, for $x \geq 0$ and $\begin{bmatrix} w(t,0) \\ v(t,0) \end{bmatrix} = \begin{bmatrix} w_{init}(t) \\ v_{init}(t) \end{bmatrix}$, for $0 < t \leq 1$. (7.4.2)

The discrete problem corresponding to (7.4.1)-(7.4.2) is given by, for $j = 1,2,\dots,N$ and $k = 1,2,\dots,K$,

$$-D_x^- W(t_j, x_k) + \epsilon\, D_t^- W(t_j, x_k) + f\left(t_j, W(t_j, x_{k-1}), V(t_j, x_{k-1})\right) = 0, \tag{7.4.3}$$

$$-D_x^- V(t_j, x_k) + D_t^- V(t_j, x_k) + g\left(t_j, W(t_j, x_{k-1}), V(t_j, x_{k-1})\right) = 0$$

with $\begin{bmatrix} W(t_0, x_k) \\ V(t_0, x_k) \end{bmatrix} = \begin{bmatrix} w(t_0) \\ v(t_0) \end{bmatrix}$, for all k (7.4.4)

and $\begin{bmatrix} W(t_j, x_0) \\ V(t_j, x_0) \end{bmatrix} = \begin{bmatrix} w_{init}(t_j) \\ v_{init}(t_j) \end{bmatrix}$, for all $j = 0,1,\dots,N$ (7.4.5) where $D_t^-\Theta(t_j, x_k) = \frac{\Theta(t_j, x_k) - \Theta(t_{j-1}, x_k)}{h_j}$, $D_x^-\Theta(t_j, x_k) = \frac{\Theta(t_j, x_k) - \Theta(t_j, x_{k-1})}{h_k}$.

The initial guess $\begin{bmatrix} w_{init}(t) \\ v_{init}(t) \end{bmatrix}$ is taken to be $\begin{bmatrix} w_{init}(t) \\ v_{init}(t) \end{bmatrix} = \begin{bmatrix} (1-t)\alpha \\ (1-t)\beta \end{bmatrix}$.

Further, $h_x = x_k - x_{k-1}$ and K are chosen as follows. Define, for $k = 1,2, \dots, K$,

$$E(k) = max_j \left(\frac{|W(t_j, x_k) - W(t_j, x_{k-1})|}{h_x}, \frac{|V(t_j, x_k) - V(t_j, x_{k-1})|}{h_x} \right). \tag{7.4.6}$$

Then h_x is chosen sufficiently small so that $E(k)$ decreases with the increasing k. Precisely,

$$E(k) \leq E(k-1) \text{ for all } k \tag{7.4.7}$$

and

$$E(K) \leq tol \tag{7.4.8}$$

where tol is a prescribed small tolerance. Algorithm given below is used to compute the numerical approximations for (7.4.1)-(7.4.2).

7.5 Algorithm

Step 1: Let $h_x = 1$.

Step 2: Suppose (7.4.7) is not satisfied for a particular k, then quit the current step and begin from x_{k-1} with h_x as $h_x/2$. Continue halving h_x until finding a h_x to which (7.4.7) is satisfied.

Step 3: If (7.4.7) is satisfied at each h_x, then continue the procedure until either (7.4.8) is satisfied or $K = 100$.

Step 4: If (7.4.8) is not satisfied, then it is assumed that the stepping process is stalled due to the choice of a large h_x. In such a case, the entire process is repeated from x_0 with $h_x/2$ instead of h_x.

Step 5: If (7.4.8) is satisfied, then $W(t_j, x_K)$ and $V(t_j, x_K)$ are taken as the numerical approximations to the solution of the non-linear system (7.2.3)-(7.2.4).

7.6 Illustrations

Different problems are presented in this section to illustrate the distinct behaviour of the nonlinear system (1)-(2). Above developed continuation method is used to solve the problems. The tolerance tol is taken to be 0.00001. The maximum point-wise error (D^N), the order of convergence (p^N) and the error constant (C_p^N) are computed using the methodology in [10].

Example:

Consider the nonlinear system

$$\epsilon \frac{dw(t)}{dt} + b\, w(t) + v(t) - a = 0$$

$$\frac{dv(t)}{dt} + \epsilon \left(\frac{(v(t))^3}{3} - v(t) - w(t) - I \right) = 0, \quad t \in (0,1]$$

For the above system with different initial conditions and for different values of the parameters a, b, I, ϵ, the values of D^N, p^N and C_p^N are presented in Tables 1-9. With aforesaid values of the parameters and for N = 128, graph of the numerical solutions for all the cases are portrayed in Figures 7.1-7.9.

The following cases are considered based on the values of α, β, a, b, I and ϵ from the literature on FitzHugh-Nagumo model.

Case 1:

$$w(0) = 0.9, v(0) = 0.7, a = 1.5, b = 0.7, I = 0.8 \;\&\; \epsilon = 12.5$$

For this case, a graph of the numerical solution is portrayed in Figure 1 and the values of D^N, p^N and C_p^N are presented in Table 7.1.

Case 2:

$$w(0) = 0.9, v(0) = 0.7, a = 1.5, b = 0.7, I = 0.8 \ \& \ \epsilon = 0.01$$

For this case, a graph of the numerical solution is portrayed in Figure 2 and the values of D^N, p^N and C_p^N are presented in Table 7.2.

Case 3:

$$w(0) = 0.9, v(0) = 0.7, a = 1.5, b = 0.7, I = 0.8 \ \& \ \epsilon = 1$$

For this case, a graph of the numerical solution is portrayed in Figure 3 and the values of D^N, p^N and C_p^N are presented in Table 7.3.

Case 4:

$$w(0) = 0.9, v(0) = 0.7, a = 0.7, b = 0.8, I = 1.5 \ \& \ \epsilon = 12.5$$

For this case, a graph of the numerical solution is portrayed in Figure 4 and the values of D^N, p^N and C_p^N are presented in Table 7.4.

Case 5:

$$w(0) = 0.9, v(0) = 0.7, a = 0.7, b = 0.8, I = 1.5 \ \& \ \epsilon = 0.01$$

For this case, a graph of the numerical solution is portrayed in Figure 5 and the values of D^N, p^N and C_p^N are presented in Table 7.5.

Case 6:

$$w(0) = 0.9, v(0) = 0.7, a = -1.45, b = -0.94, I = -0.5 \ \& \ \epsilon = 12.5$$

For this case, a graph of the numerical solution is portrayed in Figure 6 and the values of D^N, p^N and C_p^N are presented in Table 7.6.

Case 7:

$$w(0) = -1.8, v(0) = -1.8, a = -1.45, b = -0.94, I = -0.5 \,\&\, \epsilon = 12.5$$

For this case, a graph of the numerical solution is portrayed in Figure 7 and the values of D^N, p^N and C_p^N are presented in Table 7.7.

Case 8:

$$w(0) = 0.9, v(0) = 0.7, a = 1.33, b = 2.54, I = 2 \,\&\, \epsilon = 12.5$$

For this case, a graph of the numerical solution is portrayed in Figure 8 and the values of D^N, p^N and C_p^N are presented in Table 7.8.

Case 9:

$$w(0) = 0.9, v(0) = 0.7, a = -0.97, b = -0.34, I = 1.33 \,\&\, \epsilon = 12.5$$

For this case, a graph of the numerical solution is portrayed in Figure 9 and the values of D^N, p^N and C_p^N are presented in Table 7.9.

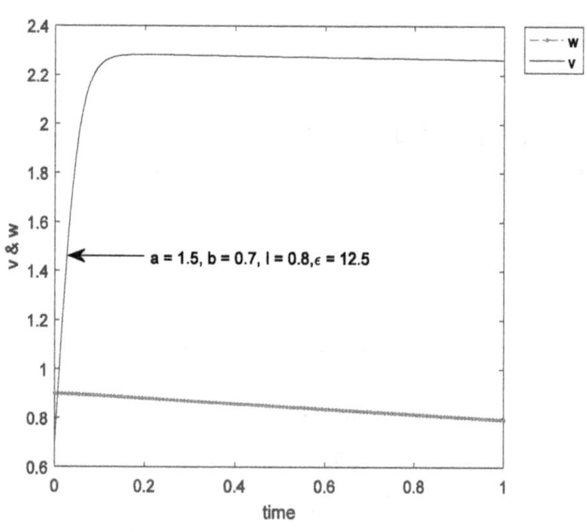

Figure 7. 1

Table 7.1

N	32	64	128	256	512	1024
D^N	6.6219e-02	3.5823e-02	1.8940e-02	9.8992e-03	5.0450e-03	2.5480e-03
p^N	8.8633e-01	9.1946e-01	9.3606e-01	9.7245e-01	9.8548e-01	
C_p^N	3.1133	3.1133	3.0426	2.9395	2.7692	2.5853

Table 7.2

N	32	64	128	256	512	1024
D^N	2.5927e-03	1.7072e-03	9.8333e-04	5.8073e-04	3.3563e-04	1.8964e-04
p^N	6.0287e-01	7.9586e-01	7.5980e-01	7.9099e-01	8.2366e-01	
C_p^N	6.1335e-02	6.1335e-02	5.3656e-02	4.8125e-02	4.2242e-02	3.6248e-02

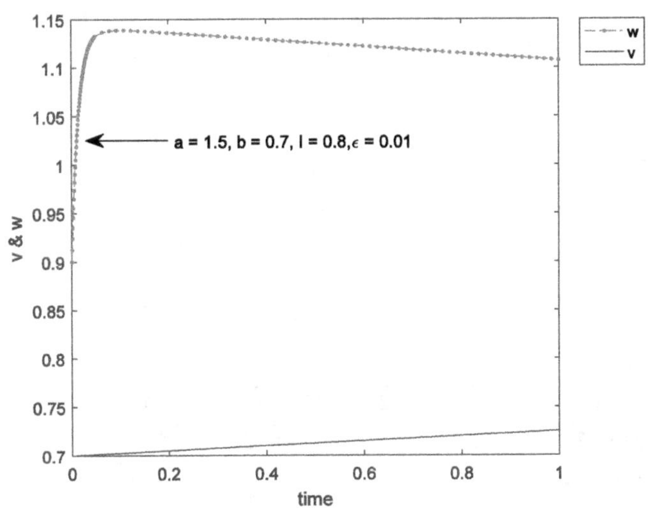

Figure 7.2

Table 7.3

N	32	64	128	256	512	1024
D^N	8.7271e-03	4.4306e-03	2.2326e-03	1.1208e-03	5.6150e-04	2.8103e-04
p^N	9.7800e-01	9.8875e-01	9.9424e-01	9.9715e-01	9.9857e-01	
C_p^N	5.2560e-01	5.2560e-01	5.2170e-01	5.1585e-01	5.0905e-01	5.0184e-01

Figure 7.3

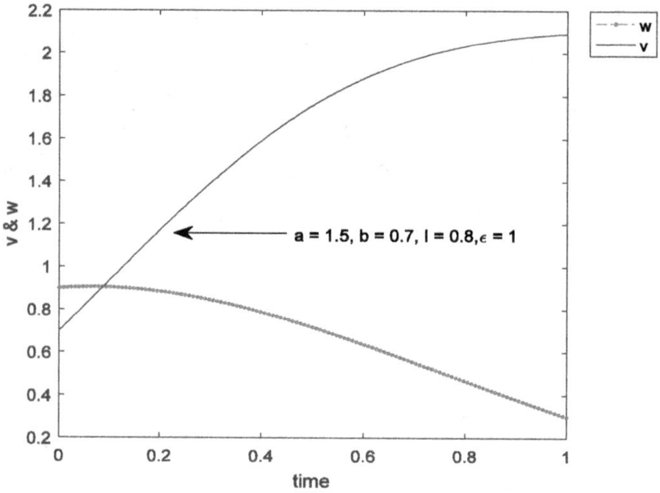

Table 7.4

N	32	64	128	256	512	1024
D^N	7.8229e-02	4.6064e-02	2.4531e-02	1.2819e-02	6.5610e-03	3.3208e-03
p^N	7.6407e-01	9.0902e-01	9.3633e-01	9.6630e-01	9.8241e-01	
C_p^N	2.6878	2.6878	2.4309	2.1573	1.8751	1.6118

Table 7.5

N	32	64	128	256	512	1024
D^N	1.0291e-02	6.6432e-03	4.1320e-03	2.4463e-03	1.4152e-03	8.0045e-04
p^N	6.3136e-01	6.8504e-01	7.5622e-01	7.8963e-01	8.2211e-01	
C_p^N	2.5894e-01	2.5894e-01	2.4948e-01	2.2880e-01	2.0503e-01	1.7963e-01

Table 7.6

N	32	64	128	256	512	1024
D^N	3.1099e-02	1.6366e-02	8.5769e-03	4.3838e-03	2.2189e-03	1.1164e-03
p^N	9.2615e-01	9.3219e-01	9.6827e-01	9.8236e-01	9.9092e-01	
C_p^N	1.6263	1.6263	1.6195	1.5729	1.5128	1.4464

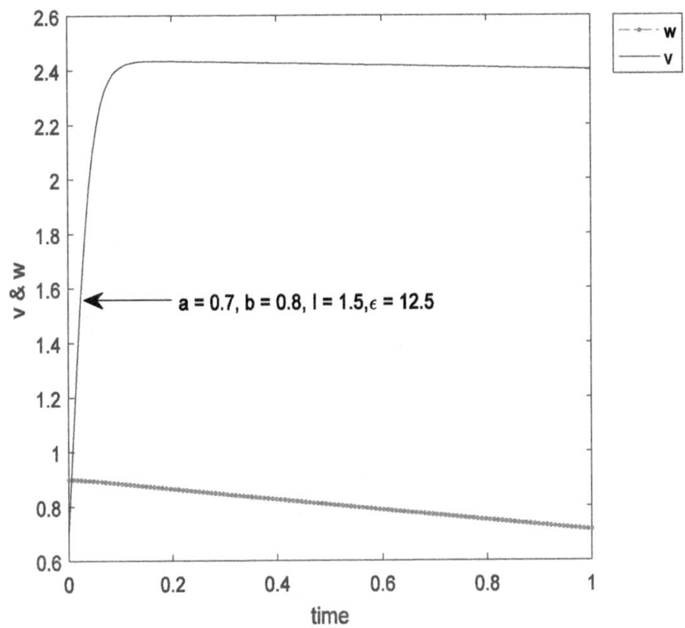

Figure 7.4

Table 7.7

N	32	64	128	256	512	1024
D^N	4.5132e-02	2.8182e-02	1.6225e-02	8.8301e-03	4.6251e-03	2.3642e-03
p^N	6.7937e-01	7.9653e-01	8.7775e-01	9.3294e-01	9.6816e-01	
C_p^N	1.2658	1.2658	1.167	1.0171	0.85315	0.69838

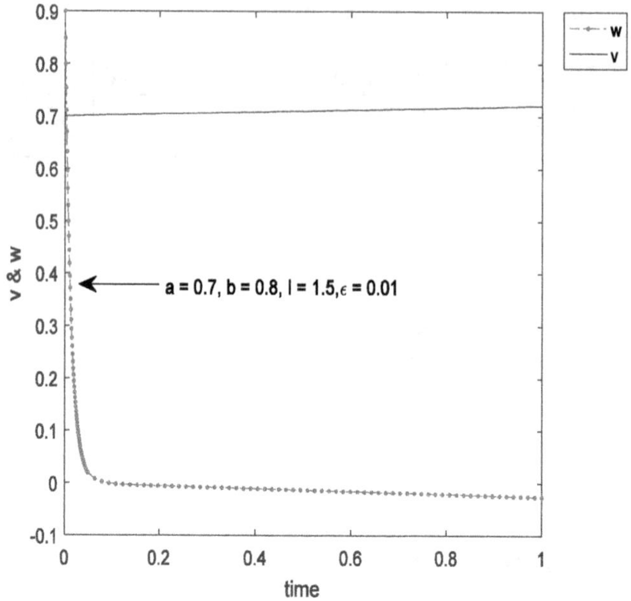

Figure 7.5

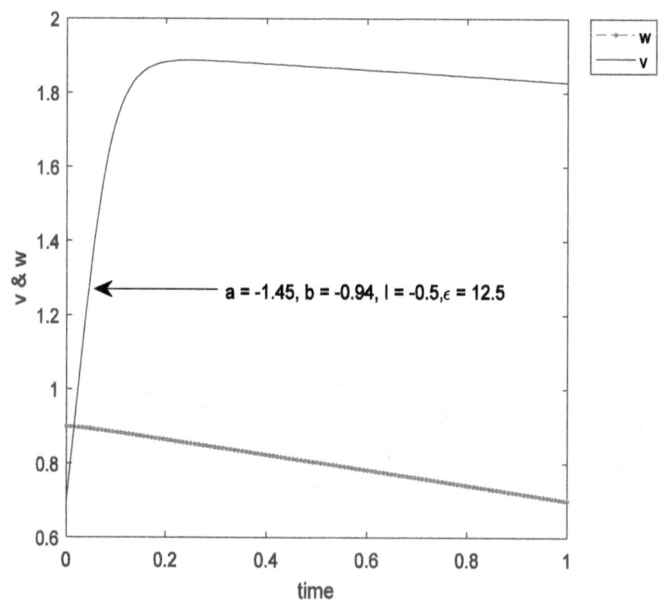

Figure 7.6

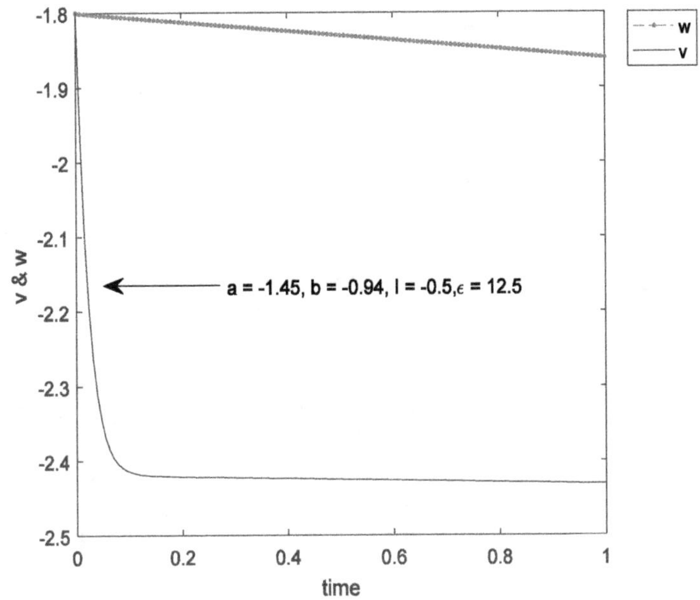

Figure 7.7

Table 7.8

N	32	64	128	256	512	1024
D^N	8.4835e-02	5.3161e-02	2.8467e-02	1.4859e-02	7.6449e-03	3.8721e-03
p^N	6.7428e-01	9.0107e-01	9.3795e-01	9.5879e-01	9.8137e-01	
C_p^N	2.3515	2.3515	2.0094	1.6738	1.3742	1.1107

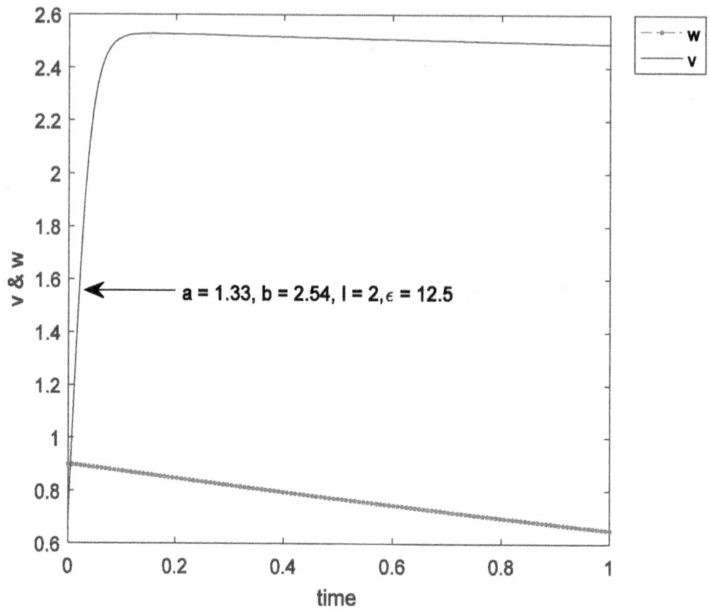

Figure 7.8

Table 7.9

N	32	64	128	256	512	1024
D^N	7.5947e-02	4.3280e-02	2.3148e-02	1.2133e-02	6.1981e-03	3.1352e-03
p^N	8.1127e-01	9.0284e-01	9.3190e-01	9.6908e-01	9.8327e-01	
C_p^N	2.9377	2.9377	2.7571	2.5359	2.2731	2.0177

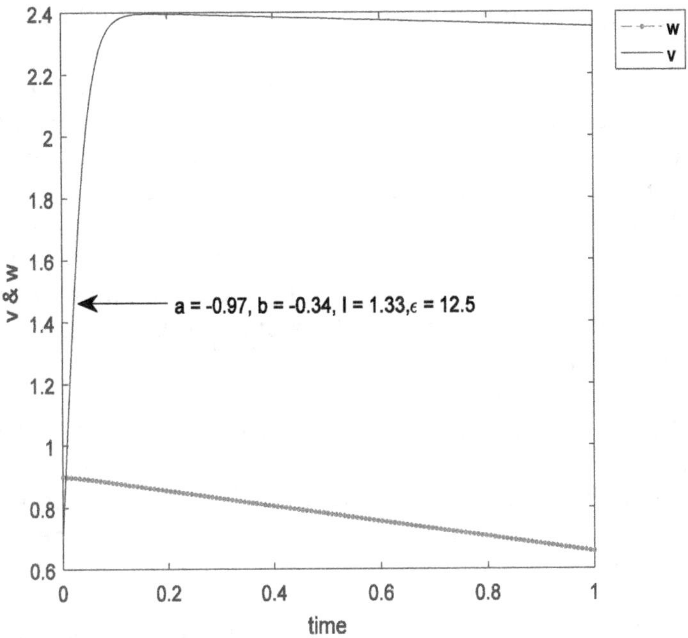

Figure 7.9

7.7 Conclusion

In this research work, a new computational technique is developed for FitzHugh-Nagumo model which works well for all values of the parameter ϵ, occurring in FitzHugh-Nagumo model. The computational technique is developed in such a way that it resolves the FitzHugh-Nagumo model when the parameter ϵ takes a large value and it traces the singular perturbation nature of a two-time scale system when the parameter ϵ takes a small value.

From Figures 7.5.1, 7.5.4, 7.5.6, 7.5.7, 7.5.8 and 7.5.9, we note that the component $v(t)$, which denotes the membrane potential in the FitzHugh-Nagumo model, changes rapidly near $t = 0$ and remains smooth elsewhere whereas the component $w(t)$, which denotes the recovery variable, is smooth and linear throughout the domain. Thus we conclude that the numerical solution obtained through the present computational technique for the nonlinear system (7.2.3)-(7.2.4) which represents the FitzHugh-Nagumo model, when the parameter ϵ takes a large value, preserves the nature of the FitzHugh-Nagumo model as mentioned in [1] and [2].

Further, from Figures 7.5.2 and 7.5.5, we note that the component $w(t)$, which denotes a fast state in a two-time scale system, changes rapidly near $t = 0$ and remains smooth elsewhere whereas the component $v(t)$, which denotes a slow state, is smooth throughout the domain. Thus we conclude that the numerical solution of the nonlinear system (7.2.3)-(7.2.4) which represents a two-time scale system, when the parameter ϵ takes a small value, preserves the singular perturbation nature of a two-time scale system as mentioned in [3], [4] and [7].

It is obvious that the nonlinear model represented in (7.2.3)-(7.2.4) becomes unperturbed when the parameter $\epsilon = 1$. From Figure 7.5.3, it is evident that the unperturbed nature of the nonlinear model represented in (7.2.3)-(7.2.4) is also preserved by the present computational technique.

From Tables 7.5.1-7.5.9, in this present work, it is evident that the maximum point-wise error (D^N) decreases monotonically and on the other hand the order of convergence (p^N) increases when the number of mesh points N is increased. This shows the consistency of the present computational technique.

Precisely, from Tables 7.5.2 and 7.5.7, we observe that all the characteristics of the numerical technique developed for a two-time scale system in [7] are preserved by the present computational technique also. This shows the efficiency of the present computational technique.

In the literature, the theoretical results for FitzHugh-Nagumo model are established when the parameter ϵ takes a large value; on the other hand the theoretical results for a two-time scale system are established when the parameter ϵ takes a small value. Thus it is highly complicated to establish the theoretical results for the nonlinear model represented in (7.2.3)-(7.2.4), irrespective of the magnitude of the parameter ϵ. So far no theoretical result for the nonlinear model represented in (7.2.3)-(7.2.4), irrespective of the magnitude of the parameter ϵ, is available in the literature. Hence establishing the theoretical result for the nonlinear model represented in

(7.2.3)-(7.2.4), irrespective of the magnitude of the parameter ϵ, is remain as a challenging open problem.

References

[1] R. Fitzhugh, Impulse and physiological states in models of nerve membrane, Biophysical Journal, 1(6), 445–466 (1961).

[2] J. S. Nagumo, S. Arimoto, and S. Yoshizawa, An active pulse transmission line simulating nurve axon, Proceedings of the Institute of Radio Engineers, 50(10), 2061–2070 (1962).

[3] X. Xu, R. M. Mathur, J. Jiang, G. J. Rogers and P. Kundur, Modeling of generators and their controls in power system simulations using singular perturbations, IEEE Transactions on Power Systems, 13(1), 109-114 (1998).

[4] Yan Zhang, D. Subbaram Naidu, Chenxiao Cai and Yun Zou, Singular perturbations and time scales in control theories and applications, International Journal of Information and System Sciences, 9(1), 1-36 (2014).

[5] C. Akkoyunlu, Compact finite differences method for fitzhugh-nagumo equation, Universal Journal of Mathematics and Applications 2, 180-187 (2019).

[6] D. P. Gui, A fourth-order compact finite difference scheme to the numerical solution of fitzhugh-nagumo equation, Applied Mechanics and Materials 873, 337-341 (2017).

[7] Manikandan Mariappan and Ayyadurai Tamilselvan, An efficient numerical method for a nonlinear system of singularly perturbed differential equations arising in a two-time scale system, Journal of Applied Mathematics and Computing, 68(2), 1069-1086 (2022).

[8] Miller, J.J.H., E. O'Riordan and Shishkin, G.I., Fitted Numerical Methods for Singular Perturbation Problems, World Scientific Publishing Co., Singapore, New Jersey, London, Hong Kong (1996).

[9] Roos H.G., Stynes M. and Tobiska L., Numerical Methods for Singularly Perturbed Differential Equations, Convection-Diffusion and Flow Problems, Springer-Verlag, New York (1996).

[10] Farrell P.A., Hegarty A.F., Miller J.J.H., O' Riordan E. and Shishkin G.I., Robust computational techniques for boundary layers, Chapman and hall/CRC, Boca Raton, Florida, USA (2000).

Chapter 8

Multiple Coexisting Attractors in Magnetoconvection

N. C. Pati [1,*], G. C. Layek [2]

[1] Department of Mathematics, Birla Institute of Technology Mesra, Ranchi-835215, Jharkhand, India
[2] Department of Mathematics, The University of Burdwan, Burdwan-713104, West Bengal, India
*Current Address of N. C. Pati: Department of Mathematics, Bidhan Chandra College, Asansol-713304, West Bengal, India.

Abstract

Multistability is the most exciting behavior in nonlinear dynamics. It basically refers to the birth of multiple coexisting attractors, viz., multi-fixed points, fixed point-periodic, multi-periodic, periodic-chaotic, multi-chaotic, etc. depending upon initial conditions of a nonlinear dynamical system for a fixed parameter setting. It has been found in almost all branches of science and technology. Exploring multistability is an active research area as the final state of a multistable system depends crucially on its initial state. A small deviation in the initial state can result in completely different asymptotic dynamics. It is wonder how multistability occurs in deterministic nonlinear systems. What is the role of nonlinearity in generating such a striking behavior? Multistability can occur in various ways, viz., bifurcation-induced, noise-induced, rate-induced, etc. In this chapter, we explore the existence of bifurcation-induced multistability by adopting low-dimensional dynamical models for magnetoconvection of electrically conducting, viscous fluid layer heated underneath subjected to uniform magnetic field imposing along horizontal and vertical directions. Analysis reveals that the oscillatory mode of convection depends strongly on the magnetic field, and hysteresis phenomenon gives birth to multistable attractors. We focus on the following contexts:

(i) determination of the existence criteria and linear stability conditions of different steady states,

(ii) conditions for boundedness of the attractors along with their bounds by constructing Lyapunov functions,

(iii) identification of various bifurcation scenarios enroute to chaotic magnetoconvection,
(iv) determination of the nature of the bifurcations and stability of the bifurcated solutions through normal-form analysis based on center manifold reduction,
(v) identification of hysteresis windows sandwiched between bifurcation-induced transitions,
(vi) characterization of the coexisting heterogeneous dynamics and their quantifications,
(vii) distinctions of the multistable attractors for the externally-imposed magnetic field in both the directions,
(viii) discussions on physical implications of multistability in magnetoconvection.

Keywords: Magnetoconvection, Generalized Lorenz system, Hopf bifurcation, Chaos, Multistability.

8.1 Introduction

Attractors are the subsets of a phase space. They have various types, viz., fixed point, periodic, non-periodic closed, and strange types. Some well-known attractors are Cantor set in chaotic logistic map, Henón attractor, Rössler attractor, etc. Their geometrical structures and basin boundaries are of special interests. For a fixed parameter set, a monostable dynamical system generally attains one of the aforementioned asymptotic state for any choice of initial conditions within its domain of attraction. However, in some dynamical systems, a small perturbation in the initial condition may change long-term behavior of the systems. Consequently, the system exhibits the coexistence of two or more attractors for a fixed set of parameters [1]. This striking phenomenon is called multistability, first time mentioned perceptionally by Attneave [2]. Subsequently, multistability has been detected in many natural phenomena and engineering systems, such as Chua's circuits [3, 4], jerk systems [5, 6, 7], lasers [8], memristor oscillators [9, 10, 11], biological systems [12, 13, 14], time delayed systems [15], acoustic gravity-wave propagation [16], and some other unusual dynamical systems [17, 18], to just name a few. Exploration of multistability is essential because there is always a final-state uncertainty in the systems. Sometimes, topologically non-equivalent attractor is born due to an infinitesimal perturbation to the initial state. In contrast to regular systems, the basins of attraction of multistable systems have mutually exclusive basin sets.

There are several constructing multistable systems with desired number of coexisting stable state attractors (see Ref. [19, 20, 21]). Depending on the mechanism, the critical transition that switches one stable state to another in a multistable system can be broadly classified into three categories, viz. bifurcation-induced transition, noise-induced transition and rate-induced transition [21, 22]. The bifurcation-induced transition occurs when the system control parameter exceeds a threshold value, which causes a qualitative change in the present stable state of the system. This type of multistability has been extensively studied in past few decades where the system exhibits a hysteresis in between two saddle-node bifurcations. Noise-induced multistability generally occurs when the basin boundaries are fractals. The third type of critical transition is not associated with bifurcation or noise. It is important in ecosystems and atmospheric processes where the environmental parameters are varied with certain speed. The speed of variation is crucial to determine whether the present stable state can be successfully tracked or not (for details see Ref. [22]).

In this chapter, hysteresis-induced multiple coexisting attractors are encountered in magnetoconvection of an electrically conducting viscous fluid heated from below subjected to externally imposed uniform magnetic field under MHD approximations. The magnetic field generally inhibits the convection depending upon the strength of the applied magnetic field. In the present study, two simple generalized Lorenz systems are considered; one for vertically imposed magnetic field and the another for horizontally imposed magnetic field. The effects of the magnetic field on the convection and its dynamic behaviors are explored theoretically. Studies disclose that both the systems exhibit windows of hysteresis in particular parameter zones. In the hysteresis regimes, different types of coexisting attractors, viz., period-periodic and periodic-chaotic, chaotic-chaotic appear depending upon the system parameters and the initial states. Interestingly, in absence of the magnetic field none of the coexisting attractors are found in convective motion.

The rest of this chapter is organized as follows. Section (8.2) presents the derivation of the physically generalized Lorenz systems.

The boundedness of the solutions, existence of different steady-states, their stability and bifurcations are discussed in Section (8.3). Section (8.4) deals with the multistability in the systems. Finally, summary and the concluding remarks are provided in Section (8.5).

8.2 The model systems

The standard two-dimensional (x -z plane) magnetoconvection in a horizontally stratified, shallow electrically conducting viscous fluid layer of depth h is taken into consideration under Rayleigh-Bénard framework. The fluid layer is heated (resp. cooled) from below (resp. top) to maintain an adverse temperature gradient through the layer. Initially we consider the magnetoconvection where a uniform magnetic field $\boldsymbol{B_0} = (0,0,B_0)$ is imposed normal to the flow plane. Then, under usual Oberbeak–Boussinesq and MHD approximations, the relevant governing magneto-hydrodynamic and heat-flow equations are written as [23, 24]

$$\frac{d\boldsymbol{v}}{dt} = -\frac{1}{\rho_0}\nabla\left(P + \frac{\boldsymbol{B}^2}{2\mu_0}\right) + \nu\nabla^2\boldsymbol{v} + \{1 - \beta(T - T_0)\}\boldsymbol{g} + \frac{1}{\rho_0\mu_0}(\boldsymbol{B}\cdot\nabla)\boldsymbol{B}, \tag{8.2.1}$$

$$\frac{dT}{dt} = \kappa\nabla^2 T, \tag{8.2.2}$$

$$\frac{d\boldsymbol{B}}{dt} = (\boldsymbol{B}\cdot\nabla)\boldsymbol{v} + \eta\nabla^2\boldsymbol{B}, \tag{8.2.3}$$

with $\nabla\cdot\boldsymbol{v} = 0$ and $\nabla\cdot\boldsymbol{B} = 0$, where $\boldsymbol{v}$ is the fluid velocity field, $\boldsymbol{B}$ the magnetic field, P the pressure, T the temperature field, and the material derivative $d/dt \equiv \partial/\partial t + \boldsymbol{v}\cdot\nabla$. The parameters μ_0, ν, β, κ, η, and ρ_0 are, respectively, magnetic permeability, the kinematic viscosity, volumetric expansion coefficient, thermal conductivity, magnetic diffusivity, fluid density at T_0, a reference temperature. Here $\boldsymbol{g} = (0,0,-g)$ represents the acceleration of gravity, and the term $\frac{1}{\rho_0\mu_0}(\boldsymbol{B}\cdot\nabla)\boldsymbol{B}$ represents the magnetic force developed in the motion.

In virtue of the two solenoidal conditions, one can introduce two functionals, one is the steam function $\psi(t,x,z)$ which is associated with the fluid velocity field $\boldsymbol{v}$ as $\boldsymbol{v} = \nabla\times(0,\psi,0)$, and the another one is a magnetic-flux function $\varphi(t,x,z)$ satisfying $\boldsymbol{B} =$

$\nabla \times (0, \varphi + B_0 x, 0)$. We assume that the two horizontal boundaries are stress-free and have fixed temperature. We also assume periodic boundary conditions in the lateral directions. We are interested to explore the evolution of perturbation to the basic (conduction) state of the thermo-magneto-fluid system under low-dimensional approach. Employing Galerkin expansion technique, we expand the three flow variables (ψ, T, φ) in truncated Fourier representation in the coordinates (x, z), following the work of Lorenz [25], Gotoda *et al.* [26] and Layek and Pati [24] as:

$$\psi(t, x, z) = \sqrt{2}\frac{1+a^2}{a}\kappa X(t) \sin\left(\frac{\pi a x}{h}\right) \sin\left(\frac{\pi z}{h}\right), \tag{8.2.4}$$

$$T(t, x, z) = T_L - (\delta T)z + h\delta T \frac{\pi^3(1+a^2)^3}{a^2 Ra}\left[\sqrt{2}Y(t) \cos\left(\frac{\pi a x}{h}\right) \sin\left(\frac{\pi z}{h}\right) - Z(t) \sin\left(\frac{2\pi z}{h}\right)\right], \tag{8.2.5}$$

$$\varphi(t, x, z) = \frac{\sqrt{2}\pi h \kappa B_0 (1+a^2)^2}{a\eta} W(t) \sin\left(\frac{\pi a x}{h}\right) \cos\left(\frac{\pi z}{h}\right), \tag{8.2.6}$$

where T_L is the temperature at $z = 0$ (the bottom surface), δT represents the temperature gradient across the fluid layer, and a is the layer's aspect ratio.

Projecting Eqs. (8.2.1)-(8.2.3) onto the modes of (8.2.4)-(8.2.6), one can easily obtain the following four-mode physically extended Lorenz system:

$$\left.\begin{aligned} \dot{X} &= \sigma(-X + Y - QW), \\ \dot{Y} &= rX - Y - XZ, \\ \dot{Z} &= -bZ + XY, \\ \dot{W} &= \sigma_m(\alpha X - W), \end{aligned}\right\} \tag{8.2.7}$$

where overdot denotes derivative w.r.t the scaled time $t' = \pi^2(1 + a^2)t/h^2$. The nonlinear system (8.2.7) consists of six dimensionless parameters, viz., the fluid Prandtl number $\sigma = \nu/\kappa$, the Chandrasekhar number $Q = B_0^2 h^2/\mu_0 \rho_0 \eta \nu$, the reduced Rayleigh number $r = Ra/Ra_c$ with $Ra = g\beta\delta T h^4/\nu\kappa$ the Rayleigh number and $Ra_c = \pi^4(1 + a^2)^3/a^4$, the diffusivity ratio $\sigma_m = \eta/\kappa (= \sigma/Pr_m$, $Pr_m = \nu/\eta$ the magnetic Prandtl number), $b = 4/(1 + a^2)$ a geometrical parameter associated with the aspect ratio of the layer, and $\alpha = 1/\pi^2(1+a^2)^2$. The parameters Q and Pr_m represent the effects of the magnetic field on the thermo-convective motion.

Similarly, in the context of magnetoconvection for horizontally applied magnetic field $\boldsymbol{B_0} = (B_0, 0{,}0)$, one can derive the following lowest-order physically extended Lorenz system (see Refs. [27, 28, 29] for detailed derivation):

$$\left.\begin{aligned} \dot{X} &= \sigma\left(Y - X - \frac{\omega_0}{\sigma}W\right), \\ \dot{Y} &= rX - Y - XZ, \\ \dot{Z} &= XY - bZ, \\ \dot{W} &= \sigma_m\left(\frac{\omega_0}{\sigma_m}X - W\right), \end{aligned}\right\} \tag{8.2.8}$$

where the parameters σ, r, b and σ_m have the same physical meaning as mentioned earlier for system (8.2.7). The dimensionless parameter ω_0 represents the magnetic frequency (proportional to B_0, the strength of the horizontally imposed uniform magnetic field). The diffusivity ratio σ_m plays a deciding role on the convective patterns, in particular astrophysical and geophysical dynamics. In context of astrophysical phenomena such as solar convection near sunspots, the diffusivity ratio $\sigma_m < 1$. On the other hand, $\sigma_m > 1$ in laboratory experiments, metallurgical applications and in the Earth's core. However, in geodynamic phenomena both $\sigma_m > 1$ and $\sigma_m < 1$ are applicable. For liquid metals, the magnetic diffusivity is very large so that $\sigma_m \gg 1$, typically of order of $10^3 - 10^6$. In this contribution, following the work of Lorenz [25], we shall study the dynamics of the above two systems (8.2.7) and (8.2.8) for fixed values of σ and b at 10 and $8/3$, respectively.

8.3 Dynamical properties of the systems

In this section, we shall study some basic dynamical properties of the systems including boundedness of solutions, determination of the region of boundedness, existence and stability of different steady states and different bifurcation patterns enroute to chaos. One can easily see that in addition to the Lorenz model [25] for Newtonian fluid convection without magnetic field, each of the systems (8.2.7) and (8.2.8) for magnetoconvection contains a new state equation representing the feedback of the induced magnetic field on the flow.

For symmetry, both the systems possess a reflectional symmetry about Z, that implies that if (X, Y, Z, W) is a solution of the systems, then so is $(-X, -Y, Z, -W)$.

Dissipativity is another important property of a dynamical system which ensures the existence of bounded solutions in the phase-space of the system. A dissipative system possesses negative flow-divergence. For our current systems, the divergence of the vector field is given by

$$\frac{\partial \dot{X}}{\partial X} + \frac{\partial \dot{Y}}{\partial Y} + \frac{\partial \dot{Z}}{\partial Z} + \frac{\partial \dot{W}}{\partial W} = -(\sigma + 1 + b + \sigma_m) < 0.$$

This implies that both the systems (8.2.7) and (8.2.8) are dissipative in nature. Hence, the systems have bounded attractors representing various dynamical scenarios, such as equilibrium (steady) state, periodic motion, quasi-periodic motion, chaos, etc.

In order to determine the bounds inside which the flow-trajectories can confine, we follow the Lyapunov function approach (see the book of Layek [30] for details).

First we construct the following Lyapunov function for the system (8.2.7):

$$V^{(v)}(X, Y, Z, W) = \frac{1}{2}\left(rX^2 + \sigma Y^2 + \sigma(Z - 2r)^2 + \frac{Pr_m rQ}{\alpha} W^2\right).$$

Then the time-rate of change of the Lyapunov function along the flow of system (8.2.7) is

$$\begin{aligned}\dot{V}^{(v)} &= \frac{\partial V^{(v)}}{\partial X}\dot{X} + \frac{\partial V^{(v)}}{\partial Y}\dot{Y} + \frac{\partial V^{(v)}}{\partial Z}\dot{Z} + \frac{\partial V^{(v)}}{\partial W}\dot{W} \\ &= -\sigma r X^2 - \sigma Y^2 - b\sigma(Z - r)^2 - \frac{\sigma r Q}{\alpha} W^2 + b\sigma r^2.\end{aligned}$$

Let us define the domain

$$\mathbb{D}^{(v)} = \left\{(X, Y, Z, W) \in \mathcal{R}^4 : rX^2 + Y^2 + b(Z - r)^2 + \frac{rQ}{\alpha} W^2 \leq br^2\right\}.$$

Then, for any point outside the domain $\mathbb{D}^{(v)}$, $\dot{V}^{(v)} < 0$, implying that all flow-trajectories outside the domain $\mathbb{D}^{(v)}$will asymptotically enter in $\mathbb{D}^{(v)}$ and will never leave it. Consequently, $\mathbb{D}^{(v)}$ is an attracting region in the solution-space of the system (8.2.7). Hence, the nonlinear system (8.2.7) has bounded global attractors.

Similarly, by constructing the following Lyapunov function

$$V^{(h)}(X,Y,Z,W) = \frac{1}{2}(rX^2 + \sigma Y^2 + \sigma(Z-2r)^2 + rW^2),$$

one can easily verify, same as above, that the system (8.2.8) also supports bounded attractors with an absorbing region

$$\mathbb{D}^{(h)} = \left\{(X,Y,Z,W) \in \mathcal{R}^4 : rX^2 + Y^2 + b(Z-r)^2 + \frac{r\sigma_m}{\sigma}W^2 \leq br^2\right\}.$$

Note that the functions chosen may not provide the optimum region of boundedness; they are merely constructed to establish that the systems cannot support unbounded attractors. Next, we shall explore different steady states, their existence criteria, stability conditions, and different associated bifurcation scenarios.

For steady states (or equilibrium states), both the systems (8.2.7) and (8.2.8) have their trivial equilibrium point at $(0,0,0,0)$. This trivial equilibrium origin represents the motionless (conduction) state of the thermo-fluid system where the fluid is at rest and the heat propagates through the fluid layer via conduction. Employing Routh–Hurwitz criteria [31], the equilibrium origin associated with the system (8.2.7) for vertically imposed magnetic field is asymptotically stable, i.e., the fluid maintains its the conduction state if

$$r < min\left\{1 + \alpha Q, 1 + \sigma_m\left(1 + \alpha Q + \frac{1}{\sigma}\right), (\sigma + \sigma_m)\left(\frac{1+\sigma_m}{\sigma} + \frac{\sigma_m \alpha Q}{(1+\sigma)}\right)\right\}.$$

Furthermore, the system undergoes a pitchfork bifurcation, leading to the birth of steady stationary convection, when the parameter r attains the critical threshold $r_e^{(v)} = 1 + \alpha Q$. Additionally, a Hopf bifurcation around the equilibrium origin (corresponding to overstable motion) comes into existence when the parameter r attains the critical threshold $r_o^{(v)} = (\sigma_m + \sigma)\left(\frac{1+\sigma_m}{\sigma} + \frac{\sigma_m \alpha Q}{1+\sigma}\right)$, provided that the Chandrasekhar number $Q > Q_o := \frac{\sigma_m(1+\sigma)}{\alpha\sigma(1-\sigma_m)}$ with the diffusivity ratio $\sigma_m < 1$ [23]. Similarly, for the system (8.2.8) for horizontally impressed magnetic field, the equilibrium origin is asymptotically stable, i.e., the conduction state persists as long as

$$r < min\left\{1 + \frac{\omega_0^2}{\sigma\sigma_m}, 1 + \sigma_m + \frac{\sigma_m + \omega_0^2}{\sigma}, (\sigma + \sigma_m)\left(\frac{\omega_0^2}{\sigma(1+\sigma)} + \frac{1+\sigma_m}{\sigma}\right)\right\}.$$

In this case, the system undergoes a pitchfork bifurcation around the origin when r attains the threshold $r_e^{(h)} = 1 + \omega_0^2/\sigma\sigma_m$, and a Hopf bifurcation around the origin takes place when r takes the threshold value $r_o^{(h)} = (\sigma + \sigma_m)\left\{\frac{1+\sigma_m}{\sigma} + \frac{\omega_0^2}{\sigma(1+\sigma)}\right\}$. In this case, the magnetic frequency parameter ω_0 must be restricted to $\omega_0^2 > \frac{\sigma_m^2(1+\sigma_m)}{(1-\sigma_m)}$ with the diffusivity ratio $\sigma_m < 1$. Therefore, for both systems, the overstable motion occurs only when $\sigma_m < 1$.

When r exceeds the critical threshold of the pitchfork bifurcation, the equilibrium origin of each system becomes unstable, giving birth to two non-trivial symmetric equilibria. For system (8.2.7), the non-zero equilibria are given by

$$E_\pm = (\pm x_v^*, \pm(1+\alpha Q)x_v^*, r - 1 - \alpha Q, \pm\alpha x_v^*),$$

and those of system (8.2.8) are

$$C_\pm = \left(\pm x_h^*, \pm\left(1 + \frac{\omega_0^2}{\sigma\sigma_m}\right)x_h^*, r - 1 - \frac{\omega_0^2}{\sigma\sigma_m}, \pm\frac{\omega_0 x_h^*}{\sigma_m}\right),$$

where

$$x_v^* = \sqrt{\frac{b(r - 1 - \alpha Q)}{1 + \alpha Q}},$$

and $x_h^* = \sqrt{\dfrac{b\left(r-1-\frac{\omega_0^2}{\sigma\sigma_m}\right)}{1+\frac{\omega_0^2}{\sigma\sigma_m}}}$.

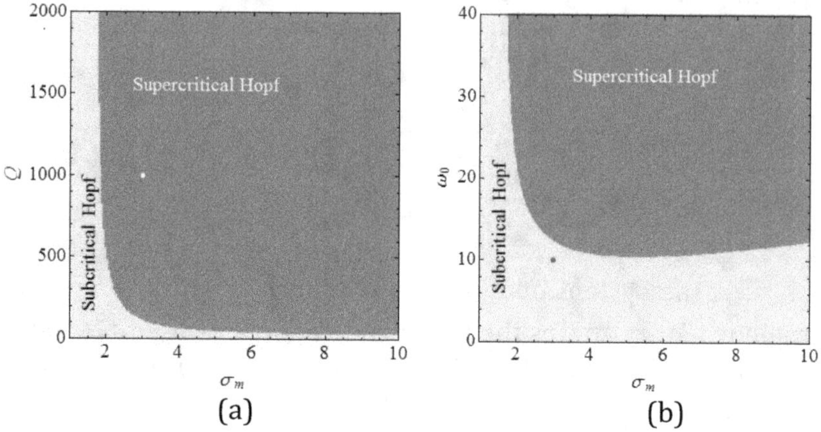

Figure 8.1 Regions of supercritical (gray color) and subcritical (cyan color) Hopf bifurcations around the nontrivial equilibria of (a) system (8.2.7) in the bi-parameter plane $(\sigma_m, Q) \in [1,10] \times [0,2000]$ and (b) system (8.2.8) in $(\sigma_m, \omega_0) \in [1,10] \times [0,40]$. Here (σ, b) are taken as (10,8/3).

Physically, the non-zero equilibrium represent the steady convection state of the magnetoconvective systems. For stability threshold of this equilibrium state, one can see after linearizing Eqs. (8.2.7) and (8.2.8) around their respective non-trivial equilibria that the steady convective state losses stability only via unsteady oscillatory convection characterized by the existence of Hopf bifurcation around the non-trivial equilibria for some certain critical threshold, $r_{\text{Hopf}}(> r_e)$, of r. The nature of the Hopf bifurcation and stability of the resultant solution can be determined by applying normal-form analysis based on center manifold reduction. This restricts the dynamics of each of the 4D nonlinear systems (8.2.7) and (8.2.8) on the center manifold to a two-dimensional reduced system of coupled equations of the form [32] $\dot{x} = -\omega y + f(x,y)$ and $\dot{y} = \omega x + g(x,y)$ with ω representing the imaginary part of the leading pair of purely imaginary eigenvalues at $r = r_{\text{Hopf}}$, and f and g are nonlinear functions of x and y such that they along with their all first-order partial derivatives w.r.t x and y are zero at

$(x, y) = (0,0)$. Based on this 2D reduced system, the nature of the Hopf bifurcation around the non-trivial equilibria of each of the systems (8.2.7) and (8.2.8) can be determined by evaluating the following index quantity (see Ref. [32]) at $r = r_{\text{Hopf}}$, $\eta_3 = \frac{1}{16}(f_{xxx} + f_{xyy} + g_{xxy} + g_{yyy}) + \frac{1}{16\omega}(f_{xy}f_{xx} + f_{xy}f_{yy} + f_{yy}g_{yy} - g_{xy}g_{xx} - g_{xy}g_{yy} - f_{xx}g_{xx})$,
where subscripts denote partial derivatives evaluated at $(x, y) = (0,0)$. With this index quantity η_3, the Hopf bifurcation is supercritical (continuous, non-catastrophic) if $\eta_3 < 0$ and it is subcritical (jump, subtle, catastrophic) when $\eta_3 > 0$.

Figure 8.1 (a) shows the parametric zones of supercritical and subcritical Hopf bifurcations for the system (8.2.7) in the bi-parameter plane $(\sigma_m, Q) \in [1,10] \times [0,2000]$, and Fig. 8.1(b) for the system (8.2.8) for $(\sigma_m, \omega_0) \in [1,10] \times [0,40]$. To distinguish the sub- and supercritical bifurcation regions, we used gray color to highlight the parameter zones for supercritical Hopf bifurcation ($\eta_3 < 0$) and cyan color for subcritical Hopf bifurcation ($\eta_3 > 0$). In absence of the magnetic field, the Hopf bifurcation is always subcritical (see Sparrow [33]). On the other hand, the magnetic field renders the subcritical bifurcation supercritical by suppressing the subcritical zone. The white dot in Fig. 8.1(a) corresponds to $(\sigma_m, Q) = (3,1000)$ at which $\eta_3 \approx -5.11 \times 10^{-5} < 0$ with $r_{\text{Hopf}}^{(v)} \approx 512.36$. Therefore, a stable limit cycle (periodic orbit) is emerged via supercritical Hopf bifurcation as r exceeds $r_{\text{Hopf}}^{(v)}$. Similarly, the red dot in Fig. 8.1(b) represents $(\sigma_m, \omega_0) = (3,10)$ at which $\eta_3 \approx 4.38173 \times 10^{-4} > 0$ with $r_{\text{Hopf}}^{(h)} \approx 43.1304$. It is well-established that the classical Lorenz system exhibits a discontinuous transition to chaos near the subcritical Hopf bifurcation. On the other hand, due to inclusion of magnetic field we found periodic attractor near subcritical Hopf bifurcation for the above parameter values. Thus, the magnetic field not only alters the Hopf bifurcation directions but also subsequently changes the attractor's structure in the subcritical regimes.

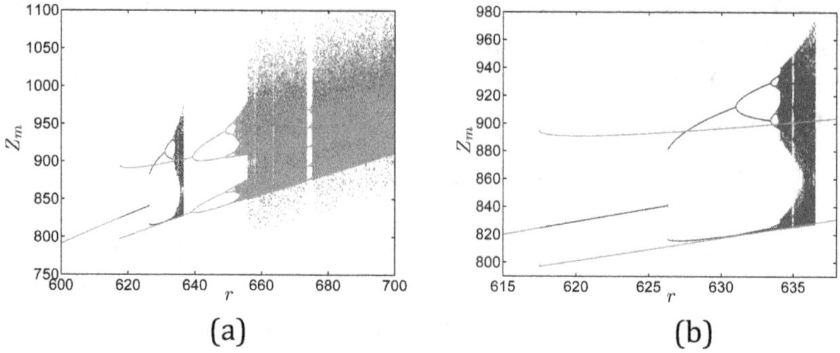

Figure 8.2 (a) Bifurcation diagrams representing local maxima, Z_m, of Z versus the reduced Rayleigh number $r \in [600,700]$ for system (8.2.7) with $\sigma = 10$, $Q = 1000$, $\sigma_m = 3$, and $b = 8/3$, (b) Enlargement of the bifurcation diagrams in (a) for $615 \leq r \leq 638$.

8.4 Multiple coexisting attractors

Our objective in this section is to explore the dynamical evolution of magnetoconvective systems beyond the Hopf bifurcation, and identification of multistability depending upon initial conditions and parameters set.

8.4.1 Multiple coexisting attractors for system (8.2.7)

We first consider the system (8.2.7) for magnetoconvective motion subject to vertically imposed magnetic field. Bifurcation diagram subject to varying a parameter keeping other parameters fixed is an important tool to elucidate the evolution of a dynamical system through different dynamical scenarios. Figure 8.2(a) displays the bifurcation diagrams for the evolution of the periodic orbit emanates from non-zero equilibrium through the supercritical Hopf bifurcation at $r_{\text{Hopf}}^{(v)} \approx 512.36$ for $\sigma_m = 3$ and $Q = 1000$. Each of the bifurcation diagrams represents variations of local maxima of the state variable Z w.r.t the parameter $r \in [600,700]$. Throughout this chapter, we use the color blue (resp. magenta) to plot the bifurcation diagram for increasing (decreasing) values of r, unless otherwise specified. We employed the '*attractor following technique*' to generate the bifurcation diagram, where the end point of each iteration of r is considered as the starting(initial) for the next iteration of r. Lyapunov exponent as another important tool to quantify different dynamical scenarios. Figure 8.3 displays the variations of the Lyapunov exponents corresponding to the dynamics obtained in the bifurcation diagrams shown in Fig. 8.2. We used the well-celebrated Wolf algorithm [34] to calculate all the Lyapunov exponent presented in this chapter. Figure 8.3(a) (resp. Fig. 8.3(b)) shows the variations of the first three Lyapunov exponents for increasing (resp. decreasing) values of r.

These bifurcation and Lyapunov exponent diagrams clearly reveal that the system (8.2.7) exhibits multiple coexisting attractors in the range $617.43 \lessapprox r \lessapprox 636.52$ for the critical parameters (see Fig. 8.2(b) for magnifications of the bifurcation diagrams in $615 \leq r \leq 638$). As disclosed by the bifurcation diagrams in Fig. 8.2, when the parameter r increases beyond $r_{\text{Hopf}}^{(v)} \approx 512.36$, the periodic attractor that emerged via the Hopf bifurcation undergoes a period-doubling cascades of chaos when r lies in the $r_{\text{Hopf}}^{(v)} < r \lessapprox 633.95$. When r increases beyond $r = 633.95$, bands of chaotic dynamics with embedded windows of periodic oscillations are observed. For further increasing r, we obtained a stable 2-periodic orbit emanates from the chaotic attractor via a tangent bifurcation (intermittency [35]) when r attains the critical threshold $r \approx 636.52$. This 2-periodic attractor then undergoes another period-doubling transition to chaos (cf. Fig. 8.2(a)) if r is increased further.

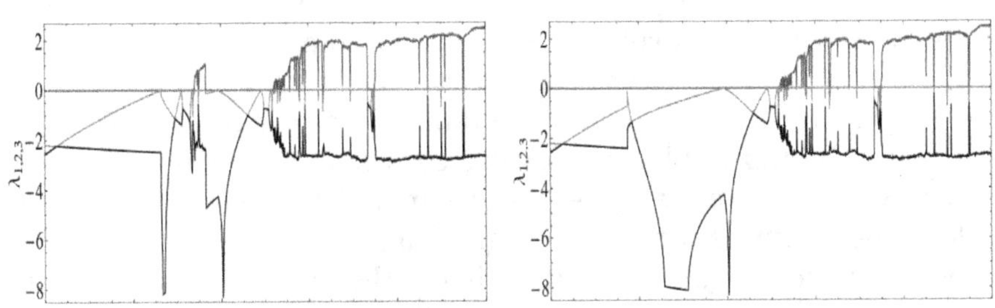

(a) (b)

Figure 8.3 Variations of the three leading Lyapunov exponents $\lambda_{1,2,3}$ ($\lambda_1 \geq \lambda_2 \geq \lambda_3$) w.r.t the parameter r for system (8.2.7). The remaining parameters are taken same as in Fig. 8.2(a). The diagram presented in subfigure (a) is obtained for increasing r and that in (b) for decreasing r.

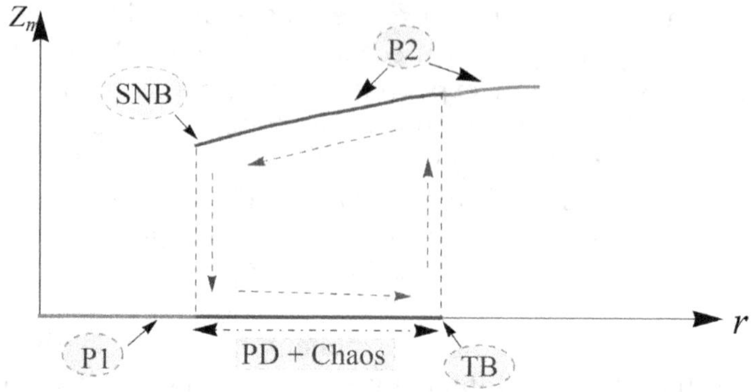

Figure 8.4 A schematic diagram of the hysteresis loop appears in the range $617.43 \lessapprox r \lessapprox 636.52$ of r for system (8.2.7). Here Pn ($n = 1{,}2$) denotes the period-n orbit, *PD* the period-doubling obits, *Chaos* the chaotic attractors, *TB* the tangent bifurcation, *SNB* the saddle-node bifurcation. The dashed arrows indicate the mechanism of formation of the hysteresis loop.

Now, if we decrease r, the periodic orbit that emerged near $r \approx 636.52$ persists until the parameter r reaches the threshold $r \approx 617.43$ at which the unstable and stable manifolds of the periodic orbit collide, resulting in a saddle-node bifurcation, and eventually, the 2-periodic attractor merges with the 1-periodic attractor, as illustrated in Fig. 8.2(b) in terms of bifurcation diagram. Thus, we have seen that in $617.43 \lessapprox r \lessapprox 636.52$ with increasing r the system exhibits chaos via period-doubling bifurcations. Next, if r is increased beyond this range and then returned back, one observes stable periodic orbit in the above parameter zone. Therefore, in the critical zone different attractors coexist for a given value of r. This interesting phenomenon is known as hysteresis [36] and was found, for example, in thermal convection with non-Fourier heat-flux [37] where the hysteresis occurs in between pitchfork and saddle-node bifurcations of equilibrium, in electrochemistry experiment of iron in sulfuric acid [38] with the hysteresis bounded in between a saddle-node bifurcation of limit cycle and a subcritical Hopf bifurcation. In the present scenario, the hysteresis is sandwiched in between saddle-node and tangent bifurcations of limit cycle (see Fig. 8.4 for a schematic representation of the hysteresis loop). In the hysteresis regime for system (8.2.7), one can find coexistence of 2-periodic attractor with 2^n-periodic attractors ($n \in \mathcal{N} \cup \{0\}$) or chaotic attractors. For instance, when $r = 620$ a

2-periodic attractor excited from the initial conditions (0,1,0,9) coexists with a pair of 1-periodic attractors excited from $(0,1,0,\pm1)$, see Fig. 8.5(a). Similarly, when $r = 635.6$ we found a 2-periodic attractor from initial conditions (0,1,0,0) coexists with two chaotic attractors excited from $(0,1,0,\pm2)$, see Fig. 8.5(b). The basin of attraction of the periodic-chaotic coexisting attractors is presented in Fig. 8.6 that shows the complex structure of the basin sets. The Lyapunov exponents of the chaotic attractors in Fig. 8.5(b) are $(0.883, 0, -2.226, -15.323)$ with the Kaplan-Yorke dimension [41] $D_{KY} \simeq 2.397$.

Now, in order to explore the effects of the magnetic parameters on the multistability behavior, we first fix (r, Q) at (635.6,1000), and vary the parameter σ_m. Since $r > 1 + \alpha Q$, the equilibrium origin is unstable. By analyzing the eigenvalues of the Jacobian matrix of the corresponding linearized system, the non-trivial equilibria are asymptotically stable as long as $\sigma_m > \sigma_m^{\text{Hopf}} \approx 3.46358$. As σ_m decreases through σ_m^{Hopf}, a limit cycle (periodic) attractor is excited non-catastrophically when σ_m attains $\sigma_m = \sigma_m^{\text{Hopf}}$. Therefore, the system can exhibit multiple coexisting attractors when $\sigma_m < \sigma_m^{\text{Hopf}}$.

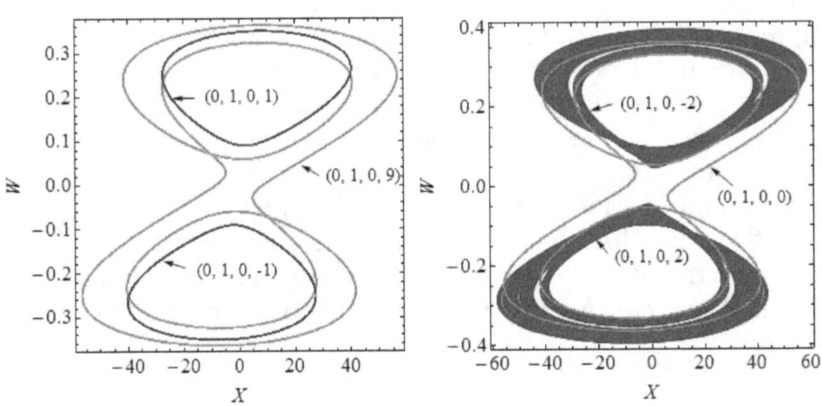

(a) (b)

Figure 8.5 Figure shows the phase portraits onto the X -W plane of the multistable attractors of the system (8.2.7): (a) a 2-periodic attractor from initial conditions (0,1,0,9) coexists with a pair of 1-periodic attractors from the initial conditions $(0,1,0,\pm1)$ for $r = 620$, and (b) a 2-periodic attractor from initial conditions (0,1,0,0) coexists with two chaotic attractors excited from the initial conditions $(0,1,0,\pm2)$ for $r = 635.6$. Here $\sigma = 10$, $b = 8/3$, $Q = 1000$ and $\sigma_m = 3.0$.

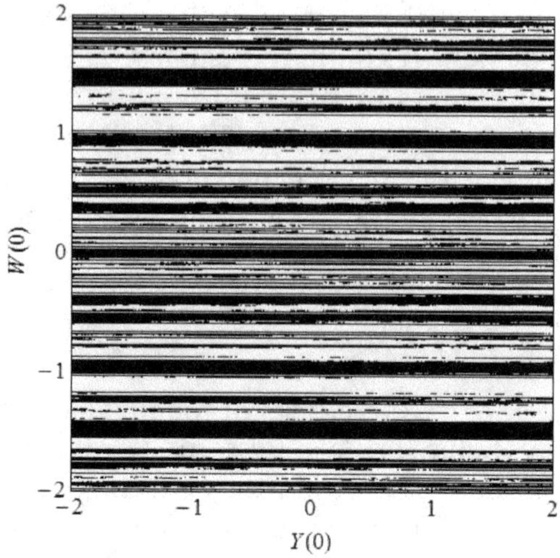

Figure 8.6 Cross-section of the attractor basin with $X(0) = Z(0) = 0$ for periodic-chaotic coexisting attractors at $r = 635.6$ (Fig. 8.5(b)). The basin sets of chaotic (resp. periodic) attractors are highlighted with yellow (resp. black) color.

Figure 8.7(a) displays the bifurcation diagrams as a function of $\sigma_m \in [0,3.8]$. It shows different ranges of σ_m where hysteresis phenomena with multiple stable states come into existence. For example, in $2.997 \leq \sigma_m \leq 3.04$, the hysteresis occurs in between a saddle-node and a tangent bifurcation of limit cycle. On the other hand, the hysteresis loops in $1.24 \leq \sigma_m \leq 1.254$ and $2.31 \leq \sigma_m \leq 2.34$ are bounded in between two saddle-node bifurcations of limit cycles. In $2.997 \leq \sigma_m \leq 3.04$, the system exhibits periodic-periodic and periodic-chaotic coexisting attractors, similar as shown in Fig. 8.5. In the other two regions, only periodic-periodic coexisting attractors are found. Since σ_m is very large for liquid metals, multiple coexisting attractors may not occur in convection of such fluids.

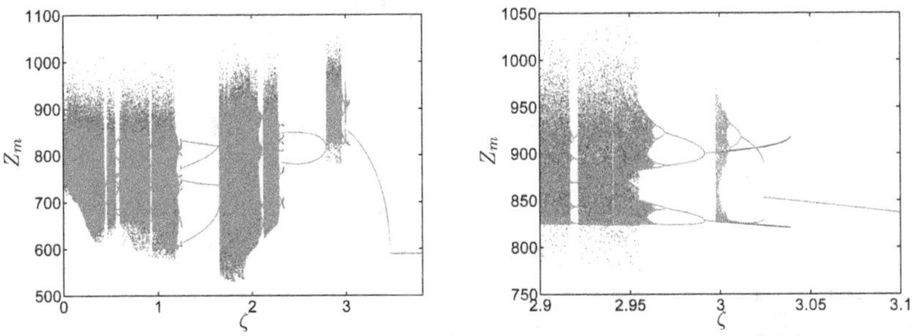

(a) (b)

Figure 8.7 (a) Bifurcation diagrams showing the variation of Z_m, local maxima of Z, as a function of $\sigma_m \in [0,3.8]$ for system (8.2.7) with $r = 635.6$, $\sigma = 10$ and $Q = 1000$. (b) Magnification of figure (a) in $2.9 \leq \sigma_m \leq 3.1$.

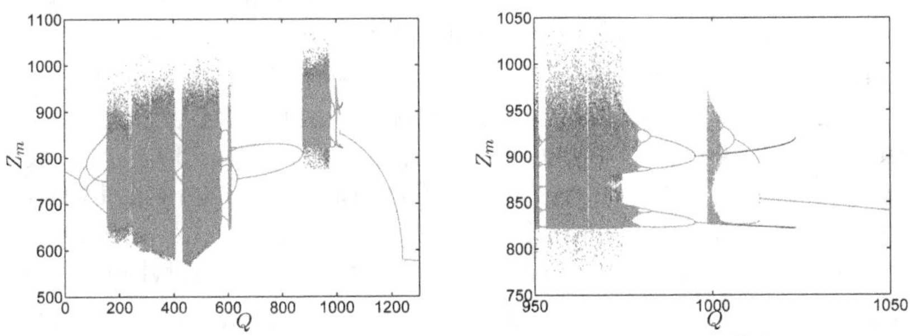

(a) (b)

Figure 8.8 (a) Bifurcation diagrams for variation of local maxima of Z w.r.t the parameter $Q \in [0,1300]$ for system (8.2.7) with $r = 635.6$, $\sigma = 10$ and $\sigma_m = 3$. (b) Magnification of the diagram (a) for $950 \leq Q \leq 1150$.

Similarly, for the effects of the applied magnetic field on the multistability behavior, we fix $r = 635.6$, $\sigma_m = 3$, and vary the Chandrasekhar number Q. Linear stability analysis reveals that the system is asymptotically stable to fixed point attractors if $Q > Q_{\text{Hopf}} \approx 1238.51$, and undergoes a non-catastrophic Hopf bifurcation giving birth to a periodic attractor when $Q = Q_{\text{Hopf}}$. Therefore, multistability is possible only when $Q < Q_{\text{Hopf}}$. The bifurcation diagrams presented in Fig. 8.8 show that multiple coexisting attractors occur when Q takes values in the range $998.5 \leq Q \leq 1013$. Surprisingly, this multistability occurs in a hysteresis regime bounded in between saddle-node and tangent bifurcations, same as illustrated in Figs. 8.2(b) and 8.7(b).

8.4.2 Coexisting multiple attractors for system (8.2.8)

We now consider the system (8.2.8). As mentioned earlier, for $\sigma_m = 3$ and $\omega_0 = 10$ system (8.2.8) exhibits a jump to stable periodic orbit near the subtle Hopf bifurcation when $r = r_{\text{Hopf}}^{(h)} \approx 43.1304$. For the evolution of the 1-periodic dynamics past the bifurcation, we plot bifurcation diagrams (see Fig. 8.9(a)) in terms of the variation of local maxima of Z w.r.t r. As previous, we use blue (resp. magenta) color to highlight the bifurcation diagrams for increasing (resp. decreasing) r. From the bifurcation diagrams we see that r is increased beyond $r_{\text{Hopf}}^{(h)}$ the periodic orbit undergoes chaos through period-doubling cascades. Thereafter, the system evolves through chaotic regimes with windows of regular oscillations. Finally, at $r = r_c^{(h)} \approx 103.77$, the two-scroll chaotic attractor switches into a three-band asymmetric chaotic attractor via a boundary crisis. Now, if r is decreased, the three-band chaotic attractor undergoes $3 \times 2^n (n \in \mathcal{N} \cup \{0\})$ period-halving bifurcations to an asymmetric period-3 orbit and ultimately merges into a two-scroll chaotic attractor through tangent bifurcation when r is decreased to $r_t^{(h)} \approx 102.77$. Thus, a hysteresis window can be obtained in the range $r_t^{(h)} \le r \le r_c^{(h)}$ (see Fig. 8.9(b) for enlargement of the bifurcation diagrams in the range $102.5 \le r \le 104.5$. Figure 8.10 for the Lyapunov exponents). In the hysteresis regime, coexisting periodic-chaotic and chaotic-chaotic attractors can be obtained depending upon the parameter r and the chosen initial conditions. For instance, when $r = 103$ we obtained two period-3 asymmetric periodic attractors coexist with a two-scroll chaotic attractor, as displayed in Fig. 8.11. For the chaotic attractor presented in Fig. 8.11(a), the all four Lyapunov exponents are computed as $(1.120, 0, -4.947, -12.840)$ with the Kaplan-Yorke dimension $D_{KY} \approx 2.226$. Similarly, when $r = 103.7$ a two-scroll chaotic attractor obtained from the initial conditions $(13,10,94,2)$ coexists with a pair of asymmetric chaotic attractors excited from $(\pm 14, \pm 10, 94, 2)$, see Fig. 8.12. For the two-scroll chaotic attractor, the Lyapunov exponents are obtained as $(1.127, 0, -5.188, -12.606)$ with the Kaplan-Yorke dimension $D_{KY} \approx 2.217$, and those for the asymmetric weak chaotic attractors in Fig. 8.12(b) the Lyapunov exponents are $(0.202, 0, -2.605, -14.264)$ with $D_{KY} \approx 2.078$.

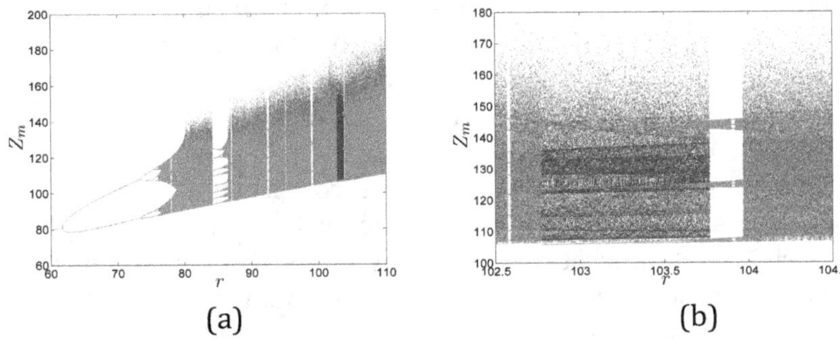

Figure 8.9 (a) Bifurcation diagrams showing the evolution of the system (8.2.8) through different dynamic regimes w.r.t $r \in [60,110]$ for $\sigma = 10$, $\omega_0 = 10$, $\sigma_m = 3$, and $b = 8/3$. The diagram in (b) is an enlargement of the bifurcation diagrams in (a) for $102.5 \le r \le 104.5$.

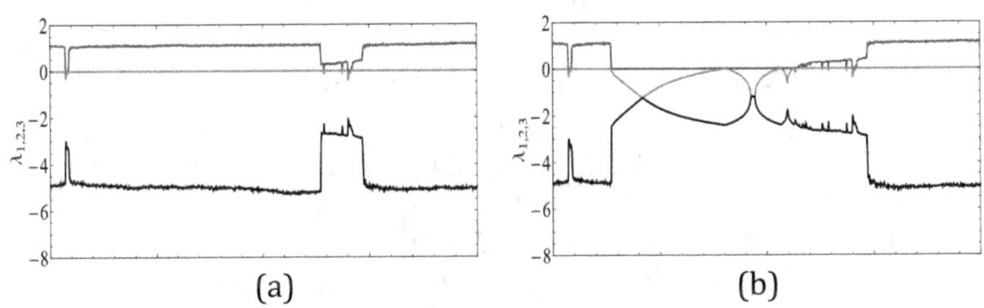

Figure 8.10 Variations of the three leading Lyapunov exponents $\lambda_{1,2,3}$ $(\lambda_1 \ge \lambda_2 \ge \lambda_3)$ with respect to the reduced Rayleigh number $r \in [102.5,104.5]$ for system (8.2.8). We have taken the remaining parameters as the same as in Fig. 8.9. Sub-figure (a) is obtained for increasing r and sub-figure (b) is for decreasing r.

8.5 Summary and conclusions

The interactive dynamical behaviors and multiple coexisting attractors depending on initial states and the change in control parameters are complex in nature. We explore theoretically coexistence of multiple stable states in magnetoconvection of electrically conducting fluid heated from below. Two low-dimensional dynamical systems are considered. One of them represents the magneto-convection for vertically imposed magnetic field and the other for horizontally imposed magnetic field. A comparative study of the bifurcation sequences enroute to multistability in the two systems is performed. The role of magnetic field on convection patterns is very crucial. The hysteresis phenomenon appeared which is intricate in magnetoconvective motion.

For primary instability, the bifurcation through which the conduction state losses stability is either pitchfork (non-oscillatory) or Hopf (oscillatory) bifurcation depending upon the magnetic parameters. On the other hand, for secondary instability the bifurcation that leads steady convection to unsteady is always Hopf type. However, the type of Hopf bifurcation depends strongly on the magnetic parameters. In absence of magnetic field, both the systems reduce to the Lorenz system. The Hopf bifurcation in the Lorenz system is always subcritical. The magnetic field renders the subcritical bifurcation to supercritical. The systems are found to exhibit multistable behaviors in the regime of unsteady magnetoconvection. For example, for system (8.2.7) when $Q = 1000$ and $\sigma_m = 3$ the unsteady convection occurs at $r = r^{(v)}_{\text{Hopf}} \approx 512.36$. As r exceeds $r^{(v)}_{\text{Hopf}}$, a stable limit cycle is excited via the Hopf bifurcation. For further increase in r, the periodic orbit undergoes a hysteresis occurs in between tangent and saddle-node bifurcations when r attains the critical range $617.43 \lessapprox r \lessapprox 636.52$. In the hysteresis regime, coexisting periodic-periodic and periodic-chaotic attractors are obtained depending on the control parameter and the initial conditions. Similar dynamics occurs if the applied magnetic field is varied. For example, when $r = 635.6$ and $\sigma_m = 3$, the system is asymptotically stable when the Chandrasekhar number Q is larger than $Q_{Hopf} \approx 1238.51$. At $Q = Q_{Hopf}$, a stable periodic orbit emerged via supercritical Hopf bifurcation, so that multistability occurs when $Q < Q_{Hopf}$. A critical range of Q, viz., $998.5 \le Q \le 1013$ is found where hysteresis-induced multistability comes into existence. As earlier, the hysteresis is sandwiched between saddle-node and tangent bifurcations, and the system exhibits periodic-periodic and periodic-chaotic coexisting attractors in this range. For the effects of the diffusivities on the emergence of coexisting attractors, we have taken $r = 635.6$, $Q = 1000$, and vary the diffusivity ratio σ_m. Then qualitative analysis reveals that the convection is stable as long as $\sigma_m > \sigma_m^{Hopf} \approx 3.46358$. As σ_m decreases through σ_m^{Hopf}, oscillatory convection occurs and a stable limit cycle appears via supercritical Hopf bifurcation. Multistability occurs when $\sigma_m < \sigma_m^{Hopf}$.

Three distinct ranges of σ_m are identified where multistability occurs via hysteresis. In $1.24 \leq \sigma_m \leq 1.254$ and $2.31 \leq \sigma_m \leq 2.34$, the hysteresis are bounded between two saddle-node bifurcations of limit cycles so that only periodic-periodic coexisting attractors exist in these ranges. On the other hand, in $2.997 \leq \sigma_m \leq 3.04$ the hysteresis happens in between saddle-node and tangent bifurcations and coexisting periodic-periodic and periodic-chaotic attractors are obtained in this range.

For system (8.2.8), when $(\sigma_m, \omega_0) = (3,10)$ the non-trivial equilibrium points become unstable via subcritical Hopf bifurcation occurs at $r_{Hopf}^{(h)} \approx 43.1304$. Note that the classical Lorenz system exhibits a discontinuous transition to chaos in the neighborhood of subcritical Hopf bifurcation when $\sigma = 10$, $b = 8/3$. On the other hand, due to inclusion of magnetic field the system exhibits a discontinuous transition to stable periodic orbit near the subcritical Hopf bifurcation at $r = r_{Hopf}^{(h)}$ when, for example, $(\sigma_m, \omega_0) = (3,10)$. When $r > r_{Hopf}^{(h)}$, there exists a range $102.77 \lessapprox r \lessapprox 103.77$ where a window of hysteresis in between a crisis and a tangent bifurcation is noted. In this hysteresis regime, coexisting chaotic-periodic and chaotic-chaotic attractors are obtained depending on the chosen initial conditions. Surprisingly, these unusual behaviors are not encountered in the absenceof applied magnetic field (i.e., in the classical Lorenz system [25]) or for sufficiently large magnetic field applicable in electrically conducting liquid metal flow. Finally, this is a theoretical study to be needed experimental evidence. Also, one can focus on dynamics in bi-parameter space for identifying different organized scenarios and transitional patterns for distribution of periodic windows in chaos regime [14, 39, 40].

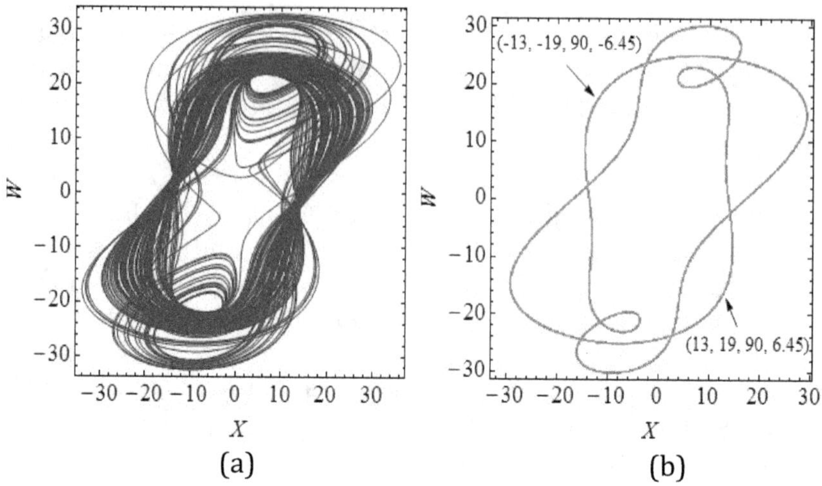

Figure 8.11 Phase-diagrams showing chaotic-periodic coexisting attractors for system (8.2.8) when $r = 103$, $\omega_0 = 10$, $\sigma_m = 3$. The diagram in (a) represents the chaotic attractor obtained from the initial conditions (12,19,90,6.45) and that in (b) is a period-3 periodic attractors excited from $(\pm 13, \pm 19, 90, \pm 6.45)$.

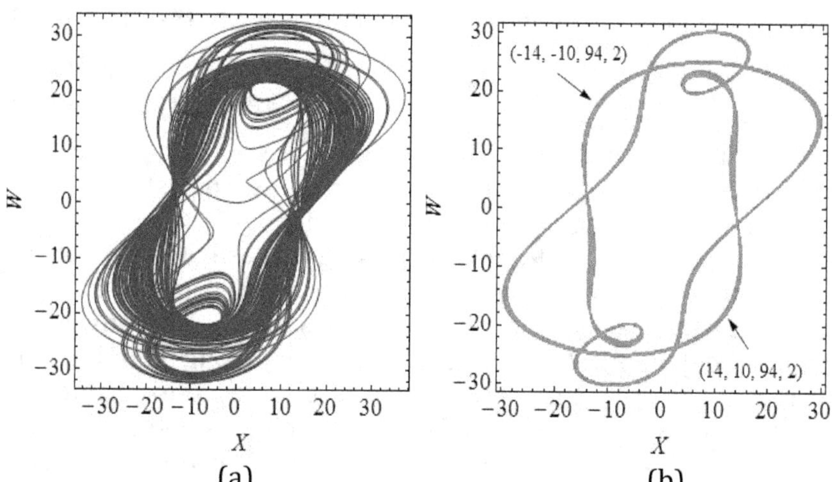

Figure 8.12 Figures show phase-portraits in the X -W plane of coexisting multi-chaotic attractors for system (8.2.8) when $r = 103.7$, $\omega_0 = 10$, $\sigma_m = 3$: (a) a two-scroll chaotic attractor obtained from the initial conditions (13,10,94,2) and (b) two asymmetric chaotic attractors excited from $(\pm 14, \pm 10, 94, 2)$.

References

[1]Feudel, U.:"Complex dynamics in multistable systems," Int. J. Bifurcation Chaos **18**, 1607–1626 (2008).

[2]Attneave, F.: "Multistability in perception," Scientific American **225**, 62–71 (1971).

[3]Bao, B. C., Li, Q. D., Wang, N. & Xu, Q.: "Multistability in Chua's circuit with two stable node-foci," Chaos **26**, 043111 (2016).

[4]Chen, M., Xu, Q., Lin, Y., Bao, B.: "Multistability induced by two symmetric stable node-foci in modified canonical Chua's circuit," Nonlinear Dyn. **87**, 789-802 (2017).

[5]Kengne, J., Njitacke, Z.T., Nguomkam Negou, A., Fouodji Tsotsop, M. & Fotsin, H. B.:"Coexistence of multiple attractors and crisis route to chaos in a novel chaotic jerk circuit," Int. J. Bifurcation Chaos **25**, 1550052 (2015).

[6]Kengne, J., Njitacke, Z. T., Fotsin, H. B.:"Dynamical analysis of a simple autonomous jerk system with multiple attractors," Nonlinear Dyn. **83**, 751–765 (2016).

[7]Kengne, J., Njikam, S.M., Signing, V. R. F.: "A plethora of coexisting strange attractors in a simple jerk system with hyperbolic tangent nonlinearity," Chaos, Solitons & Fractals **106**, 201-213 (2018).

[8]Gelens, L., Beri, S., Van der Sande, G., Mezosi, G., Sorel, M., Danckaert, J. & Verschaffelt, G.:"Exploring multistability in semiconductor ring lasers: theory and experiment," Phys. Rev. Lett. **102**, 193904 (2009).

[9]Kengne, J., Tabekoueng, Z. N., Tamba, V. K. & Negou, A. N.: "Periodicity, chaos, and multiple attractors in a memristor-based Shinriki's circuit, " Chaos **25**, 103126 (2015).

[10] Bao, B., Jiang, T., Xu, Q. et al.: "Coexisting infinitely many attractors in active band-pass filter-based memristive circuit, " Nonlinear Dyn. **86**, 1711-1723 (2016).

[11] Wang, G., Yuan, F., Chen, G. & Zhang, Y.:"Coexisting multiple attractors and riddled basins of a memristive system," Chaos **28**, 013125 (2018).

[12] Cordo, P., Inglis, J. T., Verschueren,S., Collins, J. J., Merfeld, D. M., Rosenblum, S., Buckley, S. & Moss, F.:"Noise in human muscle spindles," Nature **383**, 769 (1996).

[13] Laurent, M. & Kellershohn, N.:"Multistability: a major means of differentiation and evolution in biological systems," Trends Biochem. Sci. **24**, 418–422 (1999).

[14] Layek, G. C. & Pati, N. C.: "Organized structures of two bidirec-

tionally coupled logistic maps," Chaos **29**,093104 (2019).
[15] Kim, S., Park, S.H. & Ryu, C.S.:"Multistability in coupled oscillator systems with time delay," Phys. Rev. Lett. **79**, 2911–2914 (1997).
[16] Pati, N. C., Rech, P. C., & Layek, G. C.:"Multistability for nonlinear acoustic-gravity waves in a rotating atmosphere,", Chaos **31**, 023108 (2021).
[17] Li, C. & Sprott, J. C.:"Multistability in the Lorenz system: a broken butterfly," Int. J. Bifurcation Chaos **24**, 1450131 (2014).
[18] Lai,Q., Akgul, A., Zhao, X.W.& Pei, H.: "Various types of coexisting attractors in a new 4D autonomous chaotic system," Int. J. Bifurcation Chaos **27**, 1750142 (2017).
[19] Feudel, U., Grebogi, C., Hunt, B.R. & Yorke, J.A.: "Map with more than 100 coexisting low-period periodic attractors," Phys. Rev. E **54**, 71–81 (1996).
[20] Feudel, U., Grebogi, C.: "Multistability and the control of complexity," Chaos **7**, 597 (1997).
[21] Feudel, U., Pisarchik, A. N., Showalter, K.: "Multistability and tipping: from mathematics and physics to climate and brain- mini-review and preface to the focus issue," Chaos **28**, 033501 (2018).
[22] Ashwin, P., Wieczorek, S., Vitolo, R., Cox, P.: "Tipping points in open systems: bifurcation, noise-induced and rate-induced examples in the climate system," Phils. Trans. R. Soc. A **370**, 1166–1184 (2012).
[23] Weiss, N.O. & Proctor, M.R.E.: Magnetoconvection, Cambridge University Press, Cambridge (2014).
[24] Layek, G. C., & Pati, N.C.: "Bifurcations and hyperchaos in magnetoconvection of non-Newtonian fluids," International Journal of Bifurcation and Chaos **28**, 1830034 (2018).
[25] Lorenz, E. N.: "Deterministic nonperiodic flow," J. Atmos. Sci. **20**, 130–141 (1963).
[26] Gotoda, H., Takeuchi, R., Okuno, Y., Miyano, T.: "Low-dimensional dynamical system for Rayleigh-Bénard convection subjected to magnetic field, J. Appl. Phys. **113**, 124902 (2013).
[27] Macek, W.M. & Strumik, M.: "Model for hydromagnetic convection in a magnetized fluid," Phys. Rev. E **82**, 027301 (2010).
[28] Macek, W.M. & Strumik, M.: "Hyperchaotic intermittent convection in a magnetized viscous fluid," Phys. Rev. Lett. **112**, 074502 (2014).
[29] Macek, W.M.: "Nonlinear dynamics and complexity in the generalized Lorenz system," Nonlinear Dyn. **94**, 2957–2968 (2018).

[30] Layek, G. C.: An Introduction to Dynamical Systems and Chaos, Springer, 2ed, 2024.

[31] Salivahanan, S. and Rengaraj, R. and Venkatakrishnan, G. R.: Control systems engineering, Pearson (2014).

[32] Guckenheimer, J. & Holmes, P.: Nonlinear Oscillations, Dynamical Systems, and Bifurcations of Vector Fields, Springer-Verlag, New York (1983).

[33] Sparrow, C.: The Lorenz equations: bifurcations, chaos, and strange attractors, Springer-Verlag, New York (1982).

[34] Wolf, A., Swift, J. B., Swinney, H. L. & Vastano, J. A.: "Determining Lyapunov exponents from a time series," Physica D **16**, 285–317 (1985).

[35] Schuster, H. G., Just, W.: Deterministic Chaos: an introduction, Wiley-VCH Verlag GmbH & Co. KGaA (2005).

[36] Alligood, K. T., Sauer, T. D. & Yorke, J. A.: Chaos: an introduction of dynamical systems, Springer-Verlag, New York (1996).

[37] Layek, G. C. & Pati, N. C.: "Bifurcations and chaos in convection taking non-Fourier heat-flux," Phys. Lett. A **381**, 3568–3575 (2017).

[38] Sazou, D., Karantonis, A. & Pagitsas, M.: "Generalized Hopf, saddle-node infinite period bifurcations and excitability during the electrodissolution and passivation of iron in a sulfuric acid solution," Int. J. Bifurcations Chaos **3**, 981–997 (1993).

[39] Layek, G. C. & Pati, N. C.: "Period-bubbling transition to chaos in thermo-viscoelastic fluid systems," Int. J. Bifurcations Chaos **30**, 2030013 (2020).

[40] Silva, C. F. & Rech, P. C. "Chaos suppression, hyperchaos, period-adding, and discontinuous spirals in a bidirectional coupling of Lorenz systems,", The European Physical Journal B **96**, 4 (2023).

[41] Kaplan, J. L. & Yorke, J. A.: Chaotic behavior of multidimensional difference equations. In: Peitgen, H. O., Walther, H. O. (eds), Functional Differential Equations and Approximation of Fixed Points Lecture Notes in Mathematics, vol **730**, Springer, Berlin, pp. 204–227 (1979).

Chapter 9

Sharing Text in a Secret Way using System of Equations and Cayley Hamilton Theorem

Nagadevi Bala Nagaram[1], Kala Raja Mohan[2], R. Narmada Devi [3] and Suresh Rasappan [4]

[1,2,3] Department of Mathematics, Vel Tech Rangarajan Dr. Sagunthala R&D Institute of Science and Technology, Avadi, Chennai – 600 062, Tamil Nadu, India
[4] Mathematics Section, University of Technology and Applied Sciences, Ibri. Ibri 516, Al Dhahirah Governorate, Sultanate of Oman

Abstract

Secret Information sharing is an immense power in this internet world to secure the information. It is widely used in system security, like mobile cloud computing network, transaction etc. The mathematical technique is approached to maintain the secrecy of the information. In sending secret code or text from sender to receiver without knowing unauthorized person's knowledge the sender uses the key where the mathematics play an vital role to retain the secrecy between them. Keys, which are mathematical values plugged into the algorithm, are used in the majority of cryptographic algorithms. The algorithm's secret is represented by the number 'T' if it instructs you to encipher a message by giving each letter a numerical value (A = 1, B = 2, and so on.) and multiplying the outputs by that number. There are many areas in mathematics is utilized for constructing keys in secret way. The process to define the secret code in unknown way to others which need to utilize the simple way or difficult one who knows the secret key they can only reveal the secret code. The reverse process where applicable in this technique is adapted to do this process. This paper construct system of equations to share the secret message. The conversion of the given information into secret message is known as encryption. It kept the original information secretly when have reached unauthorized persons. With the help of secret key one can redeem the original information. For this secrecy the mathematical technique formation of system of equations is considered. To reveal the original information the Cayley Hamilton theorem is applied.

Keywords: Cryptography, Data encryption, Decryption, System of Differential equations, Characteristic Polynomial

List of Notations and Abbreviations

AES-Advanced Encryption Standard
FS-Fullstop
J-Jacobian

9.1 Introduction

Encryption—the process of transforming plaintext, which is regular information, into cipher text, which is unintelligible was the only kind of cryptography that was commonly used. Decryption is the reverse process from plaintext to the unreadable cipher text. Information is concealed or coded using cryptography so that only the recipient of a communication can decipher it. Since ancient times, messages have been coded using the technique of cryptography, which is still utilized today in e-commerce, bank cards, and computer passwords.

Algorithms and ciphers used in contemporary cryptography, like 128-bit and 256-bit encryption keys, allow for the encryption and decryption of data. Modern ciphers are thought to be nearly impenetrable, such as the Advanced Encryption Standard (AES).
The practice of encoding data to ensure that only the person may read and process the information in a typical definition of cryptography. Cryptology, another name for this cyber security practice, incorporates a number of academic fields, including computer science, engineering. Information and communications are protected using codes in a process known as cryptography, ensuring that only the intended audience may access and utilize the data.

This area of cyber security, also known as cryptology, combines elements of mathematics, engineering, and computer science to produce sophisticated codes that conceal a message's true intent.

Despite its long history, stretching back to the hieroglyphics of ancient Egypt, cryptography is still essential for protecting communication and information while it is in transit and preventing unauthorized parties from reading it. It employs methods like cryptographic keys and digital signature to convert messages into hard-to-decipher codes that preserve data privacy and secure credit card, email, and web surfing transactions.

In [1], A.P. Hiwarekar adapted Laplace Transform for encryption in cryptography technique. Following that M. Tuncay Gencoglu, Hemant K. et. al., G. Nagalakshmi, et. al., Swati Dhingra, et. al., were used Laplace transform for cryptographic technique[2,3,4,5]. Mampi Saha applied Laplace – Mellin Transform for Cryptography[6]. KalaRaja Mohan et. al. Applied Bilinear transform with probability for encryption to share the information secretly, also the same author adapted Laplace transform and hyperbolic tangent function and conformal mapping for the cryptographic technique[7,9].

Nagadevi Bala Nagaram et. al. employed ODE with variable separable method in cryptography[8]. Suresh Rasappan et.al. structured the mathematical model by system of equations for malaria transformation in our body[10]. Nagadevi Bala Nagaram et. al. constructed the mathematical model for plasmodium life cycle in their work[11]. The aim of this chapter is structured the system of equations for the text and utilized the Cayley Hamilton theorem to reveal the plain text is proposed.

Many concepts are utilized to maintain the secret between the sender and receiver. In this chapter, section 2 describes the standard definitions applied in this crypto analysis. Section 3 presents the algorithm for encryption. Section 4 demonstrates the algorithm explained in section 3 with an example. Section 5 depicts the algorithm applied for decryption. With the help of the cipher text obtained in Section 4, decryption process is explained in section 6. Section 7 is about the conclusion followed by references.

9.2 Standard Definition

The following standard definitions are applied in the cryptographic analysis in this paper.

9.2.1 Plain Text: Plain text is the data which can be directly read by any person. It is the information which has to be shared secretly to the receiver.

9.2.2 Cipher Text: The transformed form of plain text, which can be read only using the key specified by the sender is called the cipher text.

9.2.3 Cipher: Cipher refers to the algorithm through which the plain text is transformed to cipher text.

9.2.4 Encryption: Encryption is the process in which the given information is converted into secret message. This hides the original information reaching unauthorized persons, using the secret key.

9.2.5 Decryption: Decryption is the reverse process of Encryption. This is the process in which the encrypted text gets converted into original text with the usage of the secret key.

9.2.6 Jacobian Matrix: One of the most important mathematical analysis tools is the Jacobian matrix, which contains all first-order partial derivatives of system equations. It is a method of linearizing a system of nonlinear equations at a specific location such that the properties of the linear system can be used to conveniently solve and analyse that nonlinear system. The principles of partial derivatives, Jacobian matrices, and linearization of non-linear equations have been covered in this article along with the pertinent examples.

9.2.7 Cayley Hamilton Theorem: Every Square matrix satisfies its own characteristic equations.

9.3 Algorithm For Encryption

The following is a step-by-step process for performing encryption.

- The assumption is done by 1 to A, 2 to B and so on for Alphabet of 26 letters. It generates the plain text's numerical equivalent.
- The system of equation is structured with parameter values of α_1, α_2, α_3 and so on. This parameter is the numerical equivalent of the secret code.

- If the desired secret code is sent to the receiver then the system of equations is framed. The letters of secret code is assumed here as states and the numeric values for the respective letters of secret code is assigned for parameter value. The FULLSTOP(FS) state is taken in all cases. Thus, the system of equations is obtained in any nxn matrix.
- The Jacobian matrix is calculated for the linear system can be used to conveniently solve and analyse. Then the characteristic polynomial is obtained for the Jacobian matrix.
- The coefficient of $(n+1)^{th}$ degree polynomial is taken as the coded message.

9.4 Encryption Process

The word "MATHS" is chosen to demonstrate the encryption procedure, and its corresponding parameter numerical values are assigned in Table 9.1.

Table 9.1 Encryption Alphabet

α_1	α_2	α_3	α_4	α_5
13	1	20	8	19

The mathematical model for the secret code “MATHS” is presented in Figure 9.1, as follows:

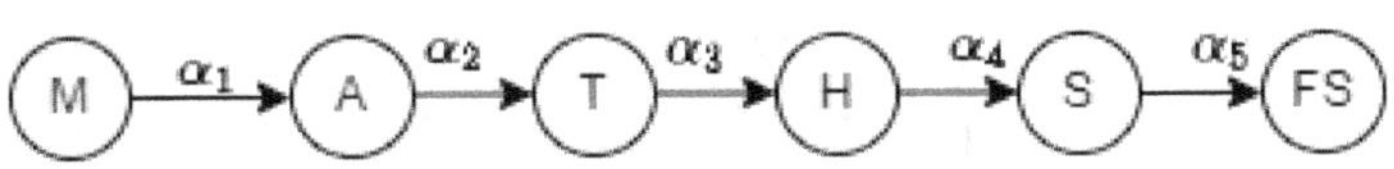

Figure 9.1 Mathematical model for the text MATHS.

The secret code “MATHS” is taken for this process. The system of equations is considered with parameters as follows:

$$\dot{M} = \alpha_1 A$$

$$\dot{A} = \alpha_2 T + \alpha_1 M$$

$$\dot{T} = \alpha_3 H + \alpha_2 A \tag{9.4.1}$$

$\dot{H} = \alpha_4 S + \alpha_3 T$

$\dot{S} = \alpha_5 S + \alpha_4 H$

$F\dot{S} = \alpha_5 S$

The letters of secret code are assumed here as states and the FS state is consider in the system of equations. The numeric values for the respective letters of secret code is assigned for parameters. Then $\dot{M}, \dot{A}, \dot{T}, \dot{H}, \dot{S}$ are considered here as moves from the present letter to successive letter respectively. Thus, the system of equations is obtained as 5x5 matrix. Then linearizing a system of linear equations (9.4.1) then obtained a Jacobian matrix (J) as follows:

$$J = \begin{bmatrix} 0 & \alpha_1 & 0 & 0 & 0 & 0 \\ \alpha_1 & 0 & \alpha_2 & 0 & 0 & 0 \\ 0 & \alpha_2 & 0 & \alpha_3 & 0 & 0 \\ 0 & 0 & \alpha_3 & 0 & \alpha_4 & 0 \\ 0 & 0 & 0 & \alpha_4 & 0 & \alpha_5 \\ 0 & 0 & 0 & 0 & \alpha_5 & 0 \end{bmatrix} \tag{9.4.2}$$

Assigned values for parameters are updated in the above matrix. Hence

$$J = \begin{vmatrix} 0 & 13 & 0 & 0 & 0 & 0 \\ 13 & 0 & 1 & 0 & 0 & 0 \\ 0 & 1 & 0 & 20 & 0 & 0 \\ 0 & 0 & 20 & 0 & 8 & 0 \\ 0 & 0 & 0 & 8 & 0 & 19 \\ 0 & 0 & 0 & 0 & 19 & 0 \end{vmatrix} \tag{9.4.3}$$

Then the characteristic polynomial is

$$\lambda^6 + 995\lambda^4 + 284250\lambda^2 + 24403600 = 0 \tag{9.4.4}$$

By Cayley Hamilton theorem, "Every square matrix satisfies its own characteristic equation". Hence, the equation (9.4.4) is written as

$$J^6 + 995J^4 + 284250J^2 + 24403600 = 0 \tag{9.4.5}$$

From the solutions obtained, the coded message is taken the coefficient of the Equation (9.4.5) as given in Table 9.2.

Table 9.2 Coded encryption message

1	0	995	0	284250	0	24403600

9.5 Algorithm For Decryption

Decryption is the reverse process of encryption. The cipher text is transformed into plain text in this step. The steps involved in the decryption process are as follows.

- The coefficient of characteristic equation is taken as cipher text.
- By Cayley Hamilton theorem, the 'J' matrix is obtained.
- By taking either super diagonal or sub diagonal values are written as in the order.
- The corresponding plain text character is marked.
- Omitting the FS at the end, the plain text is obtained.

9.6 Decryption Process

By Cayley Hamilton theorem the powers and inverse of the given matrix is calculated. Thus, to obtain J matrix Equation (9.4.4) is pre multiply by "J" as follows:

$$J = \frac{-1}{24403600}[J^7 + 995J^5 + 284250J^3] \tag{9.6.1}$$

From equation (9.6.1), the J matrix is calculated as below:

$$J = \begin{vmatrix} 0 & 13 & 0 & 0 & 0 & 0 \\ 13 & 0 & 1 & 0 & 0 & 0 \\ 0 & 1 & 0 & 20 & 0 & 0 \\ 0 & 0 & 20 & 0 & 8 & 0 \\ 0 & 0 & 0 & 8 & 0 & 19 \\ 0 & 0 & 0 & 0 & 19 & 0 \end{vmatrix}$$

Either super diagonal or subdiagonal values are arranged as in the order 13,1,20,8,19. The corresponding plain text character is marked as,

$$\mathbf{13 \rightarrow \mathit{M}, 1 \rightarrow \mathit{A}, 20 \rightarrow \mathit{T}, 8 \rightarrow \mathit{H}, 19 \rightarrow \mathit{S}}$$

Omitting the FS character at the end, the plain text is obtained AS "MATHS".

9.7 Conclusion

A new cryptographic algorithm applying Jacobian matrix and Cayley Hamilton theorem has been proposed. The plain text MATHS has been converted to cipher text using the proposed Encryption algorithm. Also, its reverse process using Decryption algorithm has been furnished to get the plain text.

References

[1] A.P. Hiwarekar, "New Mathematical Modeling for Cryptography", Journal of Information Assurance and Security, Vol. 9 (2014) pp. 027-033, 2014.

[2] M. Tuncay Gencoglu, "Cryptanalaysis of a New Method of Cryptography using Laplace Transform Hyperbolic Functions", Communications in Mathematics and Applications, Vol. 8, pp. 183-189, 2017.

[3] Dr. Hemant K. Undegaonkar, Security in Communication by Using Laplace Transform and Cryptography", International Journal of Scientific & Technology Research, Vol. 8, pp. 3207-3209, 2019.

[4] G. Nagalakshmi, A. Chandra Sekhar, D. Ravi Sankar, "Asymmetric key Cryptography using Laplace Transform", International Journal of Innovative Technology and Exploring Engineering, Vol. 9, pp. 3083-3087, 2020.

[5] Swati Dhingra, Archana A. Savalgi, Swati Jain, "Laplace Transformation based Cryptographic Technique in Network Security", International Journal of Computer Applications, Vol. 136, pp. 6-10, 2016.

[6] Mampi Saha, "Application of Laplace – Mellin Transform for Cryptography", Raj Journal of Technology Research & Innovation, Vol. V, pp. 12-17, 2017.

[7] Kala Raja Mohan, Suresh Rasappan, Regan Murugesan, Sathish Kumar Kumaravel and Ahamed A. Elngar, "Secret Infor-

mation Sharing Using Probability and Bilinear Transformation", Proceedings of 2nd International Conference on Mathematical Modeling and Computational Science, pp.115-122, 2022.

[8] Nagadevi Bala Nagaram, Kala Raja Mohan, Sathish Kumar Kumaravel, and Suresh Rasappan, "Secret Sharing of Information using First Order ODE and Variable Separable Method", Recent Trends in Computational Intelligence and Its Application, CRC Press, pp. 515-518, 2023.

[9] Kala Raja Mohan, Nagadevi Bala Nagaram, Regan Murugesan, Suresh Rasappan, "A Cryptographic Technique Using Conformal Mapping", Recent Trends in Computational Intelligence and Its Application, CRC Press, pp. 519-523, 2023.

[10] Suresh Rasappan, Nagadevi Bala Nagaram, "Stability analysis of a novel mathematical model of plasmodium life cycle in mosquito midgut", International Journal of Innovative Technology and Exploring Engineering, 2019, 8(9), pp. 1811–1813, 2019.

[11] Nagadevi Bala Nagaram, Suresh Rasappan, "A novel mathematical technique for stability analysis of plasmodium life cycle in hepatocyte", Indian Journal of Public Health Research and Development, 2019, 10(6), pp. 1534–1544, 2019.

Chapter 10

Study on Fuzzy Differential Equations using a Double Parametric Homotopy Approach

Parthkumar P. Sartanpara[1], Lalchand Verma[2], Ramakanta Meher[3]

Department of Mathematics, Sardar Vallabhbhai National Institute of Technology, Surat-395007, India

Abstract

This chapter provides an overview of several key concepts, including the fuzzy set, triangular fuzzy number, Á -cut, and double parametric approach. Moreover, we will explore the subject matter of fuzzy differential equations. The solution to the fuzzy differential equation is obtained by employing a unique double-parametric approach within the context of the homotopy analysis method. The efficacy of the homotopy analysis method in effectively tackling nonlinear differential equations has been well-established. The homotopy analysis method employs an additional parameter that provides a direct mechanism for adjusting and controlling the convergence region of the solution series. The initial value problem of the fuzzy differential equation has been successfully solved through the application of a novel double-parametric homotopy analysis method.

Keywords: Fuzzy number, Á -cut, Homotopy analysis method, convergence analysis.

10.1 Introduction

A fuzzy simulation serves as an effective means for scholars to engage in discourse pertaining to technological matters. Fuzzy set theory finds application in a diverse range of academic disciplines, encompassing robotics and some of the areas that are covered in this concentration. Zadeh [1] introduced the concept of fuzzy set theory for the first time in 1965. Fuzzy set theory is an effective methodology for characterizing situations in which data is characterized by ambiguity, imprecision, or uncertainty.

Later on, Dubois and Prade [2] established mathematical processes for working with fuzzy numbers. The trapezoidal, triangular, and Gaussian fuzzy numbers can be categorized into three distinct types. In order to ensure comprehensiveness, we will now examine the concept of fuzzy triangular numbers. The Fuzzy number is a variable that exhibits a range of values between 0 and 1. A membership grade is assigned to each numerical value within the range, with 0 representing the lowest grade and 1 indicating the highest possible grade. The data contained within concise ODEs or PDEs models of dynamical systems may occasionally require supplementation due to incompleteness, imprecision, or ambiguity. Fuzzy differential equations (FDEs) are valuable for representing and analyzing dynamical systems that exhibit ambiguity or uncertainty. Fuzzy or fuzzy triangular numbers can be mathematically utilized to define vagueness or impreciseness. Extensive scholarly investigations have been undertaken in recent years pertaining to the concept of fuzzy differential equations. The idea of fuzzy differentiability was first introduced by Seikala [3], and Kaleva [4], then they discussed fuzzy integration and differentiation. Buckley et al. [5] investigated the solution of FDEs in terms of extension principle and fuzzy numbers. The Cauchy problem for the continuation fuzzy differential equation investigated by Nieto [6]. The initial value problem for fuzzy differential equation discussed by Lakshmikantham and Mohapatra in [7]. Park and Hyo [8] employed the successive approximation approach for the existence and uniqueness solution of the fuzzy differential equation. The geometric approach for solving a system of FDEs was devised by Gasilov et al. [9]. The system of FDEs with fuzzy triangular numbers was investigated by Nizami et al. [10]. Salahsour et al. [11] used FDEs and triangular numbers to investigate the fuzzy logistic equation and the fuzzy alley effect.

In recent years many researchers have worked on a new type of approach in the fuzzy set, i.e., double parametric approach first time investigated by Behera and Chakraverty [12] to apply the system of the linear fuzzy differential equation. Further, Verma and Meher [13] have worked in fuzzy fractional fisher's equation under double parametric form. Sartanapara and Meher [14, 15] have analyzed the fuzzy fractional differential equation using the double parametric form. Verma and Meher have investigated the fuzzy fractional biological model using the double parametric approach [16]. The fuzzy time-fractional Kortewege-de Varies equation has been solved by a novel double parametric q-HAM [17]. They have discussed the crisp and uncertain case of the KdV model.

The homotopy analysis method was initially introduced by Shijun Liao in 1992, marking its first appearance in scholarly discourse. The aforementioned mathematical tool possesses significant computational capabilities in resolving nonlinear equations. The technique creates a convergent series solution for nonlinear systems using the topological homotopy notion. Homotopy-Maclaurin series to deal with non-linearities in the system has been used to enable it. In 1997, Liao introduced a convergence-control parameter, which is a nonzero auxiliary parameter, to make further modifications to a differential system in a general form, thereby establishing a homotopy. A nonphysical variable known as the convergence-control parameter offers a quick and easy technique to confirm and enforce the convergence of a solution series. In analytical and semi-analytic methods to nonlinear PDEs, the ability of HAM to naturally demonstrate the series solution's convergence is unique. In recent years many researchers have used HAM to solve the nonlinear differential equation, FDEs, and fuzzy DEs. Verma and Meher [18] have solved the fuzzy differential equation by HAM. Sartanpara and Meher [19] have used the HAM to study the Klein-Gordan equation. Dhirawat and Meher [20] employed HAM to solve nonlinear gas dynamic equations. The nonlinear partial differential equation has been solved using HAM by Kumar, and Meher [21]. Arqub et al. [22] have investigated the solution fuzzy initial value problem by HAM. The intuitionistic fuzzy DEs with the linear differential operator have been discussed via the HAM [23]. Fallahzadeh and Araghi [24] have used the HAM to solve the fuzzy Boussinesq equation. Abbasbandy and Liao [25] have to solve the nonlinear differential equation using the HAM. The researcher has used the HAM to solve the fluid flow problems. Turkyilmazoglu [26] have solved the boundary layer flow using the HAM. Using HAM, Cheng and Liao [27] have investigated the Von Karman swirling viscous flow. The utilization of the Homotopy Analysis Method (HAM) has been employed to examine the boundary layer flows of non-Newtonian fluids in the vicinity of a forward stagnation point, as discussed in the work by Xu et al. [28]. Additionally, the HAM has been applied to investigate nano-boundary layer flows incorporating a nonlinear Navier boundary condition, as explored by Cheng et al. [29]. Numerous researchers have recently employed the HAM to address fractional differential equations. The HAM has been used to obtain the solution of linear and nonlinear fractional diffusion and wave equation [30]. Jafari and Seifi [31] have analyzed the nonlinear FDEs to solve by HAM. Mohamed [32] has used the extension Khan's homotopy analysis transformation method to solve the generalized Abel's integral equations. The time-fractional linear and nonlinear fractional DEs have been discussed using Laplace HAM [33].

The primary aim of our research is to examine the characteristics and dynamics of the fuzzy differential equation. The solution to the fuzzy differential equation has been derived utilizing a novel double-parametric Homotopy Analysis Method (HAM). This chapter has addressed the topics of method convergence and error analysis. The solution of the fuzzy differential equation has been examined in both deterministic and uncertain scenarios.

This chapter is organized as Section 10.2, has discussed the basic definition of fuzzy set theory and its properties. The basic idea of the HAM has been discussed in section 10.3. Section 10.4 has presented several examples of the fuzzy differential equation. Finally, Conclusions are eventually drawn in Section 10.5

10.2 Basic definition on fuzzy set and its properties

In this discussion, we will explore the fundamental concept of fuzzy sets and their properties, which can be utilized for subsequent computational purposes.

Definition 10.2.1 Let R be a universal set and $z:\mathrm{R}\to[0,1]$ be a function, then A fuzzy set $\tilde{B}$ can be defined as a collection of a set of ordered pairs i.e.

$\tilde{B}=\{(p_0,z(p_0)):p_0\in\mathrm{R},z(p_0)\in[0,1]\}$

where z is the membership value or grade of function.

Definition 10.2.2 A fuzzy number is a fuzzy set $\tilde{B}$ if its membership function $z:\mathrm{R}\to[0,1]$, satisfying the following properties:

- It is fuzzy convex i.e.,

$$z(kp_o+(1-k)q_o \geq \min\{z(p_o),z(q_o)\}$$

where $k \in [0,1]$

- It is a normal i.e. $\exists\, z(x_0) = 1$.
- The Membership function is piece-wise continuous.
- The closure of Supp $(\tilde{B}) = \{p \in \mathrm{R}, z(p) > 0\}$ is compact.

Definition 10.2.3 A triangular fuzzy number (TFN) is a type of fuzzy number denoted as $\tilde{B} = [N_1, N_2, N_3]$. The mathematical representation of a TFN's membership function is provided as follows:

$$z(x) = \begin{cases} 0 & z\text{Ñ}N_1 \\ \dfrac{x - N_1}{N_2 - N_1} & N_1 < x\text{Ñ}N_2 \\ \dfrac{N_3 - x}{N_3 - N_2} & N_2\text{Ñ}x < N_3 \\ 0 & x\text{Ö}N_3 \end{cases}$$

Definition 10.2.4 Let $z \in \tilde{\mathrm{B}}$ and $\text{Á} \in [0,1]$,where Á is a crisps and defined as

$\tilde{B}_{\#} = \{x \in \mathrm{R} : z(x)\text{Ö}\,\text{Ą}\,\text{Á} \in [0,1]\}$

A given triangular fuzzy number $\tilde{B} = [N_1, N_2, N_3]$ can be written in an interval form by applying the idea of Á -cut, i.e., $B_{\text{Á}} = [\underline{B}(\text{Ą}), \overline{B}(\text{Ą})] = [(N_2 - N_1)\text{Á} + N_1, N_3 - (N_3 - N_2)\text{Ą}]$.

Definition 10.2.5 The parametric representation of the fuzzy number z is indicated by $[\underline{z}(\text{Ą}), \overline{z}(\text{Ą})]$. Let Á be a variable such that # belongs to the interval $[0,1]$. We can establish the following properties for $z(\text{Ą})$:

- $\underline{z}(\text{Ą})$ is left continuous, non-decreasing, bounded function over $[0,1]$,
- $\overline{z}(\text{Ą})$ is a left continuous, non-increasing, bounded function over $[0,1]$,

- $\underline{z}$(Á)Ñ $\overline{z}$(Á).

Definition 10.2.6 In a fuzzy, Á-cuts can be expressed as an interval form $U=[\underline{U},\overline{U}]$. The U can be expressed as follow:
U(Á ρ) $= \rho(\overline{U}$(Á) $- \underline{U}$(Á)) $+ \underline{U}$(Á), Á $\rho \in [0,1]$

The TFN may also be expressed as a discretized parametric form with two parameters, Á and ρ, referred to as the double parametric form concept.

Definition 10.2.7 Let $k \in \mathrm{R}$ be any scalar, and $\tilde{\Phi}=[\underline{\Phi},\overline{\tilde{\Phi}}]$ & $\tilde{\rho}=[\underline{\rho},\overline{\tilde{\rho}}]$ be two fuzzy numbers, which is satisfying the following properties:

$$\tilde{\rho}+\tilde{\Phi}=(\underline{\Phi}+\underline{\Phi},\overline{\tilde{\rho}}+\overline{\tilde{\Phi}}) \tag{10.2.1}$$

$$\tilde{\rho}-\tilde{\Phi}=(\underline{\rho}-\underline{\Phi},\overline{\tilde{\rho}}-\overline{\tilde{\Phi}}) \tag{10.2.2}$$

$$k\tilde{\Phi}=\begin{cases}(k\underline{\Phi},k\overline{\tilde{\Phi}}), k\text{Ö}0\\(k\overline{\tilde{\Phi}},k\underline{\tilde{\Phi}}), k<0\end{cases} \tag{10.2.3}$$

Definition 10.2.8 Let $\sigma,\rho \in \mathrm{E}$ be fuzzy number and $\mathrm{D}:\mathrm{E}\times\mathrm{E}\rightarrow\mathrm{R}$ be a fuzzy function. The H-distance is

$$\mathrm{D}(\sigma(\xi),\rho(\xi))=max\{sup_{\xi\in[0,1]}|\underline{\sigma}(\xi)-\underline{\rho}(\xi)|,sup_{\xi\in[0,1]}|\overline{\sigma}(\xi)-\overline{\rho}(\xi)|\} \tag{10.2.4}$$

Definition 10.2.9 A fuzzy mapping $\varphi:[a_1,a_2]\rightarrow\mathrm{E}$ is said to be continuous at point $x_0\in[a_1,a_2]$ if for any $\varepsilon>0$ and $\exists$ $\delta>0$ s.t.
$D(\varphi(x),\varphi(x_0))<\varepsilon$ when $|x-x_0|<\delta$

Definition 10.2.10 Let $\rho,\sigma\in\mathrm{E}$ be two fuzzy number and $\exists$ $\chi\in\mathrm{E}$ s.t. $\rho=\sigma+\chi$. Then χ is called H-difference of ρ and

σ, denoted by $\rho\Theta\sigma$.

Definition 10.2.11 Let $\varphi:(a_1,a_2)\to\mathrm{E}$ be a function which is generalized differentiable at point $y_0\in(a_1,a_2)$. If $\exists\ \varphi'(y_0)\in\mathrm{E}$ s.t.

- For any $h>0$ for sufficiently small, $\exists$ $\varphi(y_0+h)\Theta F(y_0)$ and $\varphi(y_0)\Theta\varphi(y_0-h)$, following limit hold:

$$lim_{h\to 0}\frac{\varphi(y_0+h)\Theta\varphi(y_0)}{h}=lim_{h\to 0}\frac{\varphi(y_0)\Theta\varphi(y_0-h)}{h}=\varphi'(y_0)$$

or

- For any $h>0$ for sufficiently small, $\exists$ $\varphi(y_0)\Theta\varphi(y_0+h)$ and $\varphi(y_0-h)\Theta\varphi(y_0)$, following limit hold:

$$lim_{h\to 0}\frac{\varphi(y_0)\Theta\varphi(y_0+h)}{-h}=lim_{h\to 0}\frac{\varphi(y_0-h)\Theta\varphi(y_0)}{-h}=\varphi'(y_0)$$

or

- For any $h>0$ for sufficiently small, $\exists$ $\varphi(y_0+h)\Theta\varphi(y_0)$ and $\varphi(y_0-h)\Theta F(y_0)$, following limit hold:

$$lim_{h\to 0}\frac{\varphi(y_0+h)\Theta\varphi(y_0)}{h}=lim_{h\to 0}\frac{\varphi(y_0-h)\Theta\varphi(y_0)}{-h}=\varphi'(y_0)$$

or

- For any $h>0$ for sufficiently small, $\exists$ $\varphi(y_0)\Theta\varphi(y_0+h)$ and $\varphi(y_0)\Theta\varphi(y_0-h)$, the following limit hold:

$$lim_{h\to 0}\frac{\varphi(y_0)\Theta\varphi(y_0+h)}{-h}=lim_{h\to 0}\frac{\varphi(y_0)\Theta\varphi(y_0-h)}{h}=\varphi'(y_0)$$

Definition 10.2.12 Consider a fuzzy valued function $\varphi:(a,b)\to\mathrm{E}$, where E is a set. Let $\varphi(y_0,r)=[\underline{\varphi}(y_0,\xi),\overline{\varphi}(y_0,\xi)]$ be an interval representation of the fuzzy value of φ at a point (y_0,r). Let y_0 be an element of the interval $[a,b]$ and ξ be an element of the interval $[0,1]$. Then

- If φ is (i)-differentiable function, then $\underline{\varphi}(y_0,\xi)$ and $\overline{\varphi}(y_0,\xi)$ are differentiable function, and $[\varphi'(y_0)]_\xi=[\underline{\varphi}'(y_0,\xi),\overline{\varphi}'(y_0,\xi)]$.

- If φ is (i)-differentiable function, then $\underline{\varphi}(y_0,\xi)$ and $\overline{\varphi}(y_0,\xi)$ are differentiable function, and $[\varphi'(y_0)]_\xi = [\overline{\varphi}'(y_0,\xi), \underline{\varphi}'(y_0,\xi)]$.

Remark 1: In this chapter, we assume (i)-differential exists, so there is no need to consider (ii)-differential.

Definition 10.2.13 Let us consider an n-th order fuzzy initial value problem of the form

$$\begin{cases} S^n(t) = F(t, S(t), S'(t), ..., S^{n-1}(t)) \\ S(t_0) = S_n \end{cases}$$

where $F:[t_0,T]\times \mathrm{E}^n \to \mathrm{E}$ is continuous and $X_n \in \mathrm{E}$, E is a fuzzy number, $n = 1,2,..,i$. This fuzzy IVP has a unique solution if it satisfies the following properties:

- Let $F:[t_0,T]\times \mathrm{E}^n \to \mathrm{E}$ is continuous and bounded. Then at least one solution exists to the fuzzy initial value problem on the interval $[t_0, T]$.
- The fuzzy IVP has a unique solution if the function $F:[t_0,T]\times \mathrm{E}^n \to \mathrm{E}$ satisfies the Lipschitz condition, i.e.,

$$d_\infty(F(t,S,S',..S^{n-1}), F(t,S_1,S_1',...,S_1^{n-1})) \\ \tilde{\mathrm{N}} k_1 d_\infty(S,S_1) + k_2 d_\infty(S',S_1') + ... + k_n d_\infty(S^{n-1},S_1^{n-1})$$

10.3 Fuzzy homotopy analysis method

This section discusses the brief idea of the HAM [34] which has been discussed here for the completeness of our discussion.

Let us consider fuzzy differential equation

$$M[\tilde{\Theta}(\eta, t, \acute{\mathrm{A}}\rho)] = 0 \tag{10.3.1}$$

The symbol M represents a non-linear operator, while η and t represent independent variables.

The function $\tilde{\Theta}(\eta,t,\acute{A}\rho)$ is an unknown fuzzy function. To maintain simplicity, we will exclude all initial and boundary conditions, which can be addressed similarly. In the study by Liao in 1992, the zeroth-order deformation equation was formulated by applying the standard homotopy method, which was extended to accommodate this purpose.

$$(1-p)L[\tilde{\psi}(\eta,t,\acute{A}\rho;q)-\tilde{\Theta}_0(\eta,0,\acute{A}\rho)] \\ = p\check{\ } \tilde{H}(\eta,t,\acute{A}\rho)M[\tilde{\psi}(\eta,t,\acute{A}\rho;q)] \quad (10.3.2)$$

where L stands for auxiliary linear operator, $\tilde{\Theta}_0(\eta,0,\acute{A}\rho)$ stands for an initial guess of $\tilde{\Theta}(\eta,t,\acute{A}\rho)$ and $\tilde{\psi}$ stands for unknown function. $q\in[0,1]$ $\check{\ }\neq 0$ is a non-zero auxiliary function, $\tilde{H}(\eta,t,\acute{A}\rho)\neq 0$ is a non-zero auxiliary parameter. Having a lot of freedom in HAM regarding supplementary items is critical. It holds when $q=0$ and $q=1$.

$$\tilde{\psi}(\eta,t,\acute{A}\rho;0)=\tilde{\Theta}_0(\eta,0,\acute{A}\rho),\tilde{\psi}(\eta,t,\acute{A}\rho;1)=\tilde{\Theta}(\eta,t,\acute{A}\rho) \quad (10.3.3)$$

Respectively. Thus, as q moves from 0 to 1, the solution $\tilde{\psi}(\eta,t,\acute{A}\rho;q)$ moves the initial guess from $\tilde{\Theta}_0(\eta,0,\acute{A}\rho)$ to $\tilde{\Theta}(\eta,t,\acute{A}\rho)$.

Now, let us consider the expansion of $\tilde{\psi}(\eta,t,\acute{A}\rho;q)$ in a Taylor series with respect to q, we have

$$\tilde{\psi}(\eta,t,\acute{A}\rho;q)=\tilde{\Theta}_0(\eta,0,\acute{A}\rho)+\sum_{n=1}^{+\infty}\tilde{\Theta}_n(\eta,t,\acute{A}\rho)q^n \quad (10.3.4)$$

where

$$\tilde{\Theta}_n(\eta,t,\acute{A}\rho)=\frac{1}{n!}\frac{\partial^n\tilde{\psi}(\eta,t,\acute{A}\rho;q)}{\partial q^n}\Big|_{q=0} \quad (10.3.5)$$

which implies the series (10.3.4) converges at $q=1$. Then we have

$$\tilde{\psi}_n(\eta,t,\acute{A}\rho)=\tilde{\Theta}_0(\eta,t,\acute{A}\rho)+\sum_{n=1}^{+\infty}\tilde{\Theta}_n(\eta,t,\acute{A}\rho) \quad (10.3.6)$$

Define the vector

$$\vec{\tilde{\Theta}}_m = \{\tilde{\Theta}_0(\eta,t,\acute{A}\rho),\tilde{\Theta}_1(\eta,t,\acute{A}\rho),\tilde{\Theta}_2(\eta,t,\acute{A}\rho),\ldots,\tilde{\Theta}_m(\eta,t,\acute{A}\rho)\}$$

By taking the nth derivative of Eq. (10.3.2) w.r.t. the embedding parameter q, evaluating at $q=0$, and dividing by $n!$, we obtain the equation representing the deformation of order n, denoted as n-order deformation equation.

$$L[\tilde{\Theta}_n(\eta,t,\acute{A}\rho)-\chi_n\tilde{\Theta}_{n-1}(\eta,t,\acute{A}\rho)] = ˇ\tilde{H}(\eta,t,\acute{A}\rho)r_n(\vec{\tilde{\Theta}}_{n-1}(\eta,t,\acute{A}\rho), \quad (10.3.7)$$

where

$$R_n(\tilde{\Theta}_{n-1},\zeta) = \frac{1}{n!}\frac{\partial^n\tilde{\psi}(\eta,t,\acute{A}\rho;q)}{\partial q^n}\bigg|_{q=0} \quad (10.3.8)$$

and

$$\chi_n = \begin{cases} 0, & nÑ1 \\ 1, & n>1 \end{cases} \quad (10.3.9)$$

On operating the inverse operator L^{-1} on both side of equation (10.3.7), we get

$$\tilde{\Theta}_n(\eta,t,\acute{A}\rho) = \chi_n\tilde{\Theta}_n(\eta,t,\acute{A}\rho) + ˇ\tilde{H}(\eta,t,\acute{A}\rho)L^{-1}[R^n(\overrightarrow{\tilde{\Theta}_n}(\eta,t,\acute{A}\rho)] \quad (10.3.10)$$

Thus in this process, one can easily find $\tilde{\Theta}_n(\eta,t,\acute{A}\rho)$ for $nÖ1$ and we have

$$\tilde{\Theta}(\eta,t,\acute{A}\rho) = \sum_{n=1}^{m}\tilde{\Theta}_n(\eta,t,\acute{A}\rho) \quad (10.3.11)$$

As $m \to +\infty$, the approximation solution converges to the exact solution of the equation (10.3.1).

Theorem 10.3.1 Let N be an operator defined from a Hilbert space R into R. The series solution $\sum_{m=0}^{\infty}\tilde{\Theta}_m(\eta,t,\acute{A}\rho)$,defined in Eq. (10.3.11) converges if $\ni 0<\gamma<1$ such that $\|\tilde{\Theta}_{m+1}\| \le \gamma\|\tilde{\Theta}_m\|$ for every $mÖm_0$ and for some $m_0 \in \mathrm{N}$.

Proof. We define the sequence $\{\tilde{\aleph}_m\}_{m=0}^{\infty}$

we have,

$$\begin{cases} \tilde{\aleph}_0 = \tilde{\Theta}_0 \\ \tilde{\aleph}_1 = \tilde{\Theta}_0 + \tilde{\Theta}_1 \\ \tilde{\aleph}_1 = \tilde{\Theta}_0 + \tilde{\Theta}_1 + \tilde{\Theta}_2 \\ \cdot \\ \cdot \\ \cdot \\ \tilde{\aleph}_m = \tilde{\Theta}_0 + \tilde{\Theta}_1 + \tilde{\Theta}_2 + \ldots + \tilde{\Theta}_m \end{cases} \tag{10.3.12}$$

To show that $\{\tilde{\aleph}_m\}_{m=0}^{\infty}$ is a Cauchy sequence defined in a Hilbert space R.

Now

$$\begin{aligned} ||d(\tilde{\aleph}_{m+1}, \tilde{\aleph}_m|| &= ||d(\tilde{\Theta}_{m+1}, \tilde{0})|| \leq \gamma||d(\tilde{\Theta}_m, \tilde{0})|| \\ &\leq \gamma^2||d(\tilde{\Theta}_{m-1}, \tilde{0}) \ldots + \ \leq \gamma^{m-k_0+1}||d(\tilde{\Theta}_{k_0}, \tilde{0})|| \end{aligned} \tag{10.3.13}$$

for every $m, n \in \mathrm{N}$ and $m Ö n$, we have,

$$\begin{aligned} \left|\left|d(\tilde{\aleph}_{m+1}, \tilde{\aleph}_m)\right|\right| &= \left|\left|d(\tilde{\aleph}_m, \tilde{\aleph}_{m-1}) + d(\tilde{\aleph}_{m-1}, \tilde{\aleph}_{m-2}) + d(\tilde{\aleph}_{m-2}, \tilde{\aleph}_{m-3}) + \cdots \right.\right. \\ &\quad \left.\left. + d(\tilde{\aleph}_{n+1}, \tilde{\aleph}_n)\right|\right| \\ &\leq ||d(\tilde{\aleph}_m, \tilde{\aleph}_{m-1})|| + ||d(\tilde{\aleph}_{m-1}, \tilde{\aleph}_{m-2})|| + |d(\tilde{\aleph}_{m-1}, \tilde{\aleph}_{m-2})|| + \cdots + ||d(\tilde{\aleph}_{n+1}, \tilde{\aleph}_n)|| \\ &\leq \gamma^m \left|\left|d(\tilde{\Theta}_0, \tilde{0})\right|\right| + \gamma^{m-1}\left|\left|d(\tilde{\Theta}_0, \tilde{0})\right|\right| + \cdots + \gamma^{n+1}\left|\left|d(\tilde{\Theta}_0, \tilde{0})\right|\right| \\ &= \frac{1-\gamma^{m-n}}{1-\gamma}\gamma^{m+n}||d(\tilde{\Theta}_0, \tilde{0})|| \end{aligned} \tag{10.3.14}$$

and $0 < \gamma < 1$,

So $\lim_{m,n\to\infty} || s_m - s_n || = 0$

i.e. $\{A_m\}_{m=0}^{\infty}$ is a Cauchy sequence in a Hilbert space R . Hence the series solution $\sum_{m=0}^{\infty}\tilde{\Theta}_n(\eta,t,Á\rho)$,defined in Eq. (10.3.11) converges.

Theorem 10.3.2 The "maximum absolute truncation error of the series" $\tilde{\Theta}(Á\rho,\eta,t)=\sum_{l=0}^{m}\tilde{\Theta}_l(Á\rho,\eta,t)$ to the example 10.1 and 10.2 by using FHAM is estimated as

$$e_l(Á\rho,\eta,t)=d(\tilde{\Theta}(Á\rho,\eta,t),\tilde{\Theta}_m(Á\rho,\eta,t))Ñ\frac{1}{1-\gamma}\gamma^{m+1}d(\tilde{\Theta}_0,\tilde{0})$$

Proof. From Theorem 10.1, According inequality (10.3.14), we get

$$d(\tilde{\aleph}_m,\tilde{\aleph}_n)=\frac{1-\gamma^{m-n}}{1-\gamma}\gamma^{m+1}d(\tilde{\Theta}_{k_0},\tilde{0})$$

for $mÖn$ then $\lim_{m,n\to\infty}\tilde{s}_m=\tilde{\Theta}(Á\rho,\zeta)$. So,

$$d(\tilde{\Theta}(Á\rho,\zeta),\sum_{l=0}^{m}\tilde{\Theta}_l(Á\rho,\zeta))Ñ\frac{1-\gamma^{m-n}}{1-\gamma}\gamma^{m+1}d(\tilde{\Theta}_{k_0},\tilde{0})$$

Also, since $0<\gamma<1$ we have $(1-\gamma^{m-n})$ so we obtained the result. Hence proof theorem 10.2.

10.4 Examples of fuzzy differential equation

Example 10.4.1 Let us consider the fuzzy differential equation is

$$\tilde{\Theta}'(t)=t^2\otimes\tilde{\Theta}(t)\oplus(1,2,3)\otimes t^2 \quad (10.4.1)$$

and the initial condition is

$$\tilde{\Theta}(0)=(1,2,3) \quad (10.4.2)$$

Equations (10.4.1) and (10.4.2) written as Á-cut, we get

$$\begin{cases} \underline{\Theta}'(t) = t^2\underline{\Theta}(t) + (\acute{A}+1)t^2 \\ \overline{\Theta}'(t) = t^2\overline{\Theta}(t) + (3-\acute{A})t^2 \end{cases} \tag{10.4.3}$$

and initial condition

$$\begin{cases} \underline{\Theta}(0) = (\acute{A}+1) \\ \overline{\Theta}(0) = (3-\acute{A}) \end{cases} \tag{10.4.4}$$

We define the nonlinear operator as follows

$$\begin{cases} N(\underline{P}(\acute{A}t,q)) = \underline{P}'(\acute{A}t,q) - t^2\underline{P}(\acute{A}t,q) - (\acute{A}+1)t^2 \\ N(\overline{P}(\acute{A}t,q)) = \overline{P}'(\acute{A}t,q) - t^2\overline{P}(\acute{A}t,q) - (\acute{A}+1)t^2 \end{cases} \tag{10.4.5}$$

where $\underline{P}(\acute{A}t,q), \overline{P}(\acute{A}t,q)$ are unknown functions and q is embedding parameters, q increases 0 to 1 $\underline{P}(\acute{A}t,q), \overline{P}(\acute{A}t,q)$ varies from $\underline{\Theta}_0(\acute{A}t,q), \overline{\Theta}_0(\acute{A}t,q)$ to $\underline{\Theta}(\acute{A}t), \overline{\Theta}(\acute{A}t)$ i.e.

$$\begin{cases} \underline{P}(\acute{A}t,0) = \underline{\Theta}_0(\acute{A}t),\ \overline{P}(\acute{A}t,0) = \overline{\Theta}_0(\acute{A}t) \\ \underline{P}(\acute{A}t,1) = \underline{\Theta}(\acute{A}t),\ \overline{P}(\acute{A}t,1) = \overline{\Theta}(\acute{A}t) \end{cases} \tag{10.4.6}$$

Expanding $\underline{P}(\acute{A}t,q), \overline{P}(\acute{A}t,q)$ in Taylor series with respect to q, we get

$$\begin{cases} \underline{P}(\acute{A}t,q) = \underline{\Theta}_0(\acute{A}t) + \sum_{n=1}^{\infty}\underline{\Theta}_n(\acute{A}t)q^n \\ \overline{P}(\acute{A}t,q) = \overline{\Theta}_0(\acute{A}t) + \sum_{n=1}^{\infty}\overline{\Theta}_n(\acute{A}t)q^n \end{cases} \tag{10.4.7}$$

where

$$\underline{\Theta}_n(\acute{A}t) = \frac{1}{n!}\frac{\partial^n \underline{P}}{\partial q^n} \overline{\Theta}_n(\acute{A}t) = \frac{1}{n!}\frac{\partial^n \overline{P}}{\partial q^n} \tag{10.4.8}$$

The linear operator can be defined as

$$L(\underline{P}(Ą t)) = \underline{P}'(Ą t),\ L(\overline{P}(Ą t)) = \overline{P}'(Ą t) \tag{10.4.9}$$

with property

$$L(c) = 0,\ L(d) = 0 \tag{10.4.10}$$

where c and d are constant. To satisfy the initial condition, we must guess the initial value of $\underline{\Theta}(Ą t), \overline{\Theta}(Ą t)$. According to the principles of solution expression and initial condition, the selection of the initial guess is

$$\underline{\Theta}_0(Ą t) = (Á-1),\ \overline{\Theta}_0(Ą t) = (3-Ą) \tag{10.4.11}$$

The zero deformation equation is

$$\begin{cases} (1-q)L(\underline{P}(Ą t,q) - \Theta_0(Ą t)) = qh\underline{H}(Ą t)N(\underline{P}(Ą t,q)) \\ (1-q)L(\overline{P}(Ą t,q) - \Theta_0(Ą t)) = qh\overline{H}(Ą t)N(\overline{P}(Ą t,q)) \\ \underline{P}(Ą 0,q) = 0,\ \overline{P}(Ą 0,q) = 0 \end{cases} \tag{10.4.12}$$

Differentiating equation (10.4.12) n times with respect to q, and dividing by $n!$ and from equation (10.4.5) to (10.4.11) and properly chosen $\underline{H}(Ą t) = \overline{H}(Ą t) = 1$. We have "the n-th order deformation equation for $n Ö 1$" represented as follows.

$$\begin{cases} \underline{\Theta}_n(Ą t) = \chi_n \underline{\Theta}_{n-1} + h\int_0^t R_n(\overrightarrow{\Theta}_{n-1})dt + c \\ \overline{\Theta}_n(Ą t) = \chi_n \overline{\Theta}_{n-1} + h\int_0^t R_n(\overrightarrow{\Theta}_{n-1})dt + d \\ \underline{\Theta}_n(Ą 0) = 0,\ \overline{\Theta}_n(Ą 0) = 0 \end{cases} \tag{10.4.13}$$

where

$$\begin{cases} R_n(\vec{\Theta}_{n-1}(\text{Ą}t)) = \underline{\Theta}'(\text{Ą}t) - t^2\sum_{i=0}^{n-1}\underline{\Theta}_i(\text{Ą}t) - (\text{Á}+1)t^2 \\ R_n(\vec{\Theta}_{n-1}(\text{Ą}t)) = \overline{\Theta}'(\text{Ą}t) - t^2\sum_{i=0}^{n-1}\overline{\Theta}_i(\text{Ą}t) - (\text{Á}+1)t^2 \end{cases} \tag{10.4.14}$$

and

$$\chi = \begin{cases} 0 & n\text{Ñ}1 \\ 1 & n > 1 \end{cases} \tag{10.4.15}$$

and following procedure of HAM, $\underline{\Theta}, \overline{\Theta}$ can expressed in terms of series form as

$$\begin{cases} \underline{\Theta}(\text{Ą}t) = \underline{\Theta}_0(\text{Ą}t) + \sum_{n=1}^{\infty}\underline{\Theta}_n(\text{Ą}t) \\ \overline{\Theta}(\text{Ą}t) = \overline{\Theta}_0(\text{Ą}t) + \sum_{n=1}^{\infty}\overline{\Theta}_n(\text{Ą}t) \end{cases} \tag{10.4.16}$$

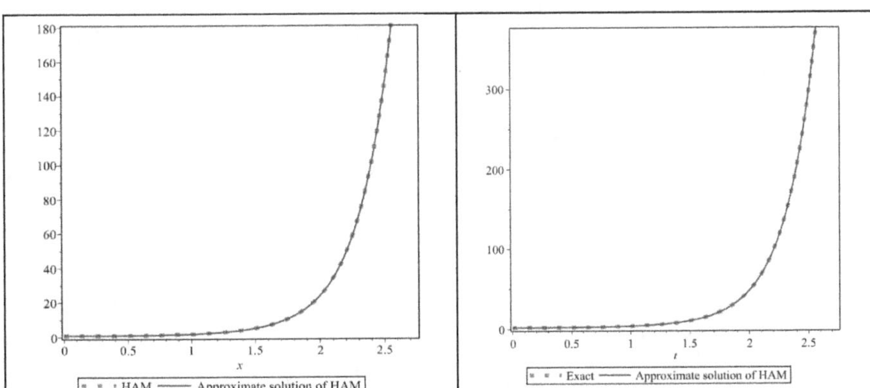

(a) $\text{Á}=0, h=-1$	(b) $\text{Á}=1, h=-1$

Figure 10.1 Comparison between exact and approximate solution of HAM of Lower bound for the value $\text{Á}=0$ and $\text{Á}=1$ of example 10.1.

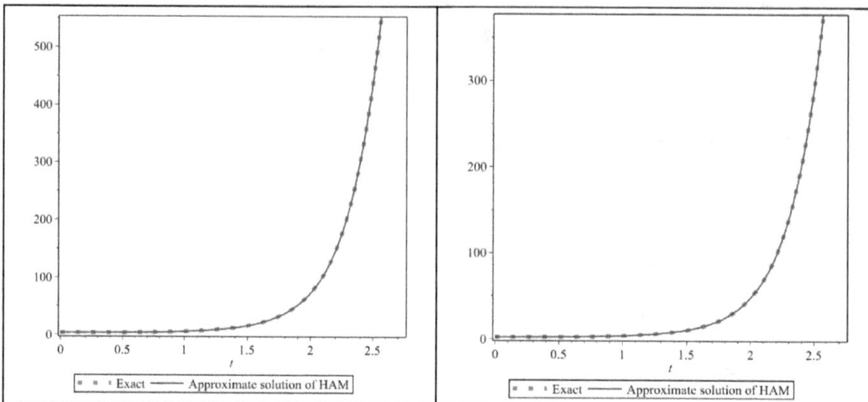

(a) $\acute{A}=0, h=-1$ (b) $\acute{A}=1, h=-1$

Figure 10.2 Comparison between exact and approximate solution of HAM of Upper bound for the value $\acute{A}=0$ and $\acute{A}=1$ of example 10.4.1.

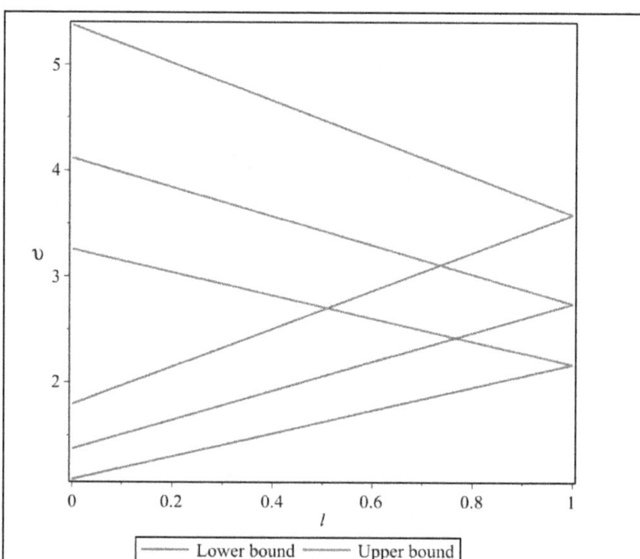

$t=0.5, 0.8, 1 \quad h=-1$

Figure 10.3 The lower bound and upper bound solution for the fixed value $t=0.5, 0.8, 1$ and different values of $\acute{A}$ of example 10.4.1.

In this context, Figures 10.1 and 10.2 illustrate the comparison between the exact and approximate solutions of the lower and upper bound solutions. These comparisons are made for various values of t while keeping the values of Á fixed at 0 and 1 in example 10.1. The graph illustrates that the HAM solution exhibits high proximity to the exact solution. Figure 10.3 depicts the lower and upper bound solutions for fixed values of $t = 0.5, 0.8, 1$ and varying values of Á in Example 10.1.

Example 10.4.2 Let us consider non-linear differential equation is given by

$$\begin{cases} \tilde{\Theta}'(t) = \tilde{\Theta}^2(t) + \tilde{\vartheta}(t) \\ \tilde{\vartheta}'(t) = \tilde{\vartheta}(t)cos(\tilde{\Theta}(t)) \end{cases} \tag{10.4.17}$$

subject to fuzzy initial condition is

$$\tilde{\Theta}(0) = (-1,0,1),\ \ \tilde{\vartheta}(0) = (0,1,2) \tag{10.4.18}$$

from above equation (10.4.17) can be written as Á-cut, we get

$$\begin{cases} [\underline{\Theta}'(t), \overline{\Theta}'(t)] = [\underline{\Theta}^2(t), \overline{\Theta}^2(t)] + [\underline{\vartheta}(t), \overline{\vartheta}(t)] \\ \quad = [\underline{\vartheta}(t), \overline{\vartheta}(t)][cos(\underline{\Theta}(t)), cos(\overline{\Theta}(t))] \end{cases} \tag{10.4.19}$$

and

$$\underline{\Theta}(0), \overline{\Theta}(0) = (Á+1, 1-Á),\ \ \underline{\vartheta}(0), \overline{\vartheta}(0) = (Á, 2-Á) \tag{10.4.20}$$

Equation (10.4.19) and (10.4.20) can be written as another parameter ρ, we have

$$\begin{cases} \{\rho[\overline{\Theta}'(t) - \underline{\Theta}'(t)] + \underline{\Theta}'(t)\} = \ \{\rho[\overline{\Theta}^2(t) - \underline{\Theta}^2(t)] + \underline{\Theta}^2(t)\} + \{\rho[\overline{\vartheta}(t) - \underline{\vartheta}(t)] + \underline{\vartheta}(t)\} \\ \{\rho[\overline{\vartheta}(t) - \underline{\vartheta}(t)] + \underline{\vartheta}(t)\} = \ \{\rho[\overline{\vartheta}(t) - \underline{\vartheta}(t)] + \underline{\vartheta}(t)\}\{\rho[cos(\overline{\Theta}(t)) - cos(\underline{\Theta}(t))] \\ \qquad + cos(\underline{\Theta}(t))\} \end{cases} \tag{10.4.21}$$

and

$$\begin{cases} \rho[\overline{\Theta}(0)-\underline{\Theta}(0)]+\underline{\Theta}(0)=\rho[-2\text{Á}]+\text{Á}+1, \\ \rho[\overline{\vartheta}(0)-\underline{\vartheta}(0)]+\underline{\vartheta}(0)=\rho[2-2\text{Á}]+\text{Á} \end{cases} \tag{10.4.22}$$

where $\text{Ą}\rho\in[0,1]$ be unknown parameters which control the fuzziness. Now we can write the according definition 10.5, we get

$$\{\rho[\overline{\Theta}'(t)-\underline{\Theta}'(t)]+\underline{\Theta}'(t)\}=\tilde{\Theta}'(\text{Ą}\rho,t) \tag{10.4.23}$$

$$\{\rho[\overline{\vartheta}'(t)-\underline{\vartheta}'(t)]+\underline{\vartheta}'(t)\}=\tilde{\vartheta}'(\text{Ą}\rho,t) \tag{10.4.24}$$

$$\{\rho[\overline{\Theta}(t)-\underline{\Theta}(t)]+\underline{\Theta}(t)\}=\tilde{\Theta}(\text{Ą}\rho,t) \tag{10.4.25}$$

$$\{\rho[\overline{\vartheta}(t)-\underline{\vartheta}(t)]+\underline{\vartheta}(t)\}=\tilde{\vartheta}(\text{Ą}\rho,t) \tag{10.4.26}$$

$$\{\rho[\overline{\Theta}^2(t)-\underline{\Theta}^2(t)]+\underline{\Theta}^2(t)\}=\tilde{\Theta}^2(\text{Ą}\rho,t) \tag{10.4.27}$$

$$\{\rho[cos(\overline{\Theta}(t))-cos(\underline{\Theta}(t))]+cos(\underline{\Theta}(t))\}=cos(\tilde{\Theta}(\text{Ą}\rho,t)) \tag{10.4.28}$$

$$\begin{cases} \rho[\overline{\Theta}(0)-\underline{\Theta}(0)]+\underline{\Theta}(0)=\tilde{\Theta}(\text{Ą}\rho,0), \\ \rho[\overline{\vartheta}(0)-\underline{\vartheta}(0)]+\underline{\vartheta}(0)=\tilde{\vartheta}(\text{Ą}\rho,0) \end{cases} \tag{10.4.29}$$

Using equation (10.4.23) to (10.4.29) in equation (10.4.21) and (10.4.22), we get

$$\begin{cases} \tilde{\Theta}'(\text{Ą}\rho,t)=\tilde{\Theta}^2(\text{Ą}\rho,t)+\tilde{\vartheta}(\text{Ą}\rho,t) \\ \tilde{\vartheta}'(\text{Ą}\rho,t)=\tilde{\vartheta}(\text{Ą}\rho,t)cos(\tilde{\Theta}(\text{Ą}\rho,t)) \end{cases} \tag{10.4.30}$$

and

$$\tilde{\Theta}(\text{Ą}\rho,0)=\rho[-2\text{Á}]+\text{Á}+1,\ \ \tilde{\vartheta}(\text{Ą}\rho,0)=\rho[2-2\text{Á}]+\text{Á} \tag{10.4.31}$$

Now we can define the two linear operators as

$$L_1(\text{Ą}\rho,t,q)=L_2(\text{Ą}\rho,t,q) \tag{10.4.32}$$

We define the nonlinear operator as follows:

$$\begin{cases} N_1(\tilde{K}_1(\text{Ą}\rho,t,q),\tilde{K}_2(\text{Ą}\rho,t,q))=\tilde{K}_1(\text{Ą}\rho,t,q)-\tilde{K}_1^2(\text{Ą}\rho,t,q)-\tilde{K}_2(\text{Ą}\rho,t,q) \\ N_2(\tilde{K}_1(\text{Ą}\rho,t,q),\tilde{K}_2(\text{Ą}\rho,t,q))=\tilde{K}_2(\text{Ą}\rho,t,q)-\tilde{K}_2(\text{Ą}\rho,t,q)cos(\tilde{K}(\text{Ą}\rho,t,q)) \end{cases}$$

(10.4.33)

and two linear operators as

$$L_1(Á\rho,t,q)=\tilde{\Theta}'(Á\rho,t,q); L_2(Á\rho,t,q)=\tilde{\vartheta}'(Á\rho,t,q) \qquad (10.4.34)$$

with property

$$L_1(c)=0,\ \ L_2(d)=0 \qquad (10.4.35)$$

where c, d is constant. The initial guess is chosen such that it satisfies the initial condition of problem; that is,

$$\tilde{\Theta}_0(Á\rho,t)=\rho[-2Á]+Á+1,\ \ \tilde{\vartheta}_0(Á\rho,t)=\rho[2-2Á]+Á \qquad (10.4.36)$$

The zero-order deformation equation is

$$\begin{cases}(1-q)L_1(\tilde{K}_1(Á\rho,t,q)-\Theta_0(Á\rho,t))=qh_1\tilde{H}(Á\rho,t)N_1(\tilde{K}_1(Á\rho,t,q),\tilde{K}_2(Á\rho,t,q))\\(1-q)L_2(\tilde{K}_2(Á\rho,t,q)-\vartheta_0(Á\rho,t))=qh_2\tilde{H}(Á\rho,t)N_2(\tilde{K}_1(Á\rho,t,q),\tilde{K}_2(Á\rho,t,q))\\ \Theta_0(Á\rho,0,q)=\vartheta_0(Á\rho,0,q)=0\end{cases}$$

(10.4.37)

Differentiating equation (10.4.37) n times with respect of q, and dividing by $n!$, and from equation (10.4.32) to (10.4.36), setting q=0, and properly chosen $\tilde{H}(Á\rho,t)=1$. "The n-th order deformation equation for nÖ1" can represented as follows

$$\begin{cases}L_1[\Theta_n(Á\rho,t)-\chi_n\Theta_{n-1}(Á\rho,t)]=h_1\vec{R}_1(\tilde{\Theta}_{n-1}(Á\rho,t),\tilde{\vartheta}_{n-1}(Á\rho,t))\\ L_2[\vartheta_n(Á\rho,t)-\chi_n\vartheta_{n-1}(Á\rho,t)]=h_1\vec{R}_2(\tilde{\Theta}_{n-1}(Á\rho,t),\tilde{\vartheta}_{n-1}(Á\rho,t))\end{cases}$$

(14.4.38)

where

$$\begin{cases}\vec{R}_1(\tilde{\Theta}_{n-1}(Á\rho,t),\tilde{\vartheta}_{n-1}(Á\rho,t))=\tilde{\Theta}'_{n-1}(Á\rho,t)-\sum_{i=1}^{n-1}\tilde{\Theta}_i\tilde{\Theta}_{n-1-i}(Á\rho,t)-\tilde{\vartheta}_{n-1}(Á\rho,t)\\ \vec{R}_2(\tilde{\Theta}_{n-1}(Á\rho,t),\tilde{\vartheta}_{n-1}(Á\rho,t))=\tilde{\vartheta}'_{n-1}(Á\rho,t)-\sum_{i=0}^{n-1}\tilde{\vartheta}(Á\rho,t)cos(\tilde{\Theta}_{n-1-i}(Á\rho,t))\end{cases}$$

(10.4.39)

and

$$\chi_n = \begin{cases} 0 & n\tilde{N}1 \\ 1 & n > 1 \end{cases} \tag{10.4.40}$$

Taking the inverse operator on both side of equation (10.4.38), we get

$$\begin{cases} \tilde{\Theta}_n(\acute{A}\rho,t) = \chi_n \tilde{\Theta}_{n-1}(\acute{A}\rho,t) + h_1 \int_o^{\infty} \overrightarrow{R_1}(\tilde{\Theta}_{n-1}(\acute{A}\rho,t), \tilde{\vartheta}_{n-1}(\acute{A}\rho,t)) \\ \tilde{\vartheta}_n(\acute{A}\rho,t) = \chi_n \tilde{\vartheta}_{n-1}(\acute{A}\rho,t) + h_2 \int_o^{\infty} \overrightarrow{R_2}(\tilde{\Theta}_{n-1}(\acute{A}\rho,t), \tilde{\vartheta}_{n-1}(\acute{A}\rho,t)) \\ \tilde{\Theta}_n(\acute{A}\rho,0) = \tilde{\vartheta}_n(\acute{A}\rho,0) = 0 \end{cases}$$

(10.4.41)

Hence the approximate solution of give fuzzy differential equation is

$$\tilde{\Theta}(\acute{A}\rho,t) = \sum_{n=0}^{\infty} \tilde{\Theta}_n(\acute{A}\rho,t)$$
$$\tilde{\vartheta}(\acute{A}\rho,t) = \sum_{n=0}^{\infty} \tilde{\vartheta}_n(\acute{A}\rho,t) \tag{10.4.42}$$

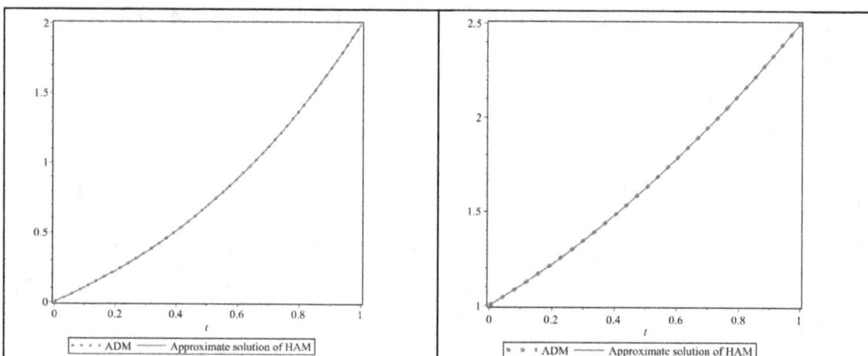

(a) $\acute{A}=0, h_1=-1, h_2=-1$	(b) $\acute{A}=1, h_1=-1, h_2=-1$

Figure 10.4 Comparison between ADM and approximate solution of HAM in the crisp case for different values of t and fixed values of $\acute{A}=1, \rho=1$ of $\tilde{\tilde{\Theta}}(\acute{A}\rho,t)$ and $\tilde{\vartheta}(\acute{A}\rho,t)$, respectively of example 10.2.

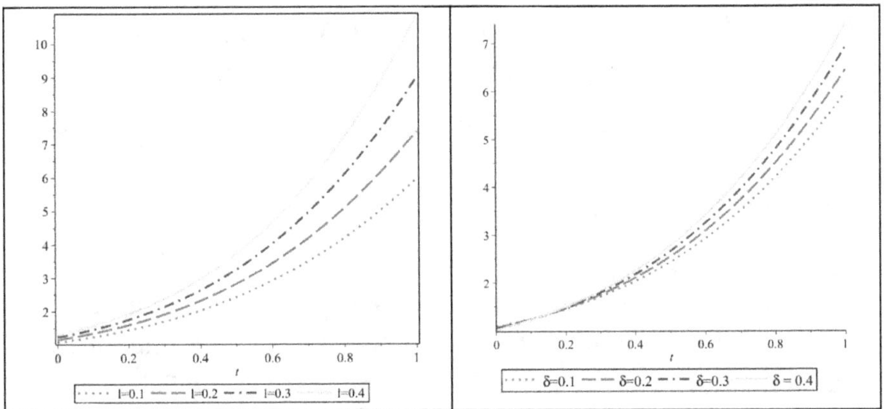

(a) $\rho = 0.1, h_1 = -1, h_2 = -1$ (b) $\# = 0.1, h_1 = -1, h_2 = -1$

Figure 10.5 The approximate solution of $\tilde{\tilde{\Theta}}(\acute{A}\rho, t)$ in uncertain case for different values of Á and ρ of example 10.4.2.

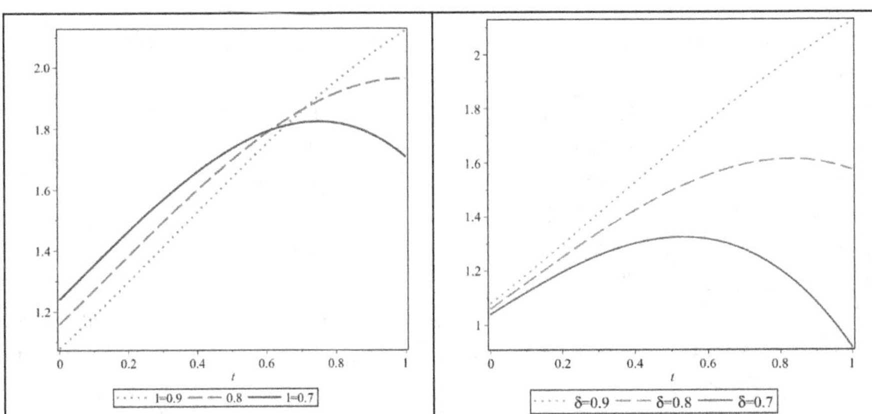

(a) $\rho = 0.9, h_1 = -1, h_2 = -1$ (b) $\acute{A} = 0.9, h_1 = -1, h_2 = -1$

Figure 10.6 The approximate solution of $\tilde{\vartheta}(\acute{A}\rho, t)$ in uncertain case for different values of Á and ρ of example 10.4.2.

Figure 10.4 represents the comparison between the ADM and HAM in crisp case, i.e., $\acute{A} = 0, \rho = 0$ and different values of t for the $\tilde{\tilde{\Theta}}(\acute{A}\rho, t) \,\&\, \tilde{\vartheta}(\acute{A}\rho, t)$, respectively of example 10.2. We have observed that the solution of the applied method is very close to the

ADM solution. The approximate solution of applied method of $\tilde{\tilde{\Theta}}(Á,\rho,t)$ and $\tilde{\vartheta}(Á,\rho,t)$ in uncertain case for different values of $Á$ and ρ are represents in the figure 10.5 and 10.6 respectively.

10.5 Conclusion

The HAM has been applied successfully to solve nonlinear differential equations under single and double parametric approach for crisp $Á\in\{0,1\}$ and uncertain case $Á\in(0,1)$ in this chapter. The obtained result has been compared with exact or available results in crisp and uncertain cases. This concludes that the HAM can be directly applied without linearizing or discretizing the nonlinear fuzzy differential equation.

References

[1] Z. Lotfi, et al., Fuzzy sets, Information and control 8(3), 338 (1965)

[2] D. Dubois, H. Prade, Operations on fuzzy numbers, International Journal of systems science 9(6), 613 (1978)

[3] S. Seikkala, On the fuzzy initial value problem, Fuzzy sets and systems 24(3), 319 (1987)

[4] O. Kaleva, Fuzzy differential equations, Fuzzy sets and systems 24(3), 301 (1987)

[5] J. Buckley, T. Feuring, Y. Hayashi, Linear systems of first order differential equations: fuzzy differential equations, Soft Comput 6, 415 (2002)

[6] J.J. Nieto, The Cauchy problem for continuous fuzzy differential equations, Fuzzy sets and systems 102(2), 259 (1999)

[7] V. Lakshmikantham, R. Mohapatra, Basic properties of solutions of fuzzy differential equations., Nonlinear Studies 8(1) (2001)

[8] J.Y. Park, H.K. Han, Existence and uniqueness theorem for a solution of fuzzy differential equations, International Journal of Mathematics and Mathematical Sciences 22(2), 271 (1999)

[9] N. Gasilov, S.G. Amrahov, A.G. Fatullayev, A geometric approach to solve fuzzy linear systems of differential equations, arXiv preprint arXiv:0910.4307 (2009)

[10] N.A. Gasilov, A.G. Fatullayev, S.E. Amrahov, Solution method for a non-homogeneous fuzzy linear system of differential equations, Applied Soft Computing 70, 225 (2018)
[11] S. Salahshour, A. Ahmadian, A. Mahata, S.P. Mondal, S. Alam, The behavior of logistic equation with alley effect in fuzzy environment: fuzzy differential equation approach, International Journal of Applied and Computational Mathematics 4(2), 1 (2018)
[12] D. Behera, S. Chakraverty, New approach to solve fully fuzzy system of linear equations using single and double parametric form of fuzzy numbers, Sadhana 40(1), 35 (2015)
[13] L. Verma, R. Meher, Solution for generalized fuzzy time-fractional Fisher's equation using a robust fuzzy analytical approach, Journal of Ocean Engineering and Science (2022). DOI https://doi.org/10.1016/j.joes.2022.03.019
[14] P.P. Sartanpara, R. Meher, A robust fuzzy-fractional approach for the atmospheric internal wave model, Journal of Ocean Engineering and Science, 8(3), 308-22, (2023)
[15] P.P. Sartanpara, R. Meher, Solution of generalised fuzzy fractional Kaup–Kupershmidt equation using a robust multi parametric approach and a novel transform, Mathematics and Computers in Simulation 205, 939 (2023)
[16] L. Verma, R. Meher, Z. Avazzadeh, O. Nikan, Solution for generalized fuzzy fractional Kortewege-de Varies equation using a robust fuzzy double parametric approach, Journal of Ocean Engineering and Science (2022)
[17] L. Verma, R. Meher, Study on generalized fuzzy fractional human liver model with Atangana–Baleanu–Caputo fractional derivative, The European Physical Journal Plus 137(11), 1 (2022)
[18] L. Verma, R. Meher, Solution of fuzzy differential equation using homotopy analysis method, in AIP Conference Proceedings, vol. 2451 (AIP Publishing LLC, 2022), vol. 2451, p. 020031
[19] P. Sartanpara, R. Meher, Computational study of Klein-Gordan equation using homotopy analysis method, in AIP Conference Proceedings, vol. 2451 (AIP Publishing LLC, 2022), vol. 2451, p. 020030
[20] K. Dhirawat, R. Meher, Semi-analytical Approach to Nonlinear Partial Differential Equations Using Homotopy Analysis Technique (HAM). Contemporary Mathematics. 4(4), 721-32, (2023).
[21] A. Kumar, R. Meher, Solving Non-linear Partial Differential Equations Using Homotopy Analysis Method (HAM), in Nonlinear Dynamics and Applications (Springer, 2022), pp. 955–964

[22] O. Abu-Arqub, A. El-Ajou, S. Momani, N. Shawagfeh, Analytical solutions of fuzzy initial value problems by HAM, Applied Mathematics & Information Sciences 7(5), 1903 (2013)

[23] S. Melliani, Z. Belhallaj, M. Elomari, L.S. Chadli, Approximate solution of intuitionistic fuzzy differential equations with the linear differential operator by the homotopy analysis method, Advances in Fuzzy Systems 2021 (2021)

[24] A. Fallahzadeh, M.A.F. Araghi, Homotopy analysis method for fuzzy Boussinesq equation, Mathematical Sciences 9(3), 145 (2015)

[25] S. Abbasbandy, Y. Tan, S. Liao, Newton-homotopy analysis method for nonlinear equations, Applied Mathematics and Computation 188(2), 1794 (2007)

[26] M. Turkyilmazoglu, A homotopy treatment of analytic solution for some boundary layer flows, International Journal of Nonlinear Sciences and Numerical Simulation 10(7), 885 (2009)

[27] C. Yang, S. Liao, On the explicit, purely analytic solution of von k´arm´an swirling viscous flow, Communications in Nonlinear Science and Numerical Simulation 11(1), 83 (2006)

[28] H. Xu, S.J. Liao, I. Pop, Series solution of unsteady boundary layer flows of non-Newtonian fluids near a forward stagnation point, Journal of non-newtonian fluid mechanics 139(1-2), 31 (2006)

[29] J. Cheng, S. Liao, R. Mohapatra, K. Vajravelu, Series solutions of nano boundary layer flows by means of the homotopy analysis method, Journal of Mathematical Analysis and Applications 343(1), 233 (2008)

[30] H. Jafari, S. Seifi, Homotopy analysis method for solving linear and nonlinear fractional diffusion-wave equation, Communications in Nonlinear Science and Numerical Simulation 14(5), 2006 (2009)

[31] H. Jafari, S. Seifi, Solving a system of nonlinear fractional partial differential equations using homotopy analysis method, Communications in Nonlinear Science and Numerical Simulation 14(5), 1962 (2009)

[32] M.S. Mohamed, K.A. Gepreel, F.A. Al-Malki, M. Al-Humyani, Approximate solutions of the generalized Abel's integral equations using the extension khan's homotopy analysis transformation method, Journal of Applied Mathematics 2015 (2015)

[33] V. Gupta, P. Kumar, Approximate solutions of fractional linear and nonlinear differential equations using laplace homotopy analysis method, Int. J. Nonlinear Sci 19(2), 113 (2015)

[34] S. Liao, Series solution of nonlinear eigenvalue problems by means of the homotopy analysis method, Nonlinear Analysis: Real World Applications 10(4), 2455 (2009)

Chapter 11

Non-conventional Cryptographic Approach based on PMF and CDF

R. Narmada Devi[1], Nagadevi Bala Nagaram[2], Kala Raja Mohan[3] and Suresh Rasappan[4]

[1,2,3] Department of Mathematics, Vel Tech Rangarajan Dr. Sagunthala R&D Institute of Science and Technology, Avadi, Chennai – 600 062, Tamil Nadu, India
[4] Mathematics Section, University of Technology and Applied Sciences, Ibri, Al Dhahirah Governorate, Sultanate of Oman

Abstract

Secret Information sharing is all time requirement in this internet world, especially in electronic communications such as system security, smart card, mobile communications etc. Cryptography is based on transformation of multiple rounds of transformation of messages in the form of plain text as input into encrypted text message. Through suitable mathematical technique, secrecy of the information is maintained. Many researchers have shown their interest in making use of mathematical techniques in Cryptography.
Cryptography plays a vital role to maintain a secrecy between the sender and the receiver by using different methods. Mathematics remains the backbone in bringing out various cryptographic methods. This chapter proposes a cryptographic technique using PMF and CDF. In this technique, plain text message is transformed as cipher text using a single key is used for both encryption and decryption. Using this proposed method, the information is also shared in safe manner and an illustration is done.

Keywords: Cryptography, Encryption, Decryption, PMF and CDF Matrix

11.1 Introduction

Cryptography is the study and practice of techniques for secure communication in the presence of third parties called adversaries. It deals with developing and analyzing protocols that prevents malicious third parties from retrieving information being shared between two entities thereby following the various aspects of information security. Secure Communication refers to the scenario where the message or data shared between two parties can't be accessed by an adversary. Cryptography is the practice and study of hiding information from all but those with the means or key to decode the message. Also the area of cryptography employs many different means of transforming normal data in to unreadable form. Cryptography mainly consists of encryption and decryption. Encryption is the transformation of data in to some unreadable form its purpose is to ensure privacy by keeping the information hidden from anyone for whom it is not intended even those who can see the encrypted data. Decryption is the reverse of encryption it is the transformation of encrypted data back in to some intelligible form. Encryption and decryption require the use of some secret information, usually referred to as a key depending on the encryption mechanism used, the same key might be used for both encryption and decryption, while for other mechanisms, the keys used for encryption and decryption might be different. This paper describes an activity build around one of the techniques that illustrates an application of probability function to cryptography.

A. P. Hiwarekar in 2014 proposed two cryptographic technique applying Laplace Transform and Hyperbolic functions [1,3]. M. Tuncay Gencoglu in 2017 introduced a cryptographic process involving Laplace Transform with Hyperbolic functions [2]. Dr. K. Hemant K. Undegaonkar introduced a secured communication method involving Laplace Transform [4]. S. Sujatha in 2013 made use of the application of Laplace Transform in the field of cryptography [5]. G. Nagalakshmi et al in 2020 involved Laplace Transform Laplace Transform using Asymmetric key for secured communication [7]. S. Dhingra et al proposed a network security method involving Laplace Transform [8]. M. Saha in 2017 utilized Laplace Mellin Transform in forming a cryptographic method for secured information sharing [9]. A. K. H. Sedeeg et al in 2016 formulated a new cryptographic algorithm applying Aboodh Transform [10]. Kala Raja Mohan et al in 2022 applied Bilinear Transform with Probability in identifying a secured information sharing algorithm and Laplace transform and hyperbolic tangent function in cryptography [11,13]. A. Meenakshi et al applied graph network in designing a cryptographic algorithm [12]. This chapter work aims at developing a cryptographic algorithm using probability functions.

In section 2, the methodology made use of in this crypto analysis are described. Section 3 and 4 represents the algorithm which is applied for encryption and decryption. The process of encryption and decryption are demonstrated with an example in section 5 and Section 6. The conclusion followed by references are given in the Section 7.

11.2 Types of Cryptographic Algorithms

There are several ways of classifying cryptographic algorithms based the number of keys that are employed for encryption and decryption, they are:

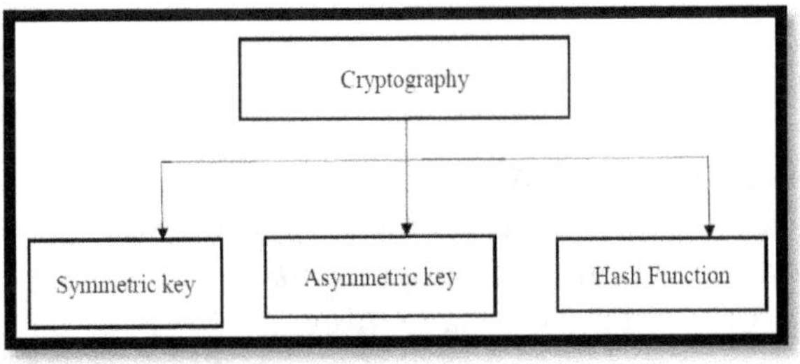

11.2.1 Secret Key Cryptography (SKC)

- Cryptography in which a single key is used for both encryption and decryption.
- It is also symmetric encryption.
- Primarily used for privacy and confidentiality.
- The sender uses the key to encrypt the plain text message and sends the cipher text message to the receiver.
- The receiver applies the same key to decrypt the message and recover the plain text message.

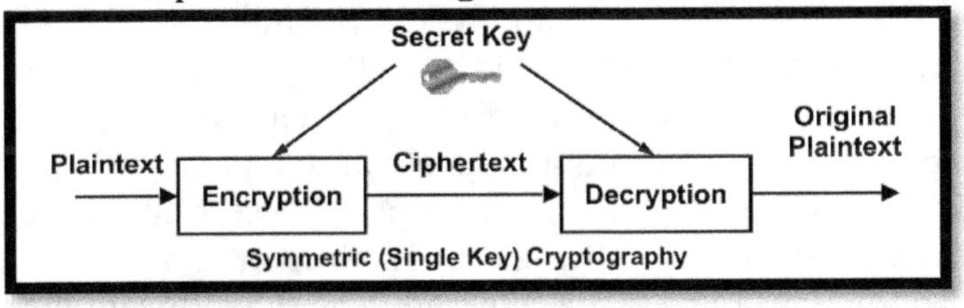

11.2.2 Public Key Cryptography (PKC)

- Cryptography in which a one key is used for encryption and another key used for decryption.
- It is also called an asymmetric encryption.
- Primarily used for authentication, non-repudiation, and key exchange.

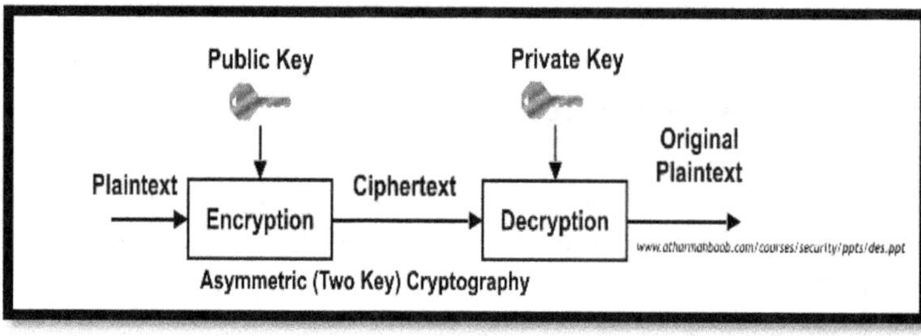

11.2.3 Hash Functions

- Cryptography has no key
- It uses a mathematical transformation to irreversibly "encrypt" information, providing a digital fingerprint.
- Primarily used for message integrity.

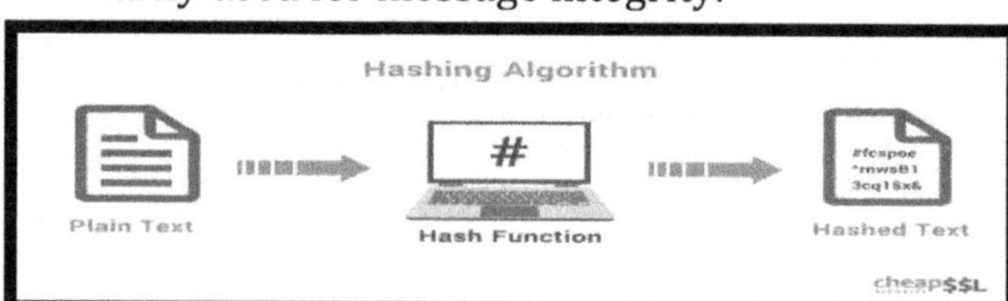

11.2.4 Probability Mass Function (PMF) and Cumulative Distribution Function (CDF)

Probability is a branch of mathematics that deals with the occurrence of a random event. It has played a pivotal role in the lives of data analysts and data scientists. Probability has been an important aspect when it comes to the field of Data Science. It has played a pivotal role in the lives of data analysts and data scientists. Probability gives information about occurring of a certain event under some assumptions i.e. It indicates the likelihood of an event occurring. A random variable can be referred to as the different outcomes which are possible in a given situation. A random variable is a rule that assigns a numerical value to each outcome in a sample space. Random variables may be either discrete or continuous. A random variable is said to be discrete which take only a limited number of values within a range. Continuous distributions are for variables that can take an infinite number of values within a range.

A probability mass function (pmf) is a function over the sample space of a discrete random variable X which gives the probability that X is equal to a certain value. Let $\boldsymbol{X}$ be a discrete random variable on a sample space S. Then the probability mass function $p(x)$ is defined as $p(x) = P[X = x]$.

The probability mass function is said to be valid if it obeys the following conditions:

(i) $p(x) \geq 0$ for all $x \in S$,
(ii) $\sum_{x \in S} p(x) = 1$
(iii) $0 \leq p(x) \leq 1$

The formula to calculate a probability mass function of a discrete random variable is given as

(i) $P(-\infty < X < \infty) = \sum_{-\infty}^{\infty} p(x)$
(ii) $P(X > k) = 1 - P(X \leq k)$

The cumulative distribution function (CDF) F(x) describes the probability that a discrete random variable X with a given probability distribution will be found at a value less than or equal to x. This function is given as $F(x) = P(X \leq x) = \sum_{-\infty}^{x} p(x)\, dx$. Let X be a random variable with cdf $F(x)$. Then $P[a < X \leq b] = F(b) - F(a)$.

11.3 Methodology

In this proposed cryptographic algorithm, the idea of Secret Key Cryptography (SKC) is used and having the following procedure:

- The information which is to be shared to the other person secretly is the plain text.
- The encrypted message making use of the key specified for the process is the cipher text.

- ➢ The process by which the plain text gets transformed into the cipher text is the cipher.
- ➢ The process involved in converting plain text into secret message is encryption.
- ➢ The reverse process of encryption is decryption, which convert secret message to plain text.
- ➢ Find the Probability mass function, cumulative distribution which are express as matrix.

11.4 Algorithm for encryption Using PMF and CDF

The steps to be followed in the process of encryption is as given below.

Step 1: Convert the plain text message of length (n) in to a stream of numerals using the following above table.
Step 2: Define a discrete random variable $\mathbb{X} = \{0,1,2, \dots, n-1\}$ where n is the number of the letter in the plain text message.
Step 3: Define probability function for each random variable by multiplied each of the numerals of the plain text message with a suitable constant term k in the order of left to right direction such that $p(x)$ becomes probability mass function (pmf).

Table 11.1 Coding table

A	B	C	D	E	F	G	H	I	J
1	2	3	4	5	6	7	8	9	10
K	L	M	N	O	P	Q	R	S	T
11	12	13	14	15	16	17	18	19	20
		U	V	W	X	Y	Z		
		21	22	23	24	25	26		

Step 4: Form the probability distribution table for the discrete random variable $\mathbb{X} = \{0,1,2, \dots, n-1\}$.
Step 5: Evaluate the cumulative density function $F(x)$ from the pmf $p(x)$ corresponding to each random variable $x \in \mathbb{X}$ using the formula as
$\mathrm{F}(x) = P(\mathbb{X} \leq x) = \sum_{-\infty}^{x} p(x)$.
Step 6: Convert as matrix of order n in which the elements are the coefficients of the cumulative density function $F(x)$ which are

obtain in the step 5 for each random variable $x \in \mathbb{X}$.

Step 7: Form a diagonal matrix of order n by taking its diagonal elements as cumulative density function $F(x)$ for each random variable $x \in \mathbb{X}$ from the step 6 which is consider as a cipher text.

11.5 Algorithm For Decryption Using PMF and CDF

The decryption process involves the steps as specified below.

Step 1: Consider the diagonal matrix of order n is obtained in the encryption process of step 7.

Step 2: Each diagonal element of the matrix assumes as cumulative density function $F(x)$ for each random variable $x \in \mathbb{X}$.

Step 3: Find the probability mass function for each cumulative distribution function for corresponding random variable $x \in \mathbb{X}$ using the formulas

$$P[X = 0] = F(0)$$
$$P[X = 1] = F(1) - F(0)$$
$$P[X = 2] = F(2) - F(1)$$

In general, $$P[X = \mathrm{n}] = F(n) - F(n - 1)$$

Step 4: Consider each probability mass function $p(x)$ and after divide by the value of k in such way that $p(x)$ is pmf.

Step 5: Obtain the corresponding numerals from step 4 and convert into corresponding alphabet using the table 11.4.1. The message formed in this and it is considered as decoded message.

11.6 Illustration of Encryption Process

In this section, the encryption process is illustrated using the text message "MATHEMATICS"

1. Consider the plain text message "MATHEMATICS" convert as numerical equivalent assigned letter using table 11. 1.

Table 11.2 Plain text to numerical values

M	A	T	H	E	M	A	T	I	C	S
13	1	20	8	5	13	1	20	9	3	19

2. This message consists of 11 characters. That is, n =11.

3. Define a discrete random variable $\mathbb{X} = \{0,1,2,\ldots,n-1\}$ Then for each random variable by multiplied each of the numerals of the plain text message.

4. Find a probability function $p(x)$ for each random variable by multiplied each of the numerals of the plain text message from step 1 by choosing value of $k = \frac{1}{112}$ such that $p(x)$ is probability mass function (pmf) which is defined as

Table 11.3 Probability distribution table

$\mathbb{X}$	0	1	2	3	4	5	6	7	8	9	10
$p(x)$	$\frac{13}{112}$	$\frac{1}{112}$	$\frac{20}{112}$	$\frac{8}{112}$	$\frac{5}{112}$	$\frac{13}{112}$	$\frac{1}{112}$	$\frac{20}{112}$	$\frac{9}{112}$	$\frac{3}{112}$	$\frac{19}{112}$

5. Calculate the corresponding cumulative distribution function $F(x)$ of the above pmf function $p(x)$ as follows:

Table 11.4 Cumulative distribution table

$\mathbb{X}$	0	1	2	3	4	5	6	7	8	9	10
$p(x)$	$\frac{13}{112}$	$\frac{1}{112}$	$\frac{20}{112}$	$\frac{8}{112}$	$\frac{5}{112}$	$\frac{13}{112}$	$\frac{1}{112}$	$\frac{20}{112}$	$\frac{9}{112}$	$\frac{3}{112}$	$\frac{19}{112}$
$F(x)$	$\frac{13}{112}$	$\frac{14}{112}$	$\frac{34}{112}$	$\frac{42}{112}$	$\frac{47}{112}$	$\frac{60}{112}$	$\frac{61}{112}$	$\frac{81}{112}$	$\frac{90}{112}$	$\frac{93}{112}$	1

6. Convert into diagonal matrix of order 11 by considering the diagonal elements as its cumulative distribution function $F(x)$ for each $\in \mathbb{X}$. Hence the matrix is given as

$$\begin{pmatrix}
\frac{13}{112} & 0 & 0 & 0 & 0 & 0 & 0 & 0 & 0 & 0 & 0 \\
0 & \frac{14}{112} & 0 & 0 & 0 & 0 & 0 & 0 & 0 & 0 & 0 \\
0 & 0 & \frac{34}{112} & 0 & 0 & 0 & 0 & 0 & 0 & 0 & 0 \\
0 & 0 & 0 & \frac{42}{112} & 0 & 0 & 0 & 0 & 0 & 0 & 0 \\
0 & 0 & 0 & 0 & \frac{47}{112} & 0 & 0 & 0 & 0 & 0 & 0 \\
0 & 0 & 0 & 0 & 0 & \frac{60}{112} & 0 & 0 & 0 & 0 & 0 \\
0 & 0 & 0 & 0 & 0 & 0 & \frac{61}{112} & 0 & 0 & 0 & 0 \\
0 & 0 & 0 & 0 & 0 & 0 & 0 & \frac{81}{112} & 0 & 0 & 0 \\
0 & 0 & 0 & 0 & 0 & 0 & 0 & 0 & \frac{0}{112} & 0 & 0 \\
0 & 0 & 0 & 0 & 0 & 0 & 0 & 0 & 0 & 0 & 1
\end{pmatrix}$$

7. This diagonal matrix which is obtained from the cumulative distribution function is to be shared to the receiver as a cipher text message.

11.7 Illustration of Decryption Process

This section describes the decryption process involved with the example cited in section 11.5. Before proceeding with the decryption process,

1. Consider the diagonal matrix which is obtained in step 6 of section 11.6.
2. Form the cumulative distribution function for corresponding each $x \in \mathbb{X}$ from each diagonal elements of the matrix and it is given as

Table 11.5 CDF

$\mathbb{X}$	0	1	2	3	4	5	6	7	8	9	10
$F(x)$	$\frac{13}{112}$	$\frac{14}{112}$	$\frac{34}{112}$	$\frac{42}{112}$	$\frac{47}{112}$	$\frac{60}{112}$	$\frac{61}{112}$	$\frac{81}{112}$	$\frac{90}{112}$	$\frac{93}{112}$	1

3. Find the probability mass function using cumulative distribution function of each $x \in \mathbb{X}$.

Table 11.6 PMF

$\mathbb{X}$	0	1	2	3	4	5	6	7	8	9	10
$p(x)$	$\frac{13}{112}$	$\frac{1}{112}$	$\frac{20}{112}$	$\frac{8}{112}$	$\frac{5}{112}$	$\frac{13}{112}$	$\frac{1}{112}$	$\frac{20}{112}$	$\frac{9}{112}$	$\frac{3}{112}$	$\frac{19}{112}$

4. Consider probability function for each $x \in \mathbb{X}$ and divided by $k = \frac{1}{112}$ and convert as numerals.
5. Using the Table 11.4.1, convert the numerals as plain text message.

Table 11.7 Numerals to plain text

Numerals	13	1	20	8	5	13	1	20	9	3	19
Plain Text	M	A	T	H	E	M	**A**	T	I	C	S

11.8 Conclusion

A new cryptographic algorithm has been proposed by applying the probability mass function and cumulative distribution function. This process involves plain text is transformed to matrices which was to be sharing cipher text. Then decryption is proceeded by using the properties of probability mass function and then convert as plain text. Thus, this is a very safe procedure in cryptography. The procedure is also illustrated using the Word "MATHEMATICS".

References

[1] A. P. Hiwarekar, "New mathematical modeling for cryptography," Journal of Information Assurance and Security, 9, 027-033 (2014).

[2] M. Tuncay Gencoglu, "Cryptanalaysis of a New Method of Cryptography using Laplace Transform Hyperbolic Functions."

Communications in Mathematics and Applications, 8, 183-189 (2017).

[3] A. P. Hiwarekar, "A new method of Cryptography ussing Laplace transform of Hyperbolic functions," International Journal of Mathematical Archive, 4, 208-213 (2013).

[4] K. Hemant K. Undegaonkar, "Security in Communication By Using Laplace Transform and Cryptography," International Journal of Scientific & Technology Research, 8, 3207-3209 (2019).

[5] S. Sujatha, "Application of Laplace Transforms in Cryptography," International Journal of Mathematical Archive, 4, 67-71 (2013).

[6] C. H. Jayanthi and V. Srinivas, "Mathematical Modelling for Cryptography using Laplace Transform," International Journal of Mathematics Trends and Technology, 65, 10-15 (2019).

[7] G. Nagalakshmi, A. Chandra Sekhar and D. Ravi Sankar, "Asymmetric key Cryptography using Laplace Transform," International Journal of Innovative Technology and Exploring Engineering, 9, 3083-3087 (2020).

[8] S. Dhingra, A. A. Savalgi and S. Jain, "Laplace Transformation based Cryptographic Technique in Network Security," International Journal of Computer Applications, 136, 6-10 (2016).

[9] M. Saha, "Application of Laplace – Mellin Transform for Cryptography," Raj Journal of Technology Research & Innovation, 5, 12-17 (2017).

[10] A. K. H. Sedeeg, M. M. Abdelrahim Mahgoub, and M. A. Saif Saeed, "An Application of the New Integral "Aboodh Transform" in Cryptography," Pure and Applied Mathematics Journal, 5, 151-154 (2016).

[11] Kala Raja Mohan, Suresh Rasappan, Regan Murugesan, Sathish Kumar Kumaravel and Ahamed A. Elngar, "Secret Information Sharing Using Probability and Bilinear Transformation", Proceedings of 2nd International Conference on Mathematical Modeling and Computational Science, 115-122 (2022).

[12] A. Meenakshi, J. Senbagamalar, and A. Neel Armstrong, "Encryption on Graph Networks", Proceedings of 2nd International Conference on Mathematical Modeling and Computational Science, 123-130 (2022).

[13] Kala Raja Mohan, Suresh Rasappan and Sathish Kumar Kumaravel, "Secret Information sharing Using Laplace Transform and

Hyperbolic Tangent Function", AIP Conference Proceedings 2516, 12003, 1-6, (2022).

Chapter 12

Analysis of Stability for a Rabies Model with Fractional Order

S. Priyadharsini

Sri Krishna Arts and Science College, BK Pudur, Sugunapuram East, Kuniyamuthur, Coimbatore– 641008, Tamil Nadu, India

Abstract

In this study, the fractional-order rabies model is the main topics of discussion. Considered is a general nonlinear differential equation of fractional order. Solutions' existence and originality are established. Study is done on the stability of equilibrium points. To examine the asymptotic stability of the model under consideration, the model's spectral values are principally used. These numerical solutions of the models are provided. An illustration of the equilibrium point is provided for systems with integer orders and of fractional orders.

Keywords: Fractional-order, Predator-prey system, equilibrium points, Stability, Rabies systems

12.1 Introduction

The predator-prey mathematical model is a mathematical framework used to study the dynamics between two interacting populations: predators and their prey. This model is often employed in ecology and population biology to understand the intricate relationships and fluctuations in population sizes over time. When considering the predator-prey model in the context of rabies, we focus on the interaction between the predator species (typically a carnivorous animal like a wolf or fox) and the prey species (usually small mammals like rabbits or rodents). Rabies, a viral disease that affects the nervous system of animals, can have a significant impact on these dynamics.

The dynamics of the predator-prey model with rabies are influenced by several factors. Firstly, the presence of rabies in the predator population can affect their behavior, making them more likely to transmit the disease to their prey while hunting or through bites. This increases the mortality rate of the prey population, causing a decline in their numbers. Secondly, the disease can also affect the predators themselves. Once infected, predators may experience a decline in their hunting abilities, leading to a decrease in their reproductive success and survival. This decline can further impact the predator population size.

Mathematically, the predator-prey model with rabies can be represented using a system of differential equations, typically based on the Lotka-Volterra model. These equations describe the rates of change in the population sizes of predators and prey over time, taking into account factors such as birth rates, death rates, and disease transmission rates. Understanding the predator-prey dynamics with rabies is crucial for wildlife management and disease control efforts. By studying the mathematical model, researchers and policymakers can gain insights into the potential impact of rabies on predator-prey interactions, predict population fluctuations, and develop strategies to mitigate the spread of the disease while maintaining ecological balance. These systems are often employed in the field of biomathematics, and theoretical research into them is now underway [1,2,3,4,5]. In some circumstances, the transformation into various type systems is also used.

It is worth noting that the specific parameters and dynamics of the predator-prey model with rabies may vary depending on the particular species involved, the environment, and other factors. Therefore, accurate modeling requires consideration of these factors and empirical data to inform the model's parameters.

12.2 Main Text

12.2.1 Stability Analysis Procedure

To check the stability of a predator-prey model with rabies, several methods can be employed. These methods help determine the equilibrium points of the model and assess their stability. Here are some common techniques used:

Equilibrium Points: Begin by identifying the equilibrium points of the predator-prey model. These points represent population sizes at which the rates of change for both predator and prey populations are zero. Equilibrium points can be found by setting the differential equations of the model to zero and solving for the population sizes.

Linear Stability Analysis: Linear stability analysis is a widely used method to assess the stability of equilibrium points. It involves linearizing the differential equations around the equilibrium points and examining the eigenvalues of the resulting linear system. The stability of an equilibrium point is determined by the signs and values of these eigenvalues. If all eigenvalues have negative real parts, the equilibrium point is stable, indicating that the populations will converge to that point over time. If at least one eigenvalue has a positive real part, the equilibrium point is unstable, suggesting that the populations will not converge but exhibit oscillatory or divergent behavior.

Phase Plane Analysis: Phase plane analysis is a graphical method that visualizes the dynamics of the predator-prey model. It involves plotting the prey population size on one axis and the predator population size on the other axis. Trajectories of the populations are then plotted on this phase plane, providing insights into their dynamics, including stability, oscillations, or other patterns.

Lyapunov Functions: Lyapunov functions are mathematical functions that can be used to prove stability in dynamical systems. By constructing a Lyapunov function for the predator-prey model with rabies, one can analyze its properties and determine the stability of the equilibrium points. If a suitable Lyapunov function can be found that satisfies certain conditions, it can establish the stability of the system.

Numerical Simulations: Computer simulations can be employed to numerically solve the differential equations of the predator-prey model with rabies. By initializing the system with different initial conditions and parameters, the dynamics of the populations can be simulated over time. Analyzing the results of these simulations can provide insights into the stability and behavior of the model.

It's important to note that the stability analysis of a predator-prey model with rabies for a complex model may require advanced mathematical techniques and computational tools.

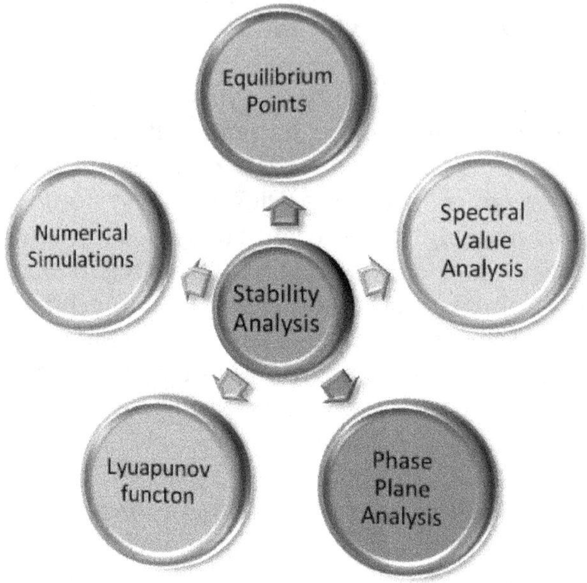

Figure 12.1 Methods of analysing stability.

12.2.2 Significance of the Model

Advantages of Predator-Prey Model with ordinary derivative:

Understanding Disease Dynamics: The model helps researchers gain insights into the complex interactions between predators, prey, and the spread of rabies. It can provide a better understanding of how the disease affects population dynamics, including the impact on both predators and prey.

Disease Control and Management: By studying the predator-prey model with rabies, scientists and policymakers can develop effective strategies for controlling and managing the disease. This may involve implementing vaccination programs, controlling the movement of infected animals, or targeting specific host populations to break the transmission cycle.

Ecological Impact Assessment: Rabies can have significant ecological consequences by altering predator-prey dynamics. The model allows scientists to assess the potential impact of rabies on the stability and structure of ecosystems. It helps identify potential cascading effects on other species and ecosystem processes.

Wildlife Conservation: Understanding the dynamics of predator-prey interactions affected by rabies is crucial for wildlife conservation efforts. The model can inform conservation strategies by highlighting the vulnerability of certain species or populations to the disease and aid in the development of targeted interventions to protect endangered or threatened species.

Veterinary and Public Health Planning: The predator-prey model with rabies provides insights into disease transmission patterns, allowing veterinarians and public health officials to plan and implement effective measures to prevent the spread of rabies among animal populations and reduce the risk of transmission to humans.
Policy and Decision Making: The model's outcomes and predictions can guide policy decisions regarding disease control, vaccination campaigns, or wildlife management. It helps policymakers assess the potential outcomes of different intervention strategies and make informed decisions to minimize the impact of rabies on both animal and human populations.

By utilizing the predator-prey model with rabies, researchers and practitioners can better understand the dynamics of the disease, evaluate the effectiveness of control measures, and develop targeted interventions to mitigate its impact on both wildlife and public health.

Advantages of the fractional predator-prey model:

Capturing Long-Term Dependencies: The fractional-order derivatives in the model allow for the consideration of memory effects and long-term dependencies in the dynamics. This can be particularly useful in understanding ecological systems where past events and interactions have a lasting impact on the current state of the populations.

Improved Realism: The fractional model provides a more realistic representation of the dynamics of predator-prey interactions with rabies by considering the fractional-order derivatives. This enables the incorporation of memory, non-local effects, and complex interactions that are not accounted for in the classical model.

Enhanced Predictability: The inclusion of fractional-order derivatives can lead to improved predictability of the system's behavior. The memory effects captured by fractional derivatives allow for a more accurate representation of historical information and can potentially improve the accuracy of predictions for population dynamics and disease spread.

Adaptability to Complex Systems: The fractional predator-prey model with rabies can be applied to complex ecological systems involving multiple predator and prey species, as well as other interacting factors. It provides a flexible framework to study and understand the dynamics of such systems in the presence of rabies.
It's important to note that analyzing fractional-order systems can be mathematically challenging, requiring specialized techniques such as fractional calculus and numerical methods designed for fractional equations. The advantages of the fractional model should be considered in relation to the specific context and objectives of the study, as well as the availability of appropriate tools for analysis.

Future Scope of the Model:

As with any mathematical model, there are several potential future extensions and variations of the fractional predator-prey model with rabies. Here are a few possible directions for future research and extensions:

Spatial Considerations: The current fractional predator-prey model with rabies assumes well-mixed populations. Future extensions could incorporate spatial effects, such as spatial diffusion, migration, or spatial heterogeneity in the environment. This would allow for a more realistic representation of population dynamics in spatially structured landscapes.

Multiple Predators and Prey Species: The current model focuses on a single predator-prey interaction. Future extensions could explore the dynamics of multiple predator and prey species interacting within the context of rabies transmission. This could involve investigating complex food web interactions and the cascading effects of rabies transmission on different species.

Nonlinear Fractional Operators: The fractional derivatives in the current model are assumed to be of the Caputo type, which is a commonly used fractional derivative definition. Future extensions could explore the dynamics and implications of using alternative fractional derivative definitions, such as the Riemann-Liouville or Grünwald-Letnikov derivatives. These nonlinear fractional operators could lead to different system behaviors and dynamics.

Epidemiological Considerations: The current model incorporates rabies transmission in the predator-prey dynamics. Future extensions could delve deeper into the epidemiological aspects of rabies, considering factors like disease transmission rates, vaccination strategies, and the impact of rabies on population dynamics and community structure.

Data-Driven Approaches: Future extensions could involve incorporating real-world data into the fractional predator-prey model with rabies. This could include fitting model parameters to observed data, validating the model's predictions against empirical data, and using data assimilation techniques to improve the accuracy of the model's predictions.

Stochastic and Uncertainty Analysis: Incorporating stochasticity and uncertainty into the fractional predator-prey model could provide insights into the effects of random fluctuations, environmental variability, and parameter uncertainty on the dynamics of the system. This could involve stochastic differential equations, Bayesian approaches, or sensitivity analysis techniques.

These are just a few potential avenues for future extensions of the fractional predator-prey model with rabies. The specific direction of research would depend on the research objectives, available data, and the complexity of the ecological and epidemiological dynamics being considered.

12.3 Examples

12.3.1 Example 1: Predator-Prey model with ordinary derivative

Let's consider a simple example of a predator-prey model with rabies involving rabbits as the prey and foxes as the predators. We'll use the Lotka-Volterra equations modified to incorporate the effects of rabies transmission. Here are the equations:

Prey equation:

$$\frac{dR}{dt} = aR - bRF \tag{12.3.1}$$

Predator equation:

$$\frac{dF}{dt} = cRF - dF \tag{12.3.2}$$

where:
R: Prey population (rabbits)
F: Predator population (foxes)
a: Prey growth rate
b: Prey death rate due to predation
c: Predator birth rate per successful predation
d: Predator death rate
To analyze the stability of this model, let's assume the following parameter values:
a = 0.3 (prey growth rate)
b = 0.02 (prey death rate due to predation)
c = 0.01 (predator birth rate per successful predation)
d = 0.4 (predator death rate)

Now, we can proceed with the stability analysis:

Equilibrium Points:

To find the equilibrium points, we set the derivatives equal to zero and solve the resulting equations. In this case, the equilibrium points occur when $\frac{dR}{dt} = 0$ and $\frac{dF}{dt} = 0$.

From the prey equation: aR - bRF = 0

We can have two equilibrium points:

Equilibrium 1: R = 0 (no prey)

Equilibrium 2: F = $\frac{a}{bR}$ (predator population dependent on prey population)

Linear Stability Analysis:

To analyze the stability of the equilibrium points, we linearize the system of equations by taking partial derivatives and evaluating them at the equilibrium points.

For Equilibrium 1 (R = 0), the linearized system becomes:

$\frac{dR}{dt}$ = -bRF

$\frac{dF}{dt}$ = cRF - dF

The eigenvalues of the Jacobian matrix at this equilibrium point will determine stability. If the real parts of the eigenvalues are negative, the equilibrium point is stable. In this case, both eigenvalues will be negative, indicating stability.

For Equilibrium 2 (F = a/(bR)), the linearized system becomes:

$\frac{dR}{dt}$ = a - bR - (acR)/d

$\frac{dF}{dt}$ = (acR)/d - dF

By evaluating the eigenvalues at this equilibrium point, we can determine stability.

12.3.2 Example 2: Predator-Prey model with fractional derivative

The fractional predator-prey model with rabies is an extension of the classical predator-prey model that incorporates fractional derivatives to capture memory effects and long-term dependencies in the dynamics. It introduces fractional-order differential equations to describe the interactions between predators, prey, and the spread of rabies. Here is a small fractional predator-prey model with rabies:

Prey equation:

$$\frac{dR(t)^{\alpha}}{dt^{\alpha}} = \text{aR(t)} - \text{bR(t)F(t)} \tag{12.3.3}$$

Predator equation:

$$\frac{dF(t)^{\beta}}{dt^{\beta}} = \text{cR(t)F(t)} - \text{dF(t)} \tag{12.3.4}$$

where:
R(t): Prey population at time t
F(t): Predator population at time t
D^{α}/Dt^{α}: Fractional derivative of order α with respect to time t for prey
D^{β}/Dt^{β}: Fractional derivative of order β with respect to time t for predators
a: Prey growth rate
b: Prey death rate due to predation
c: Predator birth rate per successful predation
d: Predator death rate
α: Fractional order for prey
β: Fractional order for predators

Let's consider a small example of analyzing a fractional predator-prey model with rabies. We'll use the following fractional-order equations:

Prey equation:

$$\frac{dR(t)^{\alpha}}{dt^{\alpha}} = 0.3R(t) - 0.02R(t)F(t)$$

Predator equation:

$$\frac{dF(t)^{\beta}}{dt^{\beta}} = 0.01R(t)F(t) - 0.4F(t)$$

where:
$\alpha = 0.8$: Fractional order for prey
$\beta = 0.9$: Fractional order for predators
To analyze the stability of this fractional predator-prey model, we can follow these steps:

Equilibrium Points:

Identify the equilibrium points by setting the derivatives equal to zero:
$0.3R - 0.02RF = 0$
$0.01RF - 0.4F = 0$

Solving these equations will give us the equilibrium points for the system.

12.4 Conclusion

In biomathematics, population models, especially predator-prey models, are a hot issue. Compared to the original historical models, several of the ones that have been presented have a highly rich dynamics, which serves as a base for all the models. The Allee effect, the fear effect, cannibalism, and immigration have all been the subject of recent studies that we have analyzed.

12.5 Acknowledgements

I gratefully thank the Editor and Reviewer for their efforts in making this work more highly developed.

References

[1] Just, W.; Nevai, A.L. A Kolmogorov-type competition model with multiple coexistence states and its applications to plant competition for sunlight. *J. Math. Anal. Appl.* **2008**, *348*, 620–636

[2] Xiao, Q.; Dai, B.; Xu, B.; Bao, L. Homoclinic bifurcation for a general state-dependent Kolmogorov type predator-prey model with harvesting. *Nonlinear Anal. Real World Appl.* **2015**, *26*, 263–273

[3] Llibre, J.; Martínez, Y.P.; Valls, C. Limit cycles bifurcating of Kolmogorov systems in R^2 and in R^3. *Commun. Nonlinear Sci. Numer. Simul.* **2020**, *91*, 10.

[4] Diz-Pita, E.; Llibre, J.; Otero-Espinar, M.V. Phase portraits of a family of Kolmogorov systems depending on six parameters. *Electron. J. Differ. Eq.* **2021**, *35*, 1–38.

[5] Diz-Pita, E.; Llibre, J.; Otero-Espinar, M.V.; Valls, C. The zero-Hopf bifurcations in the Kolmogorov systems of degree 3 in R^3. *Commun. Nonlinear Sci. Numer Simulat.* **2021**, *95*, 105621.

Chapter 13

Application of Pythagorean fuzzy soft linear space

S. Sivaramakrishnan[1], P. Balaji[2*], A. Ramesh[3]

[1] Manakula Vinayagar Institute of Technology, Kalitheerthal Kuppam, Puducherry, Tamilnadu India
[2*] Department of Mathematics, MEASI Academy of Architecture, Chennai, Tamilnadu, India (Corresponding author)
[3] Department of Mathematics, Saveetha School of Engineering, Saveetha Institute of Medical and Technical Sciences (SIMATS), Chennai, Tamilnadu, India.

Abstract

The purpose of this article is to provide the concept of Pythagorean fuzzy soft linear space by making use of the concepts of Pythagorean fuzzy linear space and soft set. The intersection of two Pythagorean fuzzy soft linear spaces results in another Pythagorean fuzzy soft linear space. However, if we look at a counter example, we can see that the union of two Pythagorean fuzzy soft linear spaces does not necessarily result in a Pythagorean fuzzy soft linear space. In addition to this, both the image and the pre image of the Pythagorean fuzzy soft linear space are presented here. Finally, an example of a problem that can be solved by applying the multi-criteria decision-making process that has been proposed.
Keywords. Fuzzy sets, soft sets, Pythagorean fuzzy sets and Pythagorean fuzzy soft linear space.

List of Notations and Abbreviations

LS-Linear space
FS-Fuzzy sets
SS-Soft sets
PFS-Pythagorean fuzzy sets
PFSS-Pythagorean fuzzy soft set
PFSLS-Pythagorean fuzzy soft linear space.
PFSM- Pythagorean fuzzy soft matrix

13.1 Introduction

The idea of fuzzy sets (FS), which later became an important aspect of research in mathematics, was initially presented by Zadeh [13]. The idea of flags and fuzzy linear space was brought to light by Lubczonok and Murali [4]. Yager [12] came up with the concept of Pythagorean fuzzy sets, sometimes known as PFS (which stipulates that the sum of the squares of the membership and non-membership functions must be less than or equal to one). PFS has found use in a wide variety of scenarios such as handling of problems involving multi criteria decision-making. Sivaramakrishnan et al. [11] proposed the concept of an interval-valued Pythagorean fuzzy linear space. Molodtsov [8] was the one who first presented the idea of Soft Sets back in 1999. Researchers have made great strides recently towards fuzzifying the theory of soft sets. The term "fuzzy soft set" was proposed by Maji et al. [5] along with several features involving fuzzy soft union, intersection, and the complement of a fuzzy soft set, among other things. In addition, Maji et al. [6] put out the idea of a fuzzy soft set, which was subsequently followed by research on Pythagorean fuzzy soft sets and a generalised Pythagorean fuzzy soft set [9].

Jun et al. [3] extended the idea of fuzzy soft sets to BCK/BCI-algebras and proposed the concept of fuzzy soft BCK/BCI-algebras as a result of their work. The authors Roy et al. [10] disclosed a few findings addressing the usefulness of fuzzy soft sets to the decision-making problem. We came up with the idea of PFSLS by combining PFSS and LS, their respective principles. We presented evidence that revealed how the intersection of two Pythagorean fuzzy soft linear spaces (PFSLSs) produces another PFSLS. In contrast, the union of two PFSLSs does not necessarily result in a PFSLS. We presented the idea of a pre-image as well as an image of a PFSLS, as well as the cartesian product of two PFSLSs. The concept of a Pythagorean fuzzy soft matrix, often known as a PFSM, as well as its many applications in the area of decision-making problem analysis are discussed. By combining the concepts of choice matrices and weighted choice matrices, the problem-solving approach known as PFSM has been utilised to provide an example of a decision-making method that has been shown.

13.2 Basic definitions

Definition 13.2.1. [12] Let $\mathcal{P}$ be the set. The Pythagorean fuzzy set (PFS) is defined as $\psi = \{< p, \mu_\psi(p), \nu_\psi(p) > | p \in \mathcal{P}\}$, such that $\mu_\psi \in [0,1]$ and $\nu_\psi \in [0,1]$ satisfy the following condition $0 \leq (\mu_\psi)^2 + (\nu_\psi)^2 \leq 1$, for every $p \in \mathcal{P}$. The set of all Pythagorean fuzzy sets over $\mathcal{P}$ is denoted by $PF^{\mathcal{P}}$.

Definition 13.2.2.[5,6] Let $\mathcal{P}$ be the Universal set and $\wp$ be the set of parameters. For Å ⊆ $\wp$, then
(i) $\mathcal{H}$: Å ⟶ P($\mathcal{P}$), where P($\mathcal{P}$) is the power set of $\mathcal{P}$ is called soft set (SS) over $\mathcal{P}$.
(ii) $\mathcal{H}$: Å ⟶ $PF^{\mathcal{P}}$, where $PF^{\mathcal{P}}$ is the is the collection of all fuzzy subsets of $\mathcal{P}$ is called is called fuzzy soft set (FSS) over $\mathcal{P}$.

Definition 13.2.3. [9] Let $\boldsymbol{\mathcal{P}}$ be the Universal set and $\wp$ be the set of parameters. For Å ⊆ $\wp$, then $\boldsymbol{\mathcal{H}}$: Å ⟶ $\boldsymbol{PF^{\mathcal{P}}}$ is called Pythagorean fuzzy soft set (PFSS) over $\boldsymbol{\mathcal{P}}$ or ($\boldsymbol{\mathcal{H}}$, Å) and we write $\boldsymbol{\mathcal{H}(e) = \{< p, \mu_{\mathcal{H}(e)}(p), \nu_{\mathcal{H}(e)}(p) > | p \in \mathcal{P}\}}$, for each e ∈ Å, the degree of membership $\boldsymbol{\mu_{\mathcal{H}(e)} \in [0, 1]}$ and degree of non-membership $\boldsymbol{\nu_{\mathcal{H}(e)} \in [0, 1]}$ satisfies the condition $0 \leq (\boldsymbol{\mu_{\mathcal{H}(e)}})^2 + (\boldsymbol{\nu_{\mathcal{H}(e)}})^2 \leq 1$.

Definition 13.2.5. [11] An IVPFS $\overline{\boldsymbol{\mathcal{P}}} = < \overline{\boldsymbol{\mu}}\ ,\ \overline{\boldsymbol{\nu}} >$ on V is known to be an IVPFLS of $\boldsymbol{\mathcal{P}}$ over F . If for all $\boldsymbol{p_1}$ and $\boldsymbol{p_2} \in \boldsymbol{\mathcal{P}}$ and $, \boldsymbol{\beta} \in$ F, it holds.
(i) $\overline{\boldsymbol{\mu}}\ (\boldsymbol{\alpha\ p_1 * \beta\, p_2}) \geq \min\{\overline{\boldsymbol{\mu}}\ (\boldsymbol{p_1}),\ \overline{\boldsymbol{\mu}}\ (\boldsymbol{p_2})\}$,
(ii) $\overline{\boldsymbol{\nu}}\ (\boldsymbol{\alpha\ p_1 * \beta\, p_2}) \leq \max\{\overline{\boldsymbol{\nu}}\ (\mathrm{x}_1),\ \overline{\boldsymbol{\nu}}\ (\boldsymbol{p_2})\}$.

Definition 13.2.6. [2] Let ($\boldsymbol{\mathcal{H}}$, Å) be a PFSS over $\boldsymbol{\mathcal{P}}$, then the subset $\boldsymbol{\mathcal{P}}$ × Å is uniquely defined by $\mathbf{R}_{Å}$ = {(p, e), e ∈ Å, p ∈ $\boldsymbol{\mathcal{P}}$}. The $\mathbf{R}_{Å}$ can be characterized by its membership function and non-membership function given by $\boldsymbol{\mu^2}_{\mathbf{R}_{Å}}$: $\boldsymbol{\mathcal{P}}$ × Å ⟶ [0, 1] and $\boldsymbol{\nu^2}_{\mathbf{R}_{Å}}$: $\boldsymbol{\mathcal{P}}$ × Å ⟶ [0, 1] respectively. If ($\boldsymbol{\mu^2}_{ij}$, $\boldsymbol{\nu^2}_{ij}$) = ($\boldsymbol{\mu^2}_{\mathbf{R}_{Å}}$($\boldsymbol{p_i}$, $\boldsymbol{e_j}$), $\boldsymbol{\nu^2}_{\mathbf{R}_{Å}}$($\boldsymbol{p_i}$, $\boldsymbol{e_j}$)) where $\boldsymbol{\mu^2}_{\mathbf{R}_{Å}}$($\boldsymbol{p_i}$, $\boldsymbol{e_j}$) is the membership of $\boldsymbol{p_i}$ the Pythagorean fuzzy set F($\boldsymbol{e_j}$) and $\boldsymbol{\nu^2}_{\mathbf{R}_{Å}}$($\boldsymbol{p_i}$, $\boldsymbol{e_j}$) is the non-membership of $\boldsymbol{p_i}$in the

Pythagorean fuzzy set F($\boldsymbol{e}_j$) respectively, then we define a matrix given by

$$[\mathrm{M}] = [m_{ij}]_{m\times n} = \begin{bmatrix} ({\mu^2}_{11}, {\nu^2}_{11}) & \cdots & ({\mu^2}_{1n}, {\nu^2}_{1n}) \\ \vdots & \ddots & \vdots \\ ({\mu^2}_{m1}, {\nu^2}_{m1}) & \cdots & ({\mu^2}_{mn}, {\nu^2}_{mn}) \end{bmatrix}$$ which

is called Pythagorean fuzzy soft matrix (PFSM) of order m × n over $\mathcal{P}$."

13.3 Pythagorean fuzzy soft linear space

Definition 13.3.1. Let $\mathcal{P}$ be a crisp linear space over a field F ($\mathcal{P}$F) with a binary operation $*$ and $\wp$ be the set of parameters and Å ⊆ $\wp$. Let $PF^{\mathcal{P}}$ be a Pythagorean fuzzy soft set (PFSS) over $\mathcal{P}$. Then ($\mathcal{H}$, Å) is said to be a Pythagorean fuzzy soft linear space (PFSLS) over $\mathcal{P}$F with respect to T = **min** and C = **max** if for each e ∈ Å, and p_1 and p_2 ∈ $\mathcal{P}$ and $\alpha, \beta \in$ F,

1. ${\mu^2}_{\mathcal{H}(e)}(\alpha\ p_1 * \ \beta\, p_2\ ) \geq \mathrm{T}\,({\mu^2}_{\mathcal{H}(e)}(p_1),\ {\mu^2}_{\mathcal{H}(e)}(p_2)\ ),$
2. ${\nu^2}_{\mathcal{H}(e)}(\alpha\ p_1 * \ \beta\, p_2\ ) \leq \mathrm{C}\,({\nu^2}_{\mathcal{H}(e)}(p_1),\ {\nu^2}_{\mathcal{H}(e)}(p_2)\ ).$

Example 13.3.2. Let $\mathcal{P} = \{i, j, k, m\}$ be the Klein 4-group defined by the binary operation $*$ as follows:

*	i	j	k	m
i	i	j	k	m
j	j	i	m	k
k	k	m	i	j
m	m	k	j	i

Let F be the field GF(2). Let $(0)h = i,\ (1)h = w$ for all $w \in \mathcal{P}$. Then $\mathcal{P}$ is a crisp linear space over F.

Let Å $= \{e_1, e_2, e_3, e_4\}$ be the parameters.

Define ($\mathcal{H}$, Å) =

$$\begin{array}{c} \\ e_1 \\ e_2 \\ e_3 \\ e_4 \end{array} \begin{array}{cccc} i & j & k & m \\ \left(\begin{array}{cccc} (0.8, 0.3) & (0.23, 0.6) & (0.23, 0.6) & (0.5, 0.7) \\ (0.23, 0.6) & (0.8, 0.3) & (0.5, 0.7) & (0.23, 0.6) \\ (0.23, 0.6) & (0.5, 0.7) & (0.8, 0.3) & (0.23, 0.6) \\ (0.5, 0.7) & (0.23, 0.6) & (0.23, 0.6) & (0.8, 0.3) \end{array}\right) \end{array}$$

Then ($\mathcal{H}$, Å) is said to be a Pythagorean fuzzy soft linear space (PFSLS) of $\mathcal{P}$F with respect to T = **min** and C = **max.**

Example 13.3.3. Let $\wp = N$ (the set of all natural numbers) be the parametric set and

$\boldsymbol{\mathcal{P}}^3$ = {p = (x, y, z) | x, y, z $\in \boldsymbol{\mathcal{P}}$} be a real vector space. Define a mapping $\mathcal{H}$: Å $\longrightarrow PF^{\mathcal{P}}$, for any p $\in \boldsymbol{\mathcal{P}}^3$ and n $\in N$, as follows:

$\mu^2{}_{F(e)}(p) = \frac{1}{n}$, $\nu^2{}_{F(e)}(p) = \frac{1}{2n}$ for all p $\in \boldsymbol{\mathcal{P}}^3$.

The t-norm T= min and the t-conorm C = max respectively. Then ($\mathcal{H}$, Å) is a Pythagorean fuzzy soft linear space (PFSLS) over $\mathcal{P}$F.

Theorem 13.3.4. Let B_1 and B_2 be two PFSLSs of $\mathcal{P}$F then $B_1 \cap B_2$ is also PFSLS of $\mathcal{P}$F.

Proof. Let p_1 and $p_2 \in \mathcal{P}, \alpha, \beta \in$ F.

$\mu^2{}_{\mathcal{H}(B_1 \cap B_2)(e)}(\alpha\ p_1 * \beta\, p_2)$

$= \mu^2{}_{\mathcal{H}B_1(e)}(\alpha\ p_1 * \beta\, p_2) \cap \mu^2{}_{\mathcal{H}B_2(e)}(\alpha\, p_1 * \beta\, p_2)$

$\geq$ T {[T ($\mu^2{}_{\mathcal{H}B_1(e)}(p_1)$, $\mu^2{}_{\mathcal{H}B_2(e)}(p_2)$] , [T ($\mu^2{}_{\mathcal{H}B_2(e)}(p_1)$, $\mu^2{}_{\mathcal{H}B_2(e)}(p_2)$]}

$\geq$ T {[T($\mu^2{}_{\mathcal{H}B_1(e)}(p_1)$, $\mu^2{}_{\mathcal{H}B_2(e)}(p_1)$], [T($\mu^2{}_{\mathcal{H}B_1(e)}(p_2)$, $\mu^2{}_{\mathcal{H}B_2(e)}(p_2)$]}

= T {$\mu^2{}_{\mathcal{H}(B_1 \cap B_2)(e)}(p_1)$, $\mu^2{}_{\mathcal{H}(B_1 \cap B_2)(e)}(p_2)$}

i.e., $\mu^2{}_{\mathcal{H}(B_1 \cap B_2)(e)}(\alpha\ p_1 * \beta\, p_2) \geq$ T {$\mu^2{}_{\mathcal{H}(B_1 \cap B_2)(e)}(p_1)$, $\mu^2{}_{\mathcal{H}(B_1 \cap B_2)(e)}(p_2)$}.

Similarly, $\nu^2{}_{\mathcal{H}(B_1 \cap B_2)(e)}(\alpha\ p_1 * \beta\, p_2) \leq$ C {$\nu^2{}_{\mathcal{H}(B_1 \cap B_2)(e)}(p_1)$, $\nu^2{}_{\mathcal{H}(B_1 \cap B_2)(e)}(p_2)$}.

This completes the proof.

Remark 13.3.5. Let B_1 and B_2 be two Pythagorean fuzzy soft linear spaces of $\mathcal{P}$F then $B_1 \cup B_2$ is not a Pythagorean fuzzy soft linear space of $\mathcal{P}$F.

Let B_1 and B_2 be two Pythagorean fuzzy soft linear spaces of $\mathcal{P}$F and $\mathcal{P}=\mathbb{R}^2$ and the parametric set $\wp = \{e_i |\ i = 1,2\}$ as following:

$$\mu^2_{\mathcal{H}B_1(e_i)}(p_1) = \begin{cases} \frac{1}{2}\ if\ the\ i^{th}\ co-ordinate\ of\ x \in \mathbb{R}^2 is\ non-zero\ only \\ 0, otherwise \end{cases}$$

$$\mu^2_{\mathcal{H}B_2(e_i)}(p_1) = \begin{cases} \frac{2}{5}\ if\ the\ i^{th}\ co-ordinate\ of\ x \in \mathbb{R}^2 is\ non-zero\ only \\ \frac{1}{10}\ otherwise \end{cases}$$

$$\nu^2_{\mathcal{H}B_1(e_i)}(p_1) = \begin{cases} \frac{2}{5}\ if\ the\ i^{th}\ co-ordinate\ of\ x \in \mathbb{R}^2 is\ non-zero\ only \\ 1, otherwise \end{cases}$$

$$\nu^2_{\mathcal{H}B_2(e_i)}(p_1) = \begin{cases} 0\ if\ the\ i^{th}\ co-ordinate\ of\ x \in \mathbb{R}^2 is\ non-zero\ only \\ \frac{1}{5}\ otherwise \end{cases}$$

Consider, T (a, b) = min {a, b} and C (a, b) = max {a, b}.

Let $B_1 \cup B_2 = B$ and we take $\alpha = \beta = 1$.

$\mu^2_{\mathcal{H}B(e_1)}(p_1 * p_2) = \mu^2_{\mathcal{H}B_1(e_1)}$ (1,1) = max {0, $\frac{1}{10}$ } = $\frac{1}{10}$

$\mu^2_{\mathcal{H}(B_1 \cup B_2)(e_1)}(\alpha\ p_1 * \beta\, p_2)$

$= \max\ \{\mu^2_{\mathcal{H}B_1(e_1)}(\alpha\ p_1 * \beta\, p_2),\ \mu^2_{\mathcal{H}B_2(e_1)}(\alpha\ p_1 * \beta\, p_2)\}$

$\geq \max\ \{\min\ [\mu^2_{\mathcal{H}B_1(e_1)}(p_1),\ \mu^2_{\mathcal{H}B_2(e_1)}(p_1)],\ \min\ [\mu^2_{\mathcal{H}B_1(e_1)}(p_2),\ \mu^2_{\mathcal{H}B_2(e_1)}(p_2)]\}$

= min {max [$\frac{1}{2}, \frac{1}{10}$], max [0, $\frac{2}{5}$]} = min {$\frac{1}{2}, \frac{2}{5}$} = $\frac{2}{5}$

$\frac{1}{10} \geq \frac{2}{5}$, which is absurd.

Hence $B_1 \cup B_2$ is not a Pythagorean fuzzy soft linear space of $\mathcal{P}$F.

Definition 13.3.6. Let B_1 and B_2 be two PFSLSs over ($\mathcal{P}_1$F, $\wp$) and ($\mathcal{P}_2$F, $\wp$) respectively. Then their cartesian product is de-

fined as $B_1 \times B_2 = B$, where $\mathcal{H}_B(\text{s, t}) = \mathcal{H}_{B_1}(\text{s}) \times \mathcal{H}_{B_2}(\text{t})$ for $(\text{s, t}) \in \wp \times \wp$.
Define $\mathcal{H}_B(\text{s, t}) = \{< (p_1, p_2), \mu^2_{\mathcal{H}_B(\text{s,t})}(p_1, p_2), \nu^2_{\mathcal{H}_B(\text{s,t})}(p_1,p_2) | (p_1, p_2) \in \mathcal{P}_1 \times \mathcal{P}_2$ with $\mu^2_{\mathcal{H}_B(\text{s,t})}(\alpha\ p_1 * \beta\, p_2)$ $= \text{T}\{ \mu^2_{\mathcal{H}_{B_1}(\text{s})}(p_1), \mu^2_{\mathcal{H}_{B_2}(\text{t})}(p_2) \}$ and
$\nu^2_{\mathcal{H}_B(\text{s,t})}(\alpha\ p_1 * \beta\, p_2) = \text{C}\{ \nu^2_{\mathcal{H}_{B_1}(\text{s})}(p_1), \nu^2_{\mathcal{H}_{B_2}(\text{t})}(p_2)\}$.

Theorem 13.3.7. Let B_1 and B_2 be two PFSLSs over $(\mathcal{P}_1\text{F}, \wp)$ and $(\mathcal{P}_2\text{F}, \wp)$ respectively. Then their cartesian product $B_1 \times B_2$ is also PFSLS over $[(\mathcal{P}_1 \times \mathcal{P}_2)\ \text{F}, \wp \times \wp]$.
Proof. Let $B_1 \times B_2 = B$, where $\mathcal{H}_B(\text{s, t}) = \mathcal{H}_{B_1}(\text{s}) \times \mathcal{H}_{B_2}(\text{t})$ for $(\text{s, t}) \in \wp \times \wp$.
Let $(p_1,q_1), (p_2, q_2) \in \mathcal{P}_1 \times \mathcal{P}_2$ and $\alpha, \beta \in \text{F}$.
Take $\alpha = \beta = 1$.
$\mu^2_{\mathcal{H}_B(\text{s,t})}[(p_1,q_1) * (p_2, q_2)] = \mu^2_{\mathcal{H}_B(\text{s,t})}[(p_1 * p_2), (q_1 * q_2)]$
$= \text{T}\{\mu^2_{\mathcal{H}_{B_1}(\text{s})}(p_1 * p_2), \mu^2_{\mathcal{H}_{B_2}(\text{t})}(q_1 * q_2)\}$
$\geq \text{T}\{[\ \text{T}(\mu^2_{\mathcal{H}_{B_1}(\text{s})}(p_1), \mu^2_{\mathcal{H}_{B_1}(\text{s})}(p_2)], [\text{T}(\mu^2_{\mathcal{H}_{B_2}(\text{t})}(q_1), \mu^2_{\mathcal{H}_{B_2}(\text{t})}(q_2)]\}$
$= \{[\ \text{T}(\mu^2_{\mathcal{H}_{B_1}(\text{s})}(p_1), \mu^2_{\mathcal{H}_{B_2}(\text{t})}(q_1)], [\text{T}(\mu^2_{\mathcal{H}_{B_1}(\text{s})}(p_2), \mu^2_{\mathcal{H}_{B_2}(\text{t})}(q_2)]\}$
$= \text{T}\{\mu^2_{\mathcal{H}_B(\text{s,t})}(p_1,q_1), \mu^2_{\mathcal{H}_B(\text{s,t})}(p_2,q_2)\}$
Hence, $\mu^2_{\mathcal{H}_B(\text{s,t})}[(p_1,q_1) * (p_2, q_2)] \geq \text{T}\{\mu^2_{\mathcal{H}_B(\text{s,t})}(p_1,q_1), \mu^2_{\mathcal{H}_B(\text{s,t})}(p_2,q_2)\}$
Similarly, $\nu^2_{\mathcal{H}_B(\text{s,t})}[(p_1,q_1) * (p_2, q_2)] \leq \text{C}\{\nu^2_{\mathcal{H}_B(\text{s,t})}(p_1,q_1), \nu^2_{\mathcal{H}_B(\text{s,t})}(p_2,q_2)\}$
Hence, the theorem is proved.

Definition 13.3.8. Let $(\mathcal{H}_1, \text{Å}_1)$ and $(\mathcal{H}_2, \text{Å}_2)$ be two PFSLSs over $(\mathcal{P}_1\text{F}, \wp)$ and $(\mathcal{P}_2\text{F}, \wp)$ respectively. A PFSS mapping ϕ is called homomorphism from $(\mathcal{H}_1, \text{Å}_1)$ to $(\mathcal{H}_2, \text{Å}_2)$ if the mapping $\phi : \mathcal{P}_1 \longrightarrow \mathcal{P}_2$ is a crisp linear space homomorphism and it satisfies the conditions; $\mu^2_{\phi(\mathcal{H}_1)(b)}(\alpha\ q_1 * \beta\, q_2) \geq \text{T}(\mu^2_{\mathcal{H}_1(a)}(p_1), \mu^2_{\mathcal{H}_1(a)}(p_2))$ and

$\nu^2{}_{\phi(\mathcal{H}_1)(b)}\ (\alpha\ q_1 * \ \beta q_2) \le C($ $\nu^2{}_{\mathcal{H}_1(a)}(p_1),\ \nu^2{}_{\mathcal{H}_1(a)}(p_2)$), for all $p_1, p_2 \in \mathcal{P}_1$, $q_1, q_2 \in \mathcal{P}_2$, a $\in$ Å$_1$ and b $\in$ Å$_2$.

Theorem 13.3.9. Let ($\mathcal{H}_1$, Å$_1$) and ($\mathcal{H}_2$, Å$_2$) be two PFSLSs over ($\mathcal{P}_1$F, ℘) and ($\mathcal{P}_2$F, ℘) and ϕ be a homomorphism from ($\mathcal{H}_1$, Å$_1$) to ($\mathcal{H}_2$, Å$_2$). Then ϕ[($\mathcal{H}_1$, Å$_1$)] is a PFSLS over ($\mathcal{P}_2$F, ℘) provided the crisp linear space homomorphism ϕ is surjective and T and C are continuous.
Proof. To prove ϕ[($\mathcal{H}_1$, Å$_1$)] = ($\phi(\mathcal{H}_1)$, Å$_1$) is a PFSLS it is enough to show that,
$\mu^2{}_{\phi(\mathcal{H}_1)(b)}\ (\alpha\ q_1 * \ \beta q_2) \ge T(\ \mu^2{}_{\mathcal{H}_1(a)}(q_1),\ \mu^2{}_{\mathcal{H}_1(a)}(q_2)\ )$ and
$\nu^2{}_{\phi(\mathcal{H}_1)(b)}\ (\alpha\ q_1 * \ \beta q_2) \le C(\ \nu^2{}_{\mathcal{H}_1(a)}(q_1),\ \nu^2{}_{\mathcal{H}_1(a)}(q_2)\ )$, for all $q_1, q_2 \in \phi(\mathcal{H}_1)$, and b $\in$ Å$_2$. Since ϕ is surjective, $p_1, p_2 \in \mathcal{P}_1$ such that $\phi(p_1) = q_1$ and $\phi(p_2) = q_2$.
Now, $\mu^2{}_{\phi(\mathcal{H}_1)(b)}\ (\alpha\ q_1 * \ \beta q_2) = \ \mu^2{}_{(\mathcal{H}_1)(b)}[\phi^{-1}(\alpha\ q_1 * \ \beta q_2)]$

$= \mu^2{}_{(\mathcal{H}_1)(b)}[\phi^{-1}(\alpha\ q_1) * \phi^{-1}(\ \beta q_2)]$,

since ϕ^{-1} is homomorphism

$= \ \mu^2{}_{(\mathcal{H}_1)(b)}\ [\ \alpha\ p_1\ * \ \beta p_2]$

$\ge T\{\mu^2{}_{(\mathcal{H}_1)(b)}(p_1),\ \mu^2{}_{(\mathcal{H}_1)(b)}(p_2)\}$

$= T\{\mu^2{}_{(\mathcal{H}_1)(b)}\ \ \phi^{-1}(q_1),\ \mu^2{}_{(\mathcal{H}_1)(b)}\ \phi^{-1}(q_2)\}$

$= T\{\mu^2{}_{\phi(\mathcal{H}_1)(b)}\ (q_1),\ \mu^2{}_{(\mathcal{H}_1)(b)}\ (q_2)\}$
In a similar way it is obtained.
$\nu^2{}_{\phi(\mathcal{H}_1)(b)}\ (\alpha\ q_1 * \ \beta q_2) \le C(\ \nu^2{}_{\mathcal{H}_1(a)}(q_1),$ $\nu^2{}_{\mathcal{H}_1(a)}(q_2)\ )$.
Hence, ϕ[($\mathcal{H}_1$, Å$_1$)] is a PFSLS over ($\mathcal{P}_2$F, ℘).

Theorem 13.3.10. Let ($\mathcal{H}_1$, Å$_1$) and ($\mathcal{H}_2$, Å$_2$) be two PFSLSs over ($\mathcal{P}_1$F, ℘) and ($\mathcal{P}_2$F, ℘) and ϕ be a homomorphism from ($\mathcal{H}_1$, Å$_1$) to ($\mathcal{H}_2$, Å$_2$). Then ϕ^{-1}[($\mathcal{H}_1$, Å$_1$)] is a PFSLS over ($\mathcal{P}_1$F, ℘) provided the crisp linear space homomorphism ϕ is surjective and T and C are continuous.

Proof. Since ϕ is surjective, $p_1, p_2 \in \mathcal{P}_1$ such that $\phi(p_1) = q_1$ and $\phi(p_2) = q_2$.

$\mu^2{}_{\phi^{-1}(\mathcal{H}_1)(a)}(\alpha\ p_1 * \beta\, p_2) = \mu^2{}_{(\mathcal{H}_1)(b)}[\phi(\alpha\ p_1 * \beta\, p_2)]$

$= \mu^2{}_{(\mathcal{H}_1)(b)}[\phi(\alpha\ p_1) * \phi(\beta\, p_2)]$,

since ϕ is homomorphism

$= \mu^2{}_{(\mathcal{H}_1)(b)}[\alpha\ q_1 * \beta\, q_2]$

$\geq T\{\mu^2{}_{(\mathcal{H}_1)(b)}(q_1), \mu^2{}_{(\mathcal{H}_1)(b)}(q_2)\}$

$= T\{\mu^2{}_{(\mathcal{H}_1)(b)}\ \phi(p_1), \mu^2{}_{(\mathcal{H}_1)(b)}\ \phi(p_2)\}$

$\mu^2{}_{\phi^{-1}(\mathcal{H}_1)(a)}(\alpha\ p_1 * \beta\, p_2) \geq T\{\mu^2{}_{\phi^{-1}(\mathcal{H}_1)(a)}(q_1), \mu^2{}_{\phi^{-1}(\mathcal{H}_1)(a)}(q_2)\}$

Similarly,

$\nu^2{}_{\phi^{-1}(\mathcal{H}_1)(a)}(\alpha\ p_1 * \beta\, p_2) \leq C\{\nu^2{}_{\phi^{-1}(\mathcal{H}_1)(a)}(q_1), \nu^2{}_{\phi^{-1}(\mathcal{H}_1)(a)}(q_2)\}$

Application in Decision Making

Definition 13.3.11. Let A = $(\mu^{2A}{}_{ij}, \nu^{2A}{}_{ij}) \in$ PFSM $(m \times n)$ be a PFSM, then the choice matrix of matrix A is given by C(A) = $\left[\left(\frac{\sum_{j=1}^{n}(\mu^{2A}{}_{ij})^2}{n}, \frac{\sum_{j=1}^{n}(\nu^{2A}{}_{ij})^2}{n}\right)\right]_{m\times 1}$ for all i when weights are equal.

Definition 13.3.12. Let A = $(\mu^{2A}{}_{ij}, \nu^{2A}{}_{ij}) \in$ PFSM $(m \times n)$ be a PFSM, then the weighted choice matrix of matrix A is given by $C(A)_w = \left[\left(\frac{\sum_{j=1}^{n} w_j\,(\mu^{2A}{}_{ij})^2}{w_j}, \frac{\sum_{j=1}^{n} w_j\,(\nu^{2A}{}_{ij})^2}{w_j}\right)\right]_{m\times 1}$ for all i where $w_j > 0$ are weights.

Algorithm for Decision Making Problem.

Step 1: Build the Pythagorean fuzzy soft matrices that correspond to PFSS.

Step 2:
Case (1)
Compute the membership and non-membership value of PFSM's choice matrix

Case (2)
Calculate the weighted choice matrix of PFSM membership and non-membership value.
Step-3: Select the option with the highest membership value.

Example 13.3.13. Assume an automobile manufacturer is looking to fill a position. There are three candidates who from the Universe $\mathcal{P} = \{ p_1, p_2, p_3 \}$, the hiring committee considers a set of parameters, $\mathring{A} = \{e_1, e_2, e_3\}$ the parameters stand for "experience", "computer knowledge" and "communication skill" on the basis of the committee will be choose the candidate for the job. Then the above problem can be model by considering the Pythagorean fuzzy soft set $(\mathcal{H}, \mathring{A})$ over $\mathcal{P}$, where $\mathcal{H}$ is mapping $\mathcal{H} : \mathring{A} \to P(\mathcal{P})$ which provides the description of candidate on the basis of different criteria.

Step 1: Construct the PFSM:

$$A = \begin{matrix} p_1 \\ p_2 \\ p_3 \end{matrix} \overset{\begin{matrix} e_1 & \quad e_2 & \quad e_3 \end{matrix}}{\begin{pmatrix} (0.8, 0.5) & (0.6, 0.5) & (0.8, 0.2) \\ (0.6, 0.6) & (0.7, 0.4) & (0.8, 0.4) \\ (0.6, 0.6) & (0.7, 0.6) & (0.9, 0.2) \end{pmatrix}}$$

Step 2:
Case 1: **Equal weights**
Evaluate the choice matrix for the Pythagorean fuzzy soft matrix A as:

$$C(A) = \begin{bmatrix} (0.5467, 0.18) \\ (0.4967, 0.2267) \\ (0.5533, 0.2533) \end{bmatrix}$$

Case 2: **Unequal weights**
If the parameter "experience" has a weight of 0.2, "computer knowledge" has a weight of 0.6, and "communication skill" has a weight of 0.2, then the weighted option matrix for A looks like this:

$$C(A)_w = \begin{bmatrix} (0.472, 0.208) \\ (0.494, 0.184) \\ (0.528, 0.296) \end{bmatrix}$$

Step 3:

Case 1: **(Equal weights):** In the case of equal weights, It is evident from the matrix that was created in Step -2 that if we offer equal preference for all of the parameters, we have 0.5467 as the highest membership value, i.e., of candidate p_1 . Therefore, in this case the most suitable candidate for the committee members would be p_1.

Case 2: **(Unequal weights):** However, it may also be observed that if the committee members give preference for the parameter "communication skill" over the other parameters, then 0.528 being the highest membership value for candidate p_3 . Therefore, in this case the most suitable candidate for the committee members would be p_3.

13.4 Conclusion

A new type of algebraic structure, which we will refer to as PFSLS, has been developed by combining the concepts of Pythagorean fuzzy linear space and soft sets. The PFSLS that results from the intersection of two other PFSLSs is also a PFSLS. However, by way of a counter example, the combination of two PFSLSs does not have to result in a PFSLS. In addition to this, we present the idea that the cartesian product of two PFSLSs is likewise a PFSLS. In addition to this, both the image and the pre image of the Pythagorean fuzzy soft linear space are presented here. In addition, the PFSM model has been used to illustrate a decision-making strategy for the purpose of problem-solving. This strategy makes use of the concepts of choice matrices as well as weighted choice matrices. In the future, we propose the algebraic structure of Pythagorean fuzzy soft rough linear space and multi criteria decision making problems.

13.5 Acknowledgements

The authors would like to express our gratitude to the Book Editor Dr. Ramkanta Meher.

References

[1] Aygünoglu, H. Aygün, 'Introduction to fuzzy soft groups', Comput. Math. Appl. 58, 1279–1286, 2009.

[2] A.Guleria, R.K.Bajaj, 'On Pythagorean fuzzy matrices, operations and their applications in decision making and medical diagnosis', Soft Comput. 23, 7889, 2018.

[3] Y. B. Jun, K. J. Lee, and C. H. Park, 'Fuzzy soft set theory applied to BCK/BCI- algebras', Computers & Mathematics with Applications, 59(9), 3180–3192, 2010.

[4] G.Lubczonok, V.Murali, 'On flags and fuzzy subspaces of vector spaces', Fuzzy Sets and Systems, 125, 201–207, 2002.

[5] P. K. Maji, R. Biswas, A.R. Roy, 'Fuzzy soft sets', The Journal of Fuzzy Mathematics, 9(3), 589–602, 2001.

[6] P. K.Maji, A. R .Roy, R.Biswas, 'An application of soft sets in a decision making Problem', Computers and Mathematics with Applications, 44(8–9), 1077–1083, 2002.

[7] P. K. Maji, R.Biswas, A.R.Roy, 'Soft set theory', Computers and Mathematics with Applications, 45(4–5), 555–562, 2003.

[8] D. Molodtsov, 'Soft set theory-first results', Computers & Mathematics with Applications, 37(4–5),19–31, 1999.

[9] X.Peng, Y.Yang, J. Song, ' Pythagorean fuzzy soft set and its application', Computer Engineering,41(7), 224–229, 2015.

[10] A.R. Roy, P.K. Maji, 'A fuzzy soft set theoretic approach to decision making problems', J. Comput. Appl. Math., 203, 412–418, 2007.

[11] Sivaramakrishnan, S, Vijayabalaji, S & Rajasekaran, R, 'Interval-valued Pythagorean fuzzy linear space', International Conference on Advanced Research in Mathematical Sciences (ICARMS-2021), Conference Proceeding, ISBN No. 938950784-7, Feb.24-25, 2021.

[12] R. R. Yager, 'Pythagorean fuzzy subsets', IFSA World Congress and NAFIPS Annual Meeting 2013 Joint, Edmonton, Canada, IEEE, 57–61, 2013.

[13] L.A. Zadeh, 'Fuzzy sets', Information Control, 8, 338-353, 1965.

Chapter 14

Strain-controlled Magnetic Domain Wall Motion in Multiferroic Devices Induced by Current and Traveling Wave Magnetic Field

Sarabindu Dolui, Sharad Dwivedi

Department of Mathematics, School of Sciences, National Institute of Technology Andhra Pradesh, Tadepalligudem - 534101, India

Abstract

This work deals with an analytical investigation of domain wall dynamics in a two-layer structure composed of a ferromagnetic nanostrip mechanically connected to a piezoelectric actuator in the presence of spin-polarized electric current and traveling wave magnetic field. The ferromagnetic layer is made of a cubic crystal material with magnetostrictive properties and linear elasticity, which causes isochoric magnetostrictive deformations. The study uses the Landau-Lifshitz-Gilbert equation to analyze how the magnetization within the ferromagnet evolves over space and time, considering the influence of magneto elasticity and magneto crystalline anisotropy. The primary objective is to understand the impact of spin-polarized current, magnetic field frequency, crystal symmetry, and magneto elasticity on the behavior of domain walls. To determine specific characteristics such as wall width, velocity, threshold, and breakdown, the analysis employs both collective coordinate and traveling wave ansatz to derive explicit expressions.

Keywords: Magnetostriction, Domain wall, Traveling wave, Strain, Landau-Lifshitz-Gilbert equation, Crystal symmetry, Walker breakdown, Frequency, Spintronics devices.

14.1 Introduction

The field of multiferroics has made significant progress and revolutionary advancements. Multiferroics are materials that exhibit both ferromagnetic and ferroelectric properties. These advancements have been spurred by the wide range of practical applications that multiferroics offer, such as magnetic resonance imaging (MRI), magnetic memories, logic, transformers, actuators, sensors, etc. [1, 2, 3, 4, 5, 6]. One key aspect of these applications involves the creation of artificial heterostructures. These heterostructures are comprised of a thin layer of magnetostrictive material (MSL) bonded to a thick layer of piezoelectric material (PEL). The ability to control the movement of domain walls within these structures is vital for their design and optimization [7, 8, 9]. The thin transition zones between the uniformly oriented magnetized regions (domains) are called the domain wall (DW).

The prime objective is to achieve robust magneto electric couplings within these structures. To be precise, when a voltage V_{PE} is applied to the PEL through lateral electrodes, it generates an electric field that induces strains in the material [10, 11, 12]. These piezo-induced strains are fully moved to the MSL via their shared surface. The presence of magnetostriction in the multiferroic heterostructures leads to the simultaneous occurrence of both direct and inverse magnetostrictive effects, as extensively studied in previous research [13, 14, 15]. In these artificial heterostructures, the crystal symmetry of the magnetostrictive material plays a crucial role in impacting anisotropy, magnetostrictive, and magneto elastic energies [16, 17, 18].

In light of advancing technologies and the growing demand for efficient data storage and processing, it becomes essential to thoroughly investigate the characteristics of DWs and understand how the key parameters that impact the DW motion vary under the influence of external stimuli. From the application viewpoint, the ability to control the position of DWs through the application of external sources is of utmost importance. By examining the dependence of these key parameters on external stimuli, we can gain a deeper understanding of the dynamical behavior of such artificial bilayer heterostructures.

The spatial and temporal variation of magnetization dynamics within the MSL can be described by the Landau-Lifshitz-Gilbert (LLG) equation [16, 17, 19, 20, 21]. When a spin-polarized electric current passes through the MSL, it applies a torque to the magnetization by transferring angular momentum to the internal structure of the DWs, resulting in the DWs moving along the direction of the applied current. This phenomenon is commonly referred to as spin-transfer torque (STT) [22, 23, 24, 25]. Also, both experimental and theoretical studies have demonstrated that the motion of DWs exhibits distinct dynamic behaviors, viz., steady-state and precessional regimes. These regimes are determined by the magnitude of the external sources. In the steady-state regime, the maximum (or minimum) value of the external stimuli required to induce DW motion corresponds to the Walker breakdown (or threshold). However, when the external stimulus surpasses the Walker breakdown (WB) limit, the DW dynamics enter into a precessional regime [26, 27].

Our study focuses on the theoretical investigation of DW dynamics in a material that exhibits magnetostriction subjected to a traveling wave-type magnetic field [28, 29, 30, 31]. Such a field offers several advantages over a static magnetic field due to enhanced control, tuneability, and versatility [32, 33, 34]. To gain a comprehensive understanding of the whole dynamics, two distinct methods are commonly employed: the traveling wave ansatz and the collective coordinate method. Each method serves a specific purpose in examining different aspects of dynamic phenomena. The traveling wave method is particularly useful when studying equations that showcase wave-like behavior, such as electromagnetic waves and acoustic waves [35, 36, 37, 38]. On the other hand, with the collective coordinate method, researchers can gain insights into the system's global behavior and macroscopic properties [39, 40, 41].
The analyses undertaken in this work build upon recent findings in the field of bilayer piezoelectric/magnetostrictive nanostructures and their interaction with traveling wave magnetic fields. Experimental observations have revealed that the coupling between the spin-polarized currents, traveling wave-type external magnetic fields, and the stress-strain response of the material can lead to a significant modification of the DW velocity and a shift in the WB limit [42, 43, 44, 45]. The results of this study demonstrate that the dynamics of DWs can be controlled by adjusting the frequency and amplitude of the magnetic field, giving rise to novel and intriguing phenomena within the system.

The manuscript is organized as follows: In Section 14.2, we detail a comprehensive explanation of the theoretical model, presenting the governing dynamics and their physical significance. In Section 14.3, we give a detailed mathematical investigation of the dynamics by employing both the traveling wave and collective coordinate techniques. Through these mathematical approaches, we gain valuable insights into the behavior and characteristics of the system under the influence of the traveling wave-type magnetic field and other key parameters. In the last section, we summarize our findings and highlight the significant issues that arise from this work.

14.2 Theoretical model

Our study investigates the propagation of magnetic DWs inside the magnetostrictive (MS) nanostrip, which possesses ferromagnetic properties and is attached above the piezoelectric (PE) actuator. The configuration is illustrated in Fig. 14.1, where the length, breadth, and thickness of the MSL are considered to be L, w, and d, positioned along $\boldsymbol{e}_x, \boldsymbol{e}_y, \boldsymbol{e}_z$ directions, respectively, subject to the condition $L \gg w > d$. Due to the thin and lengthy shape of the nanostrip, it is considered that the magnetic DWs undergo one-dimensional motion along the principal axis $\boldsymbol{e}_x$. As a result, variations of the involved physical entities along $\boldsymbol{e}_y$ and $\boldsymbol{e}_z$ directions are disregarded. A magnetic field in the form of a traveling wave $\boldsymbol{h}_{\text{app}}$ acts on MSL along with a spin-polarized electric current $\boldsymbol{J} = J\boldsymbol{e}_x$ which is uniform in space and constant in time. Moreover, the MSL experiences stress-free magnetostrictive effects and mechanical strains that are induced via the PEL. The piezoelectric (PE) material undergoes deformation when an electric voltage V_{PE} is applied using lateral electrodes. This voltage creates an electric field aligned with the $\boldsymbol{e}_y$ axis, causing the PE width to be elongated (or contracted) along that axis and resulting in contractions (or elongations). Due to the extremely thin nature of the MSL, its mechanical properties become insignificant as compared to the ones of the PEL. As the two adjacent layers are attached perfectly, piezo-induced strains produced at the PEL can be entirely transferred from the PEL to the MSL through the common interface. Also, we assumed that a $180°$ Bloch DW is nucleated in the system and that the magnetization vectors in the opposite domains are oriented along the $\pm\boldsymbol{e}_z$ axis respectively.

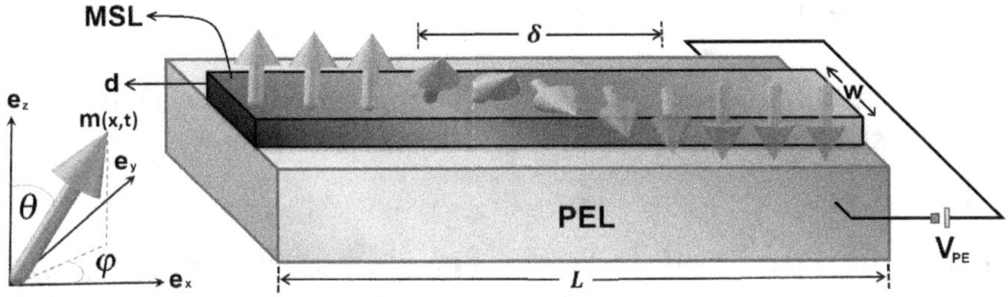

Figure 14.1 Schematics of the bilayer structure for a magnetostrictive-piezoelectric (MSL-PEL) composite, together with the magnetization orientations.

The DW motion in the MSL is governed by the Landau-Lifshitz-Gilbert (LLG) equation [16, 17, 19, 20, 26]. This equation describes the spatial and temporal variation of the magnetization vector $\boldsymbol{m} = \boldsymbol{m}(x, t)$, taking into account the combined effect of three torque contributions. These torque contributions play a vital role in delineating DW motion and are as follows:

$$\dot{\boldsymbol{m}} = \boldsymbol{t}^{\text{prec}} + \boldsymbol{t}^{\text{dis}} + \boldsymbol{t}^{\text{stt}}, \tag{14.2.1}$$

where the over dot represents partial derivative with respect to time, the unitary vector $\boldsymbol{m}(x,t) = \boldsymbol{M}(x,t)/M_S$ is normalized by saturation magnetization value M_S.

The first term on the right-hand side of Eq. (14.2.1) reflects the undamped precession of the magnetization vector around the total effective field, $\boldsymbol{h}_{\text{eff}}$, and is expressed as [26, 46, 47, 48]:

$$\boldsymbol{t}^{\text{prec}} = \gamma(\boldsymbol{h}_{\text{eff}} \times \boldsymbol{m}), \tag{14.2.2}$$

where the constant γ is defined as $\gamma = M_S\mu_0\gamma_e$, being μ_0 is the magnetic permeability of vacuum, and $\gamma_e = ge/m_e$ is the is gyromagnetic ratio (g represents the Landè factor, e represents the charge of an electron, and m_e represents the mass of an electron). The total effective field $\boldsymbol{h}_{\text{eff}}$ consists of exchange, demagnetizing, anisotropy, magneto elastic, and external field contributions.

The second term $\boldsymbol{t}^{\text{dis}}$, accounts for the dissipative phenomena associated with energy loss, commonly known as the Gilbert damping torque. It reads [17, 19, 25, 48]

$$\boldsymbol{t}^{\text{dis}} = \alpha(\boldsymbol{m} \times \dot{\boldsymbol{m}}), \tag{14.2.3}$$

being α, the dimensionless Gilbert damping coefficient characterizes the strength of the damping effect in the magnetization dynamics.

Finally, the last torque term in Eq. (14.2.1) accounts for the spin-transfer-torque t^{stt} resulting from the flow of spin current J via the MSL. It encompasses adiabatic and non-adiabatic contributions, delineating the DW distortion and motion [17, 19, 22, 23, 42], respectively.

$$\boldsymbol{t}^{\text{stt}} = -u_0 J \frac{\partial \boldsymbol{m}}{\partial x} - \eta u_0 J \left(\frac{\partial \boldsymbol{m}}{\partial x} \times \boldsymbol{m} \right), \tag{14.2.4}$$

where, η represents the phenomenological non-adiabatic coefficient, and J is the strength of electric current density. Moreover, the coefficient $u_0 = g\mu_B P/(2eM_S)$, being μ_B Bohr magneton and P the current polarization factor.

14.2.1 Governing equation

The Eq. (14.2.1) that rules the magnetization dynamics inside the ferromagnetic MSL that explains how the magnetization vector evolves over time in response to external influences can be expressed as:

$$\frac{\partial \boldsymbol{m}}{\partial t} = \gamma(\boldsymbol{h}_{\text{eff}} \times \boldsymbol{m}) + \alpha \left(\boldsymbol{m} \times \frac{\partial \boldsymbol{m}}{\partial t} \right) + \left[-u_0 J \frac{\partial \boldsymbol{m}}{\partial x} - \eta u_0 J \left(\frac{\partial \boldsymbol{m}}{\partial x} \times \boldsymbol{m} \right) \right] \tag{14.2.5}$$

where, $\boldsymbol{m}\colon \mathbb{R} \times \mathbb{R}^+ \to \mathbb{S}^2$ and $\times$ represents the vector product on $\mathbb{R}^3$.

In our analysis, we consider that all the vector quantities involved in this study have a functional relationship with both the major axis coordinate x, representing the position along the strip, and time t. This one-dimensional methodology is widely acknowledged and firmly established in the literature for thin and elongated structures like nanostrips or nanowires [7, 16, 26, 42].

Now, let's delve into the details of the total effective field $\boldsymbol{h}_{\text{eff}}$, which includes account exchange, demagnetizing, anisotropy, magneto elastic, and external field contributions.

$$\boldsymbol{h}_{\text{eff}} = \boldsymbol{h}_{\text{ex}} + \boldsymbol{h}_{\text{dmg}} + \boldsymbol{h}_{\text{ani}} + \boldsymbol{h}_{\text{me}} + \boldsymbol{h}_{\text{app}}. \tag{14.2.6}$$

The first term in Eq. (14.2.6) corresponds to the exchange field $\boldsymbol{h}_{\text{ex}}$, which plays a crucial role in aligning neighboring magnetic dipole moments parallel to each other. The expression for the exchange field can be expressed as [17, 19, 26, 44]:

$$\boldsymbol{h}_{\text{ex}} = A\frac{\partial^2 \boldsymbol{m}}{\partial x^2}, \tag{14.2.7}$$

where, $A = (2A_{\text{ex}})/(\mu_0 M_s^2)$ and A_{ex} denotes the exchange constant that depends on the material property.

The second term in the equation represents the demagnetizing (stray) field $\boldsymbol{h}_{\text{dmg}}$, which arises as a consequence of the internal magnetization distribution within the material. It is characterized by a magnetic field that opposes the original magnetization, effectively reducing the overall magnetization strength. This field strongly depends on the shape and geometry of the object and can be expressed as [9, 19, 26, 27]:

$$\boldsymbol{h}_{\text{dmg}} = -N_x m_x \boldsymbol{e}_x - N_y m_y \boldsymbol{e}_y - N_z m_z \boldsymbol{e}_z, \tag{14.2.8}$$

where N_x, N_y, and N_z represent the demagnetizing factors and satisfy the normalization constraint $N_x + N_y + N_z = 1$.

The third term represents the magneto crystalline anisotropy field $\boldsymbol{h}_{\text{ani}}$, which accounts for the energy required to deviate the magnetization away from the easy axis. The symmetry of the crystal influences this energy and can be expressed for cubic crystal class as [17, 19, 26, 42]:

$$\begin{aligned}\boldsymbol{h}_{\text{ani}} &= -\frac{2K}{\mu_0 M_S^2}\left[m_x(1-m_x^2)\boldsymbol{e}_x + m_y\left(1-m_y^2\right)\boldsymbol{e}_y + m_z(1-m_z^2)\boldsymbol{e}_z\right] \\ &= (h_x^{an}, h_y^{an}, h_z^{an}),\end{aligned} \tag{14.2.9}$$

where, K represents the cubic anisotropy coefficient.

The fourth term describes the magneto elastic field $\boldsymbol{h}_{\text{me}}$ that arises from the coupling between the magnetization and the strain within the material, leading to changes in the magnetic behavior due to mechanical deformations and vice versa. This interaction gives rise to various effects, including magnetostriction. In the case of a linear elastic material, the expression for the magneto elastic field can be described as follows [17, 19, 26, 42]:

$$\boldsymbol{h}_{\text{me}} = \frac{1}{\mu_0 M_S^2}(\boldsymbol{\varepsilon} - \boldsymbol{\varepsilon}^{\boldsymbol{m}}) : \mathbb{C} : \frac{\partial \boldsymbol{\varepsilon}^{\boldsymbol{m}}}{\partial \boldsymbol{m}} = (h_x^{\text{me}}, h_y^{\text{me}}, h_z^{\text{me}}), \tag{14.2.10}$$

where, $\boldsymbol{\varepsilon}^{\boldsymbol{m}} = \mathbb{Z} : (\boldsymbol{m} \otimes \boldsymbol{m})$ denotes the magnetostrictive strain tensor, while $\boldsymbol{\varepsilon}$ represents the total strain tensor encompassing contributions from both the elastic and magnetostrictive tensors. The strain related to the local magnetization changes is $\boldsymbol{\varepsilon}^{\boldsymbol{m}} = \lambda^{\boldsymbol{m}} \boldsymbol{m}\boldsymbol{m}^T$, where $\lambda^{\boldsymbol{m}}$ denotes the magneto-mechanical coupling tensor. In the case of a cubic crystal, $\boldsymbol{\varepsilon}^{\boldsymbol{m}}$ is given by [16, 18, 21, 26],

$$[\boldsymbol{\varepsilon}^{\boldsymbol{m}}] = \begin{cases} \frac{3}{2}\lambda_{100}\left(m_i m_j - \frac{1}{3}\right), & i = j \\ \frac{3}{2}\lambda_{111} m_i m_j, & i \neq j \end{cases},$$

$$= \frac{3}{2}\begin{bmatrix} \lambda_{100}\left(m_x^2 - \frac{1}{3}\right) & \lambda_{111} m_x m_y & \lambda_{111} m_x m_z \\ \lambda_{111} m_x m_y & \lambda_{100}\left(m_y^2 - \frac{1}{3}\right) & \lambda_{111} m_y m_z \\ \lambda_{111} m_x m_z & \lambda_{111} m_y m_z & \lambda_{100}\left(m_z^2 - \frac{1}{3}\right) \end{bmatrix} \tag{14.2.11}$$

where, λ_{100} and λ_{111} are the magnetostriction constants along the $\langle 100 \rangle$ and $\langle 111 \rangle$ directions. It is important to note that for hexagonal or uniaxial crystals, the equation describing the magnetoelastic field would require modifications [15, 16].

Moreover, the fourth-order magnetostriction tensor is denoted by $\mathbb{Z}$, and the fourth-order elasticity tensor is denoted by $\mathbb{C}$. The double contraction and tensor product are denoted by " : " and " $\otimes$ ", respectively. For the considered cubic crystal classes, the elasticity tensor can be given in terms of three independent elastic constants, which are usually referred to as c_{11}, c_{12} and c_{44} in Voigt notation. The matrix $\mathbb{C}$ can be expressed in a concise form as follows:

$$[\mathbb{C}] = \begin{bmatrix} c_{11} & c_{12} & c_{12} & & & \\ c_{12} & c_{11} & c_{12} & & [0] & \\ c_{12} & c_{12} & c_{11} & & & \\ & & & c_{44} & 0 & 0 \\ & [0] & & 0 & c_{44} & 0 \\ & & & 0 & 0 & c_{44} \end{bmatrix}. \quad (14.2.12)$$

The magnetostriction tensor $\mathbb{Z}$ has three independent components. However, if the system is considered to undergo a fully isochoric magnetostrictive deformation, then the number of independent components reduces to two. Therefore, $\mathbb{Z}$ is given by [8, 14, 21]:

$$[\mathbb{Z}] = \begin{bmatrix} Z_{1111} & -\frac{1}{2}Z_{1111} & -\frac{1}{2}Z_{1111} & & & \\ -\frac{1}{2}Z_{1111} & Z_{1111} & -\frac{1}{2}Z_{1111} & & [0] & \\ -\frac{1}{2}Z_{1111} & -\frac{1}{2}Z_{1111} & Z_{1111} & & & \\ & & & Z_{1212} & 0 & 0 \\ & [0] & & 0 & Z_{1212} & 0 \\ & & & 0 & 0 & Z_{1212} \end{bmatrix}. \quad (14.2.13)$$

The two magnetostriction coefficients can be expressed in terms of the Miller-Bravais indices as follows: $\lambda_{100} = Z_{1111}$ and $\lambda_{111} = \frac{4}{3}Z_{1212}$ [42, 47]. To properly determine the magneto elastic field $\boldsymbol{h}_{\mathrm{me}}$, it is necessary to define the six independent components of the total strain tensor ε. Since the difference in thickness between the MSL and the PEL is significant, variations in strain along the $\boldsymbol{e}_z$ axis can be considered negligible. As a result, the total strains affecting the MSL can be calculated by assuming the mechanical boundary conditions at the upper and bottom xy surfaces. Therefore, the overall strains experienced by the MSL can be determined by taking into account the mechanical boundary conditions at the upper and bottom xy surfaces.

At the shared surface between the PEL and the MSL, the planar strains $\varepsilon_{xx}, \varepsilon_{xy}$ and ε_{yy} are imposed by the PEL. These strains can be classified as constant in time and uniform in space for a fixed applied voltage. Also, previous numerical findings indicate that the in-plane shear strain ε_{xy} can be disregarded. The rest of the components of the strain tensor, namely ε_{xz}, ε_{yz}, and ε_{zz}, can be determined by solving the traction boundary conditions $\bar{\boldsymbol{\sigma}}\hat{\boldsymbol{n}} = \boldsymbol{0}$ at the top xy -surface. Here, $\bar{\boldsymbol{\sigma}} = \mathbb{C}: (\boldsymbol{\varepsilon} - \boldsymbol{\varepsilon}^m)$ represents the Cauchy stress tensor. Since the normal vector $\hat{\boldsymbol{n}}$ to the free top surface is equivalent to the unit vector $\boldsymbol{e}_z$, these aforementioned conditions yield the following set of equations:

$$\begin{aligned}\sigma_{xz} &= 0 \Rightarrow \varepsilon_{xz} = \tfrac{3}{2}\lambda_{111} m_x m_z,\\ \sigma_{yz} &= 0 \Rightarrow \varepsilon_{yz} = \tfrac{3}{2}\lambda_{111} m_y m_z,\\ \sigma_{zz} &= 0 \Rightarrow \varepsilon_{zz} = -\tfrac{c_{12}}{c_{11}}\left(\varepsilon_{xx} + \varepsilon_{yy}\right) + \tfrac{3}{2}\lambda_{100}\tfrac{c_{11}-c_{12}}{c_{11}}\left(m_z^2 - \tfrac{1}{3}\right).\end{aligned} \tag{14.2.14}$$

where taking into account Eqs. (14.2.10)-(14.2.13) and Eq. (14.2.17), the components of the magneto elastic field $\boldsymbol{h}_{\mathrm{me}}$ reads:

$$\begin{aligned}h_x^{\mathrm{me}}(\theta,\varphi) &= \frac{3\cos\varphi\sin\theta}{\mu_0 M_S^2}\left\{\lambda_{100}\frac{c_{11}-c_{12}}{c_{11}}\left[\left(c_{11}+c_{12}\right)\varepsilon_{xx} + c_{12}\varepsilon_{yy}\right]\right.\\ &\quad -3c_{44}\lambda_{111}^2\sin^2\varphi\sin^2\theta + \frac{3}{2}\lambda_{100}^2\frac{c_{12}-c_{11}}{c_{11}}\left[c_{12}\left(\sin^2\theta\sin^2\varphi - \frac{1}{3}\right)\right.\\ &\quad \left.\left. +\left(c_{11}+c_{12}\right)\left(\sin^2\theta\cos^2\varphi - \frac{1}{3}\right)\right]\right\},\\ h_y^{\mathrm{me}}(\theta,\varphi) &= \frac{3\sin\varphi\sin\theta}{\mu_0 M_S^2}\left\{\lambda_{100}\frac{c_{11}-c_{12}}{c_{11}}\left[c_{12}\varepsilon_{xx} + \left(c_{11}+c_{12}\right)\varepsilon_{yy}\right]\right.\\ &\quad -3c_{44}\lambda_{111}^2\cos^2\varphi\sin^2\theta + \frac{3}{2}\lambda_{100}^2\frac{c_{12}-c_{11}}{c_{11}}\left[c_{12}\left(\sin^2\theta\cos^2\varphi - \frac{1}{3}\right)\right.\\ &\quad \left.\left. +\left(c_{11}+c_{12}\right)\left(\sin^2\theta\sin^2\varphi - \frac{1}{3}\right)\right]\right\},\\ h_z^{\mathrm{me}}(\theta,\varphi) &= 0.\end{aligned} \tag{14.2.15}$$

Finally, the last term in Eq. (14.2.6) represents the applied magnetic field $\boldsymbol{h}_{\mathrm{app}}$ of the form:

$$\boldsymbol{h}_{\text{app}} = h_x\boldsymbol{e}_x + h_y\boldsymbol{e}_y + h_z\boldsymbol{e}_z. \tag{14.2.16}$$

Using spherical coordinates, the unitary magnetization vector can be expressed as follows:

$$\boldsymbol{m} = \sin\theta\cos\varphi\ \boldsymbol{e}_x + \sin\theta\sin\varphi\ \boldsymbol{e}_y + \cos\theta\ \boldsymbol{e}_z, \tag{14.2.17}$$

being θ and φ the polar and azimuthal angles, respectively.

By inserting Eq. (14.2.6) into Eq. (14.2.5), we obtain:

$$\begin{aligned}\frac{\partial\theta}{\partial t}+\alpha\sin\theta\frac{\partial\varphi}{\partial t} = {}& \gamma\Big\{A\sin\theta\frac{\partial^2\varphi}{\partial x^2}+2A\cos\theta\frac{\partial\theta}{\partial x}\frac{\partial\varphi}{\partial x}+\left(N_x-N_y\right)\sin\theta\cos\varphi\sin\varphi \\ &+\left(h_y+h_y^{\text{me}}+h_y^{\text{an}}\right)\cos\varphi-\left(h_x+h_x^{\text{me}}+h_x^{\text{an}}\right)\sin\varphi\Big\}-\eta u_0 J\sin\theta\frac{\partial\varphi}{\partial x}-u_0 J\frac{\partial\theta}{\partial x},\end{aligned} \tag{14.2.18}$$

$$\begin{aligned}\sin\theta\frac{\partial\varphi}{\partial t}-\alpha\frac{\partial\theta}{\partial t} = {}& \gamma\Big\{-A\frac{\partial^2\theta}{\partial x^2}+A\sin\theta\cos\theta\left(\frac{\partial\varphi}{\partial x}\right)^2+\sin\theta\left(h_z+h_z^{\text{me}}+h_z^{\text{an}}\right) \\ &-\cos\theta\cos\varphi\left(h_x+h_x^{\text{me}}+h_x^{\text{an}}\right)-\cos\theta\sin\varphi\left(h_y+h_y^{\text{me}}+h_y^{\text{an}}\right) \\ &+\sin\theta\cos\theta\left[N_x\cos^2\varphi+N_y\sin^2\varphi-N_z\right]\Big\} \\ &-u_0 J\sin\theta\frac{\partial\varphi}{\partial x}+\eta u_0 J\frac{\partial\theta}{\partial x}.\end{aligned}$$

14.3 Mathematical Formulation

In this section, we aim to obtain explicit expressions for key characteristics, including DW width, velocity, threshold, and WB limit under the action of external stimuli. More precisely, we explore how key factors, such as spin-polarized current, frequency of the magnetic field oscillation, crystal symmetry, and magnetoelasticity, impact the dynamics of DWs. We employ the collective coordinate and the traveling wave ansatz techniques to analyze DW motion. A traveling wave magnetic field is a specific type of electromagnetic wave that exhibits the propagation of electric and magnetic fields through space and time, with their magnitudes periodically changing. These waves are generated by oscillating electric or magnetic fields and propagate at the speed of light. Traveling wave magnetic fields find applications in various fields, including wireless communication, radar systems, and medical imaging [32, 33, 34, 43].

Mathematically, a traveling wave magnetic field can be described as a sinusoidal function of both time and position [28, 29, 30, 31, 45]. It follows the form:

$$\boldsymbol{h}_{\mathrm{app}}(x,t) = (h_x, h_y, h_z)\sin(nx - \omega t), \tag{14.3.1}$$

where $\boldsymbol{h}_{\mathrm{app}}$, x, t, n, and ω denote the total applied magnetic field, the position along the wave, the time variable, wave number, and the angular frequency, respectively. The magnitude of such magnetic fields varies sinusoidally in space and time, and they travel with a velocity of ω/n.

14.3.1 Traveling wave ansatz

The traveling wave ansatz is a useful technique for analyzing the physical phenomenon in which the governing equation can be written in the form of a function that travels at a constant speed along a specific direction. It is particularly useful in dealing with equations that exhibit wave-like phenomena, such as electromagnetic waves, acoustic waves, and other types of propagating disturbances [35, 36, 37, 38].

In this section, we employ the traveling wave technique within the steady-state regime of DW propagation. In this regime, the dynamics of the DW are characterized by a rigid motion maintaining a constant velocity v along the nanostrip axis $\boldsymbol{e}_x$ having a fixed azimuthal angle ϕ_0, and the polar angle adheres to the classical traveling wave ansatz, i.e., $\theta = \theta(\xi)$ where $\xi = x - vt$ [16, 19, 26, 42]. Using these assumptions, Eq. (14.2.18) and Eq. (14.2.19) reduce to:

$$\begin{aligned}(u_0 J - v)\theta' &= \gamma\{(N_x - N_y)\sin\theta\sin\varphi_0\cos\varphi_0 + (h_y + h_y^{\mathrm{me}} + h_y^{an})\cos\varphi_0 \\ &\qquad -(h_x + h_x^{\mathrm{me}} + h_x^{an})\sin\varphi_0\},\end{aligned} \tag{14.3.2}$$

$$\begin{aligned}\left[\alpha v-\eta u_0 J\right]\theta' = & -\gamma\Big\{A\theta''+\left(N_z-N_x\cos^2\varphi_0-N_y\sin^2\varphi_0\right)\sin\theta\cos\theta \\ & -\left(h_z+h_z^{\mathrm{me}}+h_z^{an}\right)\sin\theta+\Big[\left(h_x+h_x^{\mathrm{me}}+h_x^{an}\right)\cos\varphi_0 \\ & -\left(h_y+h_y^{\mathrm{me}}+h_y^{an}\right)\sin\varphi_0\Big]\cos\theta\Big\}\end{aligned} \tag{14.3.3}$$

In our analysis, the applied magnetic field is assumed to be a traveling wave type and acting only in $\boldsymbol{e}_z$ direction, i.e., $h_x = 0 = h_y$ and,

$$h_z = h\sin(nx-\omega t) = h\sin(n(x-vt)) = h\sin(n\xi). \tag{14.3.4}$$

Now, taking Eqs. (14.2.7) - (14.2.10), and Eqs. (14.2.15) -(14.2.16) into account, Eq. (14.3.2) can be recast as:

$$\theta' = \Gamma\left(\sin\theta+\Lambda\sin^3\theta\right), \tag{14.3.5}$$

where,

$$\Gamma = \frac{\gamma}{2(u_0 J-v)}\left[N_x-N_y+\frac{3\lambda_{100}}{\mu_0 M_S^2}(c_{11}-c_{12})(\varepsilon_{yy}-\varepsilon_{xx})\right]\sin 2\varphi_0, \tag{14.3.6}$$

and,

$$\Lambda = \frac{\left[9\lambda_{100}^2\left(c_{11}-c_{12}\right)-18\lambda_{111}^2 c_{44}-4K\right]\left(\cos^2\varphi_0-\sin^2\varphi_0\right)}{2\mu_0 M_S^2\left(N_x-N_y\right)+6\lambda_{100}\left(c_{11}-c_{12}\right)\left(\varepsilon_{yy}-\varepsilon_{xx}\right)}. \tag{14.3.7}$$

Then, by inserting Eq. (14.3.5) into Eq. (14.3.3) we obtain,

$$\Upsilon_1\sin\theta+\Upsilon_2\sin\theta\cos\theta+\Upsilon_3\sin^3\theta\cos\theta+\Upsilon_4\sin^3\theta+\Upsilon_5\sin^5\theta\cos\theta = 0 \tag{14.3.8}$$

where,

$$\Upsilon_1 = \Gamma(\alpha_G v-\eta u_0 J)-\gamma\, h\sin(n\xi), \tag{14.3.9}$$

$$\Upsilon_2 = \gamma \left\{ \frac{3\lambda_{100}}{\mu_0 M_S^2} \frac{c_{11} - c_{12}}{c_{11}} [\varepsilon_{xx} \left(c_{11} \cos^2\varphi_0 + c_{12} \right) + \varepsilon_{yy} \left(c_{11} \sin^2\varphi_0 + c_{12} \right) \right.$$

$$\left. + \frac{\lambda_{100} \left(c_{11} + 2c_{12} \right)}{2} \right] + \mathrm{A}\Gamma^2 + (N_z - N_y \sin^2\varphi_0 - N_x \cos^2\varphi_0) - \frac{2K}{\mu_0 M_S^2} \bigg\}, \tag{14.3.10}$$

$$\Upsilon_3 = \gamma \left\{ 4\mathrm{A}\Gamma^2 \Lambda + \frac{4K\left(1 - \cos^2\varphi_0 \sin^2\varphi_0\right)}{\mu_0 M_S^2} - \frac{18}{\mu_0 M_S^2} c_{44} \lambda_{111}^2 \cos^2\varphi_0 \sin^2\varphi_0 \right.$$

$$\left. - \frac{9}{2\mu_0 M_S^2} \lambda_{100}^2 \frac{c_{11} - c_{12}}{c_{11}} \left[c_{12} + c_{11} \left(1 - 2\cos^2\varphi_0 \sin^2\varphi_0 \right) \right] \right\}, \tag{14.3.11}$$

$$\Upsilon_4 = \Lambda\, \Gamma(\alpha_G v - \eta u_0 J), \tag{14.3.12}$$

$$\Upsilon_5 = 3A\gamma\Lambda^2\Gamma^2. \tag{14.3.13}$$

Furthermore, by considering the relevant condition $\Upsilon_2 = 0$,we can derive the expression for the DW width $\delta = 1/\Gamma$ as:

$$\delta = \left(\frac{\mathrm{A}_{\mathrm{ex}}}{K - \beta + \frac{1}{2} \mu_0 M_s^2 \left(N_x \cos^2\varphi_0 + N_y \sin^2\varphi_0 - N_z \right)} \right)^{1/2} \tag{14.3.14}$$

where, $\beta = \frac{3}{2} \lambda_{100} \frac{c_{11} - c_{12}}{c_{11}} [\varepsilon_{xx} \left(c_{11} \cos^2\varphi_0 + c_{12} \right)$

$$+ \varepsilon_{yy} \left(c_{11} \sin^2\varphi_0 + c_{12} \right) + \frac{\lambda_{100} \left(c_{11} + 2c_{12} \right)}{2} \Bigg]. \tag{14.3.15}$$

The expression (14.3.14) reveals that the DW width is independent of external fields; however, it is influenced by magnetostriction, piezo-strains, and magneto crystalline anisotropy.

In this case, Eq. (14.3.5) provides us with two types of steady-state solutions, each representing the configuration of two distant domains. The first type of solution arises when $\sin\theta = 0$. These solutions correspond to the well-known standard Walker profile of a Bloch type DW, where the polar angle θ varies between $0°$ and $180°$. On the other hand, the second type of solution occurs when $\sin^2\theta = -1/\Lambda$. It is important to note that this class of solutions exists only when $\Lambda < 0$, allowing configurations such as the $71°$ and $109°$ DWs mainly observed in materials with cubic crystal symmetry [26, 28]. By restricting the analysis to only the first scenario, we obtain the solution of Eq. (14.3.5) for $\Lambda \geq 0$:

$$\ln\left(\tan\frac{\theta}{2}\right) - \frac{1}{2}\sqrt{\frac{\Lambda}{1+\Lambda}}\ln\left(\frac{1+2\Lambda+\tan^2\left(\frac{\theta}{2}\right)-2\sqrt{\Lambda(1+\Lambda)}}{1+2\Lambda+\tan^2\left(\frac{\theta}{2}\right)+2\sqrt{\Lambda(1+\Lambda)}}\right) = \Gamma\xi + c_1, \tag{14.3.16}$$

whereas, for $-1 < \Lambda < 0$:

$$\ln\left(\tan\frac{\theta}{2}\right) + \frac{1}{2}\sqrt{\frac{-\Lambda}{1+\Lambda}}\arg\left(\frac{1+2\Lambda+\tan^2\left(\frac{\theta}{2}\right)-2i\sqrt{-\Lambda(1+\Lambda)}}{1+2\Lambda+\tan^2\left(\frac{\theta}{2}\right)+2i\sqrt{-\Lambda(1+\Lambda)}}\right) = \Gamma\xi + c_2, \tag{14.3.17}$$

where, c_1 and c_2 are the arbitrary constants whose values can be easily estimated by applying the symmetry to the DW profile with regard to its center. Upon careful examination, we can easily verify that when the magnetostriction and cubic anisotropy effects are absent i.e., $\Lambda = 0$, Eq. (14.3.17) simplifies and reduces into the classical Walker solution:

$$\begin{aligned}\theta(\xi) &= 2\arctan\{\exp(\Gamma\xi)\} \\ &= 2\arctan\left\{\exp\left(\frac{x-vt}{\delta}\right)\right\}\end{aligned} \tag{14.3.18}$$

By taking average of the Eq. (14.3.8) over the range $0 \leq \theta \leq \pi$, we obtain the following expression:

$$\alpha(3+2\Lambda)v = \eta u_0(3+2\Lambda)J + 3\gamma\delta h\sin(n\xi). \tag{14.3.19}$$

The threshold field is the minimum field required to initiate motion, and the condition $v \geq 0$ must be satisfied for DW motion to occur. Hence from Eq. (14.3.19) we obtain:

$$h^{TH} = \frac{\eta u_0(3+2\Lambda)}{3\gamma\delta\sin(kx-\omega t)} J. \tag{14.3.20}$$

Now, we evaluate the WB limit, beyond which the DW motion becomes oscillatory. To this aim, using the condition $|\sin(2\varphi_0)| \leq 1$, Eq. (14.3.6) yields:

$$v \leq u_0 J + \frac{\gamma\delta}{2}\left|N_x - N_y + \frac{3\lambda_{100}}{\mu_0 M_s^2}(c_{11}-c_{12})(\varepsilon_{yy}-\varepsilon_{xx}).\right| \tag{14.3.21}$$

Using Eq. (14.3.19) and Eq. (14.3.21), we establish the WB limit of the external magnetic field:

$$h^{\mathrm{WB}} = \frac{(3+2\Lambda)}{3\sin(nx-\omega t)}\left\{\frac{(\alpha-\eta)u_0 J}{\gamma\delta} + \frac{\alpha}{2}\left|N_x - N_y + \frac{3\lambda_{100}}{\mu_0 M_s^2}(c_{11}-c_{12})(\varepsilon_{yy}-\varepsilon_{xx})\right|\right\}. \tag{14.3.22}$$

Eq. (14.3.20) and Eq. (14.3.22) revealed that both the threshold and Walker breakdown (WB) limit depend on the behavior of the external field configuration, particularly emphasizing the influence of the wave nature.

14.3.2 Collective Coordinate Method

This section delves into the systematic investigation of DW dynamics within the framework of the collective coordinate method. By employing a small number of collective variables, this technique simplifies the study of complex magnetization dynamics. In this particular case, we kept the configuration of the applied magnetic field consistent with the previous scenario [28, 29, 30, 31, 45], i.e.,

$$\boldsymbol{h}_{\mathrm{app}}(x,t) = h\sin(nx-\omega t)\,\boldsymbol{e}_z\,. \tag{14.3.25}$$

where, h is the amplitude of the magnetic field, while $\omega = 2\pi f$ corresponds to the angular velocity, with f denoting the frequency of the traveling wave field. The frequency of the oscillation determines the electromagnetic energy carried by the field, with higher frequencies carrying higher energy.

Our objective is to determine the DW velocity by solving Equations (14.2.18) and (14.2.19). To achieve this, we carried out the analysis employing the trial functions approach introduced by Schryer-Walker [39, 40, 41]:

$$\theta(x,t) = 2\tan^{-1}\exp\left(\frac{x - X(t)}{\delta}\right), \tag{14.3.26}$$

$$\varphi(x,t) = \varphi(t),$$

where, X(t) is the position of the DW center and δ represents the its width. Eq. (14.3.24) assumes that the DW moves without any variation in the static profile as time progresses. Additionally, Eq. (14.3.24) considers that the DW excitation occurs exclusively in the yz -plane over time, independent of space. In this context, the term $\varphi(t)$ denotes the excitation angle, which corresponds to the angle between the projection of magnetization (xy -plane) and the positive x-axis direction.

From Eq. (14.2.18) and Eq. (14.2.19), we obtain:

$$\begin{aligned}\frac{\partial\theta}{\partial t} + \alpha\sin\theta\frac{\partial\varphi}{\partial t} = {} & \gamma\left[\mathrm{A}\sin\theta\frac{\partial^2\varphi}{\partial x^2} + 2\mathrm{A}\cos\theta\frac{\partial\theta}{\partial x}\frac{\partial\varphi}{\partial x} - \left(N_x - N_y\right)\sin\theta\sin\varphi\cos\varphi\right] \\ & -\eta u_0 J\sin\theta\frac{\partial\varphi}{\partial x} - u_0 J\frac{\partial\theta}{\partial x} - \frac{\gamma K}{\mu_0 M_S^2}\sin(4\varphi)\sin^3\theta \\ & + \frac{3\gamma\sin\varphi\cos\varphi\sin\theta}{\mu_0 M_s^2}\{\lambda_{100}(c_{11} - c_{12})(\varepsilon_{yy} - \varepsilon_{xx}) \\ & - \frac{3}{2}\lambda^2{}_{100}(c_{11} - c_{12})\sin^2\theta\cos 2\varphi\},\end{aligned} \tag{14.3.25}$$

and,

$$\begin{aligned}\sin\theta\frac{\partial\varphi}{\partial t}-\alpha\frac{\partial\theta}{\partial t}= & -\gamma[A\sin\theta\frac{\partial^2\theta}{\partial x^2}-A\sin\theta\cos\theta\left(\frac{\partial\varphi}{\partial x}\right)^2-h\sin(nx-\omega t)\sin\theta \\ & +(N_z-N_x\cos^2\varphi-N_y\sin^2\varphi)\sin\theta\cos\theta]+\eta u_0 J\frac{\partial\theta}{\partial x} \\ & -u_0 J\sin\theta\frac{\partial\varphi}{\partial x}+\frac{2\gamma K}{\mu_0 M_S^2}(1-2\sin^2\varphi\cos^2\varphi)\sin^3\theta\cos\theta \\ & -\frac{3\gamma\sin\theta\cos\theta}{\mu_0 M_S^2}\{\frac{3}{2}c_{44}\lambda_{111}^2\sin(2\varphi)(\sin\varphi+\cos\varphi)\sin^2\theta\cos\theta \\ & +\lambda_{100}\frac{(c_{11}-c_{12})}{c_{11}}[\varepsilon_{xx}(c_{11}\cos^2\varphi+c_{12})+\varepsilon_{yy}(c_{11}\sin^2\varphi+c_{12})]\}.\end{aligned} \tag{14.3.26}$$

On substituting $\frac{\partial\theta}{\partial t}$ from Eq. (14.3.25) into Eq. (14.3.26), we get:

$$\begin{aligned}(1+\alpha^2)\sin\theta\frac{\partial\varphi}{\partial t}= & A\gamma\left[\alpha\sin\theta\frac{\partial^2\varphi}{\partial x^2}+2\alpha\sin\theta\cos\theta\frac{\partial\theta}{\partial x}\frac{\partial\varphi}{\partial x}-\frac{\partial^2\theta}{\partial x^2}\right. \\ & \left.+\sin\theta\cos\theta\left(\frac{\partial\varphi}{\partial x}\right)^2\right]-\gamma h\sin(nx-\omega t)\sin\theta \\ & +\frac{\gamma\alpha}{2}\left[\left(N_x-N_y\right)+\frac{3\lambda_{100}}{\mu_0 M_s^2}(c_{11}-c_{12})\left(\varepsilon_{yy}-\varepsilon_{xx}\right)\right]\sin\theta\sin 2\varphi \\ & -\gamma\left[\frac{3\lambda_{100}}{\mu_0 M_s^2}\left(\frac{c_{11}-c_{12}}{c_{11}}\right)\left\{c_{11}\left(\varepsilon_{xx}\cos^2\varphi+\varepsilon_{yy}\sin^2\varphi\right)\right.\right. \\ & \left.\left.+c_{12}\left(\varepsilon_{xx}+\varepsilon_{yy}\right)\right\}-\left(N_x\cos^2\varphi+N_y\sin^2\varphi-N_z\right)\right]\sin\theta\cos\theta \\ & +\eta u_0 J\left\{\frac{\partial\theta}{\partial x}-\alpha\sin\theta\frac{\partial\varphi}{\partial x}\right\}-u_0 J\left\{\alpha\frac{\partial\varphi}{\partial x}+\sin\theta\frac{\partial\theta}{\partial x}\right\} \\ & +\frac{\gamma K}{\mu_0 M_S^2}\left\{-2(1-2\sin^2\varphi\cos^2\varphi)\cos\theta-\alpha\sin(4\varphi)\right\}\sin^3\theta \\ & +\frac{9\gamma\lambda_{100}^2}{8\mu_0 M_S^2}(c_{12}-c_{11})\sin(4\varphi)\sin^3\theta \\ & -\frac{9\gamma\lambda_{111}^2}{2\mu_0 M_S^2}c_{44}\sin(2\varphi)(\sin\varphi+\cos\varphi)\sin^3\theta\cos^2\theta.\end{aligned} \tag{14.3.27}$$

Similarly, by substituting $\frac{\partial\varphi}{\partial t}$ from Eq. (14.3.26) into Eq. (14.3.25), we get:

$$
\begin{aligned}
\left(1+\alpha^2\right)\frac{\partial\theta}{\partial t} = \; & A\gamma\left[\alpha\frac{\partial^2\theta}{\partial x^2} - \alpha\sin\theta\cos\theta\left(\frac{\partial\varphi}{\partial x}\right)^2 + \sin\theta\frac{\partial^2\varphi}{\partial x^2}\right.\\
& \left. +2\cos\theta\frac{\partial\theta}{\partial x}\frac{\partial\varphi}{\partial x}\right] - \gamma\alpha h\sin(nx-\omega t)\sin\theta\\
& +\frac{\gamma}{2}\left[\left(N_x - N_y\right) + \frac{3\lambda_{100}}{\mu_0 M_s^2}\left(c_{11}-c_{12}\right)\left(\varepsilon_{yy}-\varepsilon_{xx}\right)\right]\sin\theta\sin 2\varphi\\
& +\alpha\gamma\left[\frac{3\lambda_{100}}{\mu_0 M_s^2}\left(\frac{c_{11}-c_{12}}{c_{11}}\right)\left\{c_{11}\left(\varepsilon_{xx}\cos^2\varphi + \varepsilon_{yy}\sin^2\varphi\right)\right.\right.\\
& \left.\left. +c_{12}\left(\varepsilon_{xx}+\varepsilon_{yy}\right)\right\} - \left(N_x\cos^2\varphi + N_y\sin^2\varphi - N_z\right)\right]\sin\theta\cos\theta\\
& -\eta u_0 J\left\{\alpha\frac{\partial\theta}{\partial x} + \sin\theta\frac{\partial\varphi}{\partial x}\right\} + u_0 J\left\{\alpha\sin\theta\frac{\partial\varphi}{\partial x} - \frac{\partial\theta}{\partial x}\right\}\\
& -\frac{\gamma K}{2\mu_0 M_S^2}\left\{\sin(4\varphi) - 2\alpha(1-2\sin^2\varphi\cos^2\varphi)\cos\theta\right\}\sin^3\theta\\
& +\frac{3\gamma\lambda_{111}^2}{2}c_{44}\sin(2\varphi)(\sin\varphi+\cos\varphi)\sin^2\theta\cos\theta.
\end{aligned}
\tag{14.3.28}
$$

Now, evaluating Eq. (14.3.24) at $x = X(t)$, we derive the following identities:

$$\theta(X,t) = \pi/2, \tag{14.3.29}$$

$$\frac{\partial\theta(X,t)}{\partial x} = \frac{1}{\delta}, \tag{14.3.30}$$

$$\frac{\partial^2\theta(X,t)}{\partial x^2} = 0, \tag{14.3.31}$$

$$\frac{\partial^3\theta(X,t)}{\partial x^3} = -\frac{1}{\delta^3}, \tag{14.3.32}$$

$$\frac{\partial\theta(X,t)}{\partial t} = -\frac{1}{\delta}\left(\frac{dX}{dt}\right), \tag{14.3.33}$$

$$\varphi(X,t) = \varphi(t), \tag{14.3.34}$$

$$\frac{\partial\varphi(X,t)}{\partial t}=\frac{d\varphi(t)}{dt}, \tag{14.3.35}$$

$$\frac{\partial\varphi(X,t)}{\partial x}=\frac{\partial^2\varphi(X,t)}{\partial x^2}=\frac{\partial^2\varphi(X,t)}{\partial x\,\partial t}=0. \tag{14.3.36}$$

By substituting the identities (14.3.29)-(14.3.36) into the Eq. (14.3.27), we derive a differential equation for the excitation angle $\varphi(t)$:

$$(1+\alpha^2)\frac{d\varphi}{dt}=\gamma\alpha\left\{(N_x-N_y)+\frac{3\lambda_{100}}{\mu_0 M_S^2}(c_{11}-c_{12})(\varepsilon_{yy}-\varepsilon_{xx})\right\}\cos\varphi\sin\varphi$$
$$+\gamma\, h\sin(nx-\omega t)+(\eta-\alpha)\frac{u_0 J}{\delta}+\frac{\gamma\sin(4\varphi)}{\mu_0 M_S^2}\{9\lambda_{100}(c_{12}-c_{11})-\alpha K\} \tag{14.3.37}$$

Similarly, from Eq. (14.3.28) we deduce the expression for instantaneous DW velocity $v(t)$:

$$v(t)=\frac{dx}{dt}=\frac{\alpha\gamma\delta\sin(nx-\omega t)}{(1+\alpha^2)}h+\frac{u_0(1+\eta\alpha)}{(1+\alpha^2)}J+\frac{\gamma\delta K}{2\mu_0 M_S^2(1+\alpha^2)}\sin(4\varphi)$$
$$-\frac{\gamma\delta}{2(1+\alpha^2)}\left[(N_x-N_y)+\frac{3\lambda_{100}}{\mu_0 M_S^2}(c_{11}-c_{12})(\varepsilon_{yy}-\varepsilon_{xx})\right]\sin(2\varphi). \tag{14.3.38}$$

Eqs. (14.3.37) and (14.3.38) provide the excitation angle and velocity over time, respectively, taking into account the influence of the electric current and external magnetic field. Notably, since Eqs. (14.3.37) and (14.3.38) involve the excitation angle $\varphi(t)$, solving Eq. (14.3.37) is necessary to determine the velocity. Furthermore, the DW position $X(t)$, at any given time t can be obtained by integrating Eq. (14.3.38).

14.4 Conclusion

In a nutshell, our theoretical investigation aimed to determine the explicit expressions for the key features of DW dynamics, including wall width, WB limit, threshold, and velocity, by employing both the traveling wave ansatz and collective coordinate approaches. Our results have revealed that the most relevant DW dynamic features, *viz.* DW width, WB limit, threshold, and DW velocity are significantly influenced by the key parameters such as spin-polarized current, the frequency of magnetic field oscillation, crystal symmetry, and magnetoelasticity. These findings shed light on the underlying mechanisms that delineate the DW motion in multiferroic devices and provide crucial insights for future research. The explicit expressions derived through our theoretical investigation offer a comprehensive understanding of the interplay between the key parameters and their effects on DW motion.

Finally, we emphasize that this analytical investigation provides a new path for manipulating magnetic domains in these multiferroic devices and plays a vital role in designing next-generation spintronic devices based on these artificial heterostructures.

14.5 Acknowledgement

Sharad Dwivedi would like to thank the Science and Engineering Research Board (SERB), Department of Science and Technology, Government of India, and the National Institute of Technology Andhra Pradesh for the financial support through Projects CRG/2019/003101 and NITAP/SDG/15/2020, respectively.

References

[1]Son, S. J., Reichel, J., He, B., Schuchman, M., Lee, S. B., Magnetic nanotubes for magnetic-field-assisted bio separation, bio interaction, and drug delivery, Journal of the American Chemical Society, 127(20), 7316-7317 (2005).

[2]Parkin, S. S., Hayashi, M., & Thomas, L., Magnetic domain-wall racetrack memory, Science, 320(5873), 190-194 (2008).

[3]Diegel, M., Mattheis, R., & Halder, E., 360/spl deg/domain wall investigation for sensor applications, IEEE transactions on magnetics, 40(4), 2655-2657 (2004).

[4]Hubert, A., and Schäfer, R., Magnetic domains: the analysis of magnetic microstructure, Springer Science & Business Media, USA (2008).

[5]Morales, R., Kovylina, M., Schuller, I. K., Labarta, A., & Batlle, X., Antiferromagnetic/ferromagnetic nanostructures for multidigit

storage units, Applied Physics Letters, 104(3), 032401 (2014).

[6]Ben-Romdhane, N., Zhao, W. S., Zhang, Y., Klein, J. O., Wang, Z. R., & Ravelosona, D., Design and analysis of racetrack memory based on magnetic domain wall motion in nanowires, In 2014 IEEE/ACM International Symposium on Nanoscale Architectures (NANO-ARCH), (pp. 71-76) (2014).

[7]Dwivedi, S., & Dubey, S., Field-driven motion of ferrofluids in ferromagnetic nanowire under the influence of inertial effects, Procedia Engineering, 127, 3-9 (2015).

[8]Chikazumi, S., Physics of ferromagnetism, Oxford university press, UK (1997).

[9]Shahu, C. K., Dwivedi, S., & Dubey, S., Dynamics of curved domain walls in hard ferromagnets with nonlinear dissipative and inertial effects, Physica D: Nonlinear Phenomena, 448, 133737 (2023).

[10] Vaz, C. A., Hoffman, J., Ahn, C. H., & Ramesh, R., Magnetoelectric coupling effects in multiferroic complex oxide composite structures, Advanced materials, 22(26-27), 2900-2918 (2010).

[11] Lei, N., Devolder, T., Agnus, G., Aubert, P., Daniel, L., Kim, J. V., ... & Lecoeur, P., Strain-controlled magnetic domain wall propagation in hybrid piezoelectric/ferromagnetic structures, Nature communications, 4(1), 1378 (2013).

[12] De Ranieri, E., Roy, P. E., Fang, D., Vehsthedt, E. K., Irvine, A. C., Heiss, D., ... & Wunderlich, J., Piezoelectric control of the mobility of a domain wall driven by adiabatic and non-adiabatic torques, Nature materials, 12(9), 808-814 (2013).

[13] Geoffroy, O., O'Brien, D., Cugat, O., & Delamare, J., Practical and theoretical investigations of a rotating coilless actuator using the inverse magnetostrictive effect, IEEE Transactions on Magnetics, 46(2), 606-609 (2010).

[14] Mballa-Mballa, F. S., Hubert, O., He, S., Depeyre, S., & Meilland, P., Micromagnetic modeling of magneto-mechanical behavior, IEEE Transactions on magnetics, 50(4), 1-4 (2014).

[15] Shu, Y. C., Lin, M. P., & Wu, K. C., Micromagnetic modeling of magnetostrictive materials under intrinsic stress, Mechanics of materials, 36(10), 975-997 (2004).

[16] Consolo, G., Federico, S., & Valenti, G., Magnetostriction in transversely isotropic hexagonal crystals. Physical Review B, 101(1), 014405 (2020).

[17] Dwivedi, S., Singh, Y. P., & Consolo, G., On the statics and dynamics of transverse domain walls in bilayer piezoelectric-magnetostrictive nanostructures. Applied Mathematical Model-

ling, 83, 13-29 (2020).

[18] Liang, C. Y., Keller, S. M., Sepulveda, A. E., Bur, A., Sun, W. Y., Wetzlar, K., & Carman, G. P., Modeling of magnetoelastic nanostructures with a fully coupled mechanical-micromagnetic model, Nanotechnology, 25(43), 435701 (2014).

[19] Consolo, G., Currò, C., Martinez, E., & Valenti, G., Mathematical modeling and numerical simulation of domain wall motion in magnetic nanostrips with crystallographic defects, Applied Mathematical Modelling, 36(10), 4876-4886 (2012).

[20] Zhang, S., & Zhang, S. S. L., Generalization of the Landau-Lifshitz-Gilbert equation for conducting ferromagnets, Physical review letters, 102(8), 086601 (2009).

[21] Baňas, L. U., A numerical method for the Landau–Lifshitz equation with magnetostriction, Mathematical methods in the applied sciences, 28(16), 1939-1954 (2005).

[22] Zhang, S., & Li, Z., Roles of nonequilibrium conduction electrons on the magnetization dynamics of ferromagnets, Physical review letters, 93(12), 127204 (2004).

[23] Thiaville, A., Nakatani, Y., Miltat, J., & Suzuki, Y., Micromagnetic understanding of current-driven domain wall motion in patterned nanowires. Euro physics Letters, 69(6), 990 (2005).

[24] Li, Z., & Zhang, S., Magnetization dynamics with a spin-transfer torque, Physical Review B, 68(2), 024404 (2003).

[25] Maity, S., Dolui, S., & Dwivedi, S., Strain-induced fast domain wall motion in hybrid piezoelectric-magnetostrictive structures with Rashba and nonlinear dissipative effects, Acta Mechanica Sinica, 40(9), 423613 (2024).

[26] Consolo, G., Federico, S., & Valenti, G., Strain-mediated propagation of magnetic domain-walls in cubic magnetostrictive materials. Ricerche di Matematica, 70(1), 81-97 (2021).

[27] Dolui, S., Maity, S., & Dwivedi, S., Strain-induced ultrafast magnetization dynamics in cubic magnetostrictive materials with inertial and nonlinear dissipative effects, Zeitschrift für angewandte Mathematik und Physik, 75(4), 149 (2024).

[28] Wu, L., Tong, P., Wang, X., & Su, R., An absolute linear displacement sensor based on orthogonal dual traveling wave magnetic field. IEEE Sensors Journal, 22(6), 6019-6026 (2022).

[29] Sudan, R. N., Interaction of a conducting fluid stream with a traveling wave of magnetic field of finite extension, Journal of applied physics, 34(3), 641-650 (1963).

[30] Zahn, M., & Wainman, P. N., Effects of fluid convection and par-

ticle spin on ferrohydrodynamic pumping in traveling wave magnetic fields, Journal of magnetism and magnetic materials, 122(1-3), 323-328 (1993).

[31] Zahn, M., & Pioch, L. L., Ferrofluid flows in AC and traveling wave magnetic fields with effective positive, zero or negative dynamic viscosity, Journal of Magnetism and Magnetic Materials, 201(1-3), 144-148 (1999).

[32] Mark, A. R., James, A. S., & William, A. H., Principles of modern radar: Basic principles, Institution of Engineering and Technology: London, UK (2010).

[33] Rappaport, T. S., Wireless Communications–Principles and Practice, Microwave Journal, 45(12), 128-129 (2002).

[34] Brown, R. W., Cheng, Y. C. N., Haacke, E. M., Thompson, M. R., & Venkatesan, R., Magnetic resonance imaging: physical principles and sequence design, John Wiley & Sons, (2014).

[35] Duran, S., Travelling wave solutions and simulation of the Lonngren wave equation for tunnel diode, Optical and Quantum Electronics, 53(8), 458 (2021).

[36] Zhang, K., Li, D., Chang, K., Zhang, K. and Li, D., Electromagnetic theory for microwaves and optoelectronics, 274-283, B Springer, Berlin (1998).

[37] Chen, M., Torres, M., & Walsh, T., Existence of traveling wave solutions of a high-order nonlinear acoustic wave equation, Physics Letters A, 373(11), 1037-1043 (2009).

[38] Sun, Y., New travelling wave solutions for Sine-Gordon equation, Journal of Applied Mathematics, (2014).

[39] Schryer, N. L., & Walker, L. R., The motion of 180 domain walls in uniform dc magnetic fields, Journal of Applied Physics, 45(12), 5406-5421 (1974).

[40] Li, Z., & Zhang, S., Domain-wall dynamics driven by adiabatic spin-transfer torques, Physical Review B, 70(2), 024417 (2004).

[41] Dolui, S., Maity, S., Dwivedi, S., & Consolo, G., Depinning of domain walls in a notched ferromagnetic nanostrip: role of inertial and nonlinear damping effects, Physica Scripta, 99(9), 095237 (2024).

[42] Maity, S., Dolui, S., Dwivedi, S., & Consolo, G., Domain wall dynamics in cubic magnetostrictive materials subject to Rashba effect and nonlinear dissipation, Zeitschrift für angewandte Mathematik und Physik, 74(1), 23 (2023).

[43] Griffiths, D. J., Introduction to Electrodynamics, Cambridge University Press, UK (2017).

[44] Dolui, S., & Dwivedi, S., Tuning domain wall dynamics in a notched ferromagnetic nanostrip with Rashba effect, Chaos: An Interdisciplinary Journal of Nonlinear Science, 34(12), 123106 (2024).

[45] Yellen, B. B., Erb, R. M., Son, H. S., Hewlin Jr, R., Shang, H., & Lee, G. U., Traveling wave magnetophoresis for high resolution chip based separations, Lab on a Chip, 7(12), 1681-1688 (2007).

[46] Halder, A., Dolui, S., & Dwivedi, S., Dynamics of transverse domain wall in bilayer ferromagnetic-heavy metal nanostructures: Interplay of spin-orbit torque, dry-friction dissipation, and the interfacial Dzyaloshinskii-Moriya interaction, Mechanics of Advanced Materials and Structures, pp. 1–15 (2024).

[47] Zhang, J. X., & Chen, L. Q., Phase-field micro elasticity theory and micro magnetic simulations of domain structures in giant magnetostrictive materials, Acta Materialia, 53(9), 2845-2855 (2005).

[48] Dolui, S., Halder, A., & Dwivedi, S., Ultrafast domain wall motion in hexagonal magnetostrictive materials: role of inertial damping, magnetostriction, and dry-friction dissipation, Acta Mechanica, 235(12), 7121-7139 (2024).

Chapter 15

Some Invariant Solutions of (3+1)-dimensional Zakharov-Kuznetsov-Burgers Equation

Vishalkumar J. Prajapati, Ramakanta Meher

Department of Mathematics, Sardar Vallabhbhai National Institute of Technology, Surat-395007, Gujarat, INDIA

Abstract

In this study, the Lie group method is employed to perform invariant solutions and similarity reduction to the $(3+1)$-dimensional Zakharov-Kuznetsov-Burgers (ZKB) equation. Using the Lie group symmetry method, the ZKB equation has been transformed into the new partial differential equation with fewer independent variables, which has then been transformed further into an ordinary differential equation. For the ZKB equation, we found the infinitesimal generators, Lie algebra's commutator table, similarity reduction, and the symmetry group. Additionally, the reduction equation is used to obtain the ZKB equation's solitary wave solutions. As a result, we find some new exact explicit solutions to the (3+1)-dimensional ZKB equation which describes the dynamics of dust ion acoustic waves.

Keywords: $(3+1)$ -dimensional Zakharov-Kuznetsov-Burgers (ZKB) equation, Lie symmetries, Infinitesimal generators, Invariant solutions.

15.1 Introduction

Nonlinear partial differential equations can be found in a variety of scientific disciplines, including physics, engineering, and chemistry. Although it is challenging, a wide variety of techniques have been developed for finding significant PDE solutions like Painleve analysis [1], the tanh method [2], the multiple exp-function method [3], the Bäcklund transformation method [4], etc. Among them the method of group analysis [5, 6, 7], which is first introduced by Sophus Lie, provides a systematic approach to understanding the symmetries of differential equations and using them to simplify and solve the equations. It has had a profound impact on various branches of mathematics and physics, including mechanics, electromagnetism, quantum mechanics, and field theory. The symmetry transformation method [8, 9, 10, 11] is indeed an important tool for solving higher-order differential equations and physical models. This method relies on the concept of symmetries, which are transformations that leave the equations or models invariant. The symmetry transformation method is particularly useful for finding exact solutions to differential equations, especially when the equations are nonlinear or of higher order.

In this study, we examine a (3+1)-dimensional Zakharov-Kuznetsov-Burgers equation [12, 13, 14] from the perspective of Lie symmetries

$$\Psi_T + A\Psi\Psi_X + B\Psi_{XXX} + C\Psi_{XYY} + C\Psi_{XZZ} + D\Psi_{XX} = 0, \qquad (15.1.1)$$

where Ψ is a dependent function X,Y,Z,T and A,B,C,D are arbitrary constants. The ZKB equation arises in various areas of physics, including plasma physics, fluid dynamics, and nonlinear optics. It describes the propagation of nonlinear waves in dispersive and dissipative media. The three-dimensional Zakharov-Kuznetsov-Burgers (ZKB) equation for dust ion acoustic waves involves investigating the behavior of nonlinear waves in plasmas where dust particles and ions interact.

The goal of this study is to apply the symmetry approach to produce infinitesimal generators, Lie algebra commutator tables, symmetry groups, similarity reduction, and, most crucially, some invariant solutions to the (3+1)-dimensional Zakharov-Kuznetsov-Burgers problem.

The following is the sketch of the current paper: Section 2 describes the Lie symmetry method's algorithm. In Section 3, we discuss the general infinitesimal generator and the symmetries accepted by Eq. (15.1.1) using the Lie symmetry analysis. In Section 4, we find the Lie symmetry transformations group of Eq. (15.1.1). In Section 5, we derive reduction equations by employing similarity variables, and by solving reduction equations, we gain accurate solutions to Eq. (15.1.1). Section 6 contains a brief conclusion.

15.2 Procedure of Lie Symmetries

In this section, we will look at how to derive the symmetries for any system of partial differential equations. For that let us consider the system of m partial differential equations of order n in r-independent variables and s-dependent variables as follows:

$$F_i(X, w^{(n)}) = 0, \quad i = 1, 2, ..., m, \tag{15.2.1}$$

where $z = (z^1, ..., z^r)$, $w = (w^1, ..., w^s)$ and $w^{(n)}$ is the $n-th$ derivative of w.

Now let us consider the Lie group of transformations in one-parameter as follows:

$$\overline{z}^i = z^i + \varepsilon\xi^i(w, z) + O(\varepsilon^2), \tag{15.2.2}$$

$$\overline{w}^j = w^j + \varepsilon\eta^j(w, z) + O(\varepsilon^2), \tag{15.2.3}$$

where $i = 1, 2, ..., r$ and $j = 1, 2, ..., s$. The parameter ε is very small parameter of group of transformations and ξ^i, η^j are the infinitesimals for the independent and dependent variables, respectively.

Now the infinitesimal operator or vector field V associated with the preceding transformation group can be represented as follows:

$$V = \sum_{i=1}^{r}\xi^i(w, z)\frac{\partial}{\partial z^i} + \sum_{j=1}^{s}\eta^j(w, z)\frac{\partial}{\partial w^j} \tag{15.2.4}$$

A symmetry of a system of differential equations is a transformation that links one solution to the another solutions. This leads to the invariance condition of the system under the group of transformation as

$$Pr^{(n)}V\left[F_i(z,w^{(n)})\right]=0, \quad i=1,2,...,m, \tag{15.2.5}$$

whenever $F_i(z,w^{(n)})=0$. Here $Pr^{(n)}V$ is called the prolongation equation for vector field V of order n and it is given by

$$Pr^{(n)}V = V + \sum_{i=1}^{r}\sum_{j=1}^{s}\zeta_i^j \frac{\partial}{\partial w_i^j} + \cdots + \sum_{k_1=1}^{r}\cdots\sum_{k_n=1}^{r}\sum_{j=1}^{s}\zeta_{k_1k_2\cdots k_n}^j \frac{\partial}{\partial w_{k_1k_2\cdots k_n}^j} \tag{15.2.6}$$

where the coefficients $\zeta_{k_1k_2\cdots k_t}^j$, $t=1,2,...,n$ depend on the infinitesimals $\xi(w,z)$ and $\eta(w,z)$ by the following relations

$$\zeta_i^j = D_i(\eta^j) - \sum_d^r w_d^j D_i(\xi^d), \tag{15.2.7}$$

$$\zeta_{k_1k_2\cdots k_t}^j = D_{k_t}(\zeta_{k_1k_2\cdots k_{t-1}}^j) - \sum_d^r w_k^j D_{k_1k_2\cdots k_{t-1}d}(\xi^k) D_{k_t}(\xi^d), \tag{15.2.8}$$

where D_i is the total derivative operator with respect to the variable z^i. Since ξ^i, η^j are independent of the derivatives of w^j, so the coefficients like derivatives of w^j in (6) can be equated to obtained the following over determined system of equations. After solving the over determined system of differential equations, we will get the infinitesimals ξ^i, η^j. And from these infinitesimals we can find infinitesimal generators. To understand the whole procedure, lets apply this algorithm to the (3+1) dimensional Zakharov-Kuznetsov-Burgers equation.

15.3 Lie Symmetry analysis of (3+1)-dimensional Zakharov-Kuznetsov Burgers Equation

The following is a $(3+1)$ -dimensional Zakharov–Kuznetsov–Burgers(ZKB) equation

$$\frac{\partial\Psi}{\partial T}+A\Psi\frac{\partial\Psi}{\partial X}+B\frac{\partial^3\Psi}{\partial X^3}+C\frac{\partial^3\Psi}{\partial X\partial Y^2}+C\frac{\partial^3\Psi}{\partial X\partial Z^2}+D\frac{\partial^2\Psi}{\partial X^2}=0, \tag{15.3.1}$$

where Ψ is a dependent function X,Y,Z,T and A,B,C,D are arbitrary constants.

Now we considering the one-parameter Lie symmetry group of infinitesimal transformation on X,Y,Z,T and Ψ as follows:

$$\bar{X}=X+\varepsilon\xi^1+O(\varepsilon^2), \tag{15.3.2}$$
$$\bar{Y}=Y+\varepsilon\xi^2+O(\varepsilon^2), \tag{15.3.3}$$
$$\bar{Z}=Z+\varepsilon\xi^3+O(\varepsilon^2), \tag{15.3.4}$$
$$\bar{T}=T+\varepsilon\xi^4+O(\varepsilon^2), \tag{15.3.5}$$
$$\bar{\Psi}=\Psi+\varepsilon\eta^1+O(\varepsilon^2), \tag{15.3.6}$$

where $\varepsilon(<<1)$ is a very small parameter of group and ξ^1,ξ^2,ξ^3,ξ^4 and η^1 are functions of X,Y,Z,T which are determined further. Here ξ^1,ξ^2,ξ^3,ξ^4 and η^1 are called the infinitesimals of the transformations for the independent variables X,Y,Z,T and the dependent variable Ψ, respectively. Now the the Lie algebra of ZKB equation is spanned by vector fields

$$\tilde{V}=\xi^1\frac{\partial}{\partial X}+\xi^2\frac{\partial}{\partial Y}+\xi^3\frac{\partial}{\partial Z}+\xi^4\frac{\partial}{\partial T}+\eta^1\frac{\partial}{\partial\Psi}. \tag{15.3.7}$$

Now the third order prolongation equation $Pr^{(3)}$ for 3-D ZKB equation (10) is calculated by

$$Pr^{(3)}\tilde{V}=\tilde{V}+\zeta^T\frac{\partial}{\partial\Psi_T}+\zeta^X\frac{\partial}{\partial\Psi_X}+\zeta^{XX}\frac{\partial}{\partial\Psi_{XX}}+\zeta^{XXX}\frac{\partial}{\partial\Psi_{XXX}}+\zeta^{XYY}\frac{\partial}{\partial\Psi_{XYY}}+\zeta^{XZZ}\frac{\partial}{\partial\Psi_{XZZ}}, \tag{15.3.8}$$

where the coefficients $\zeta^{T}, \zeta^{X}, \zeta^{XX}, \zeta^{XXX}, \zeta^{XYY}$ and ζ^{XZZ} are determined by following formulae:

$$\zeta^{T} = D_T(\eta^1) - \Psi_X D_T(\xi^1) - \Psi_Y D_T(\xi^2) - \Psi_Z D_T(\xi^3) - \Psi_T D_T(\xi^4), \tag{15.3.9}$$

$$\zeta^{X} = D_X(\eta^1) - \Psi_X D_X(\xi^1) - \Psi_Y D_X(\xi^2) - \Psi_Z D_X(\xi^3) - \Psi_T D_X(\xi^4), \tag{15.3.10}$$

$$\zeta^{XX} = D_X(\zeta^X) - \Psi_{XX} D_X(\xi^1) - \Psi_{XY} D_X(\xi^2) - \Psi_{XZ} D_X(\xi^3) - \Psi_{XT} D_X(\xi^4), \tag{15.3.11}$$

$$\zeta^{XY} = D_Y(\zeta^X) - \Psi_{XX} D_Y(\xi^1) - \Psi_{XY} D_Y(\xi^2) - \Psi_{XZ} D_Y(\xi^3) - \Psi_{XT} D_Y(\xi^4), \tag{15.3.12}$$

$$\zeta^{XZ} = D_Z(\zeta^X) - \Psi_{XX} D_Z(\xi^1) - \Psi_{XY} D_Z(\xi^2) - \Psi_{XZ} D_Z(\xi^3) - \Psi_{XT} D_Z(\xi^4), \tag{15.3.13}$$

$$\zeta^{XXX} = D_X(\zeta^{XX}) - \Psi_{XXX} D_X(\xi^1) - \Psi_{XXY} D_X(\xi^2) - \Psi_{XXZ} D_X(\xi^3) - \Psi_{XXT} D_X(\xi^4), \tag{15.3.14}$$

$$\zeta^{XYY} = D_Y(\zeta^{XY}) - \Psi_{XYX} D_Y(\xi^1) - \Psi_{XYY} D_Y(\xi^2) - \Psi_{XYZ} D_Y(\xi^3) - \Psi_{XYT} D_Y(\xi^4), \tag{15.3.15}$$

$$\zeta^{XZZ} = D_Z(\zeta^{XZ}) - \Psi_{XZX} D_Z(\xi^1) - \Psi_{XZY} D_Z(\xi^2) - \Psi_{XZZ} D_Z(\xi^3) - \Psi_{XZT} D_Z(\xi^4), \tag{15.3.16}$$

where D_S $(S = X, Y, Z, T)$ stands for the total derivative operator with respect to the variable S and it defined by,

$$D_S = \frac{\partial}{\partial S} + \Psi_S \frac{\partial}{\partial \Psi} + \Psi_{XS} \frac{\partial}{\partial \Psi_X} + \Psi_{YS} \frac{\partial}{\partial \Psi_Y} + \Psi_{ZS} \frac{\partial}{\partial \Psi_Z} + \Psi_{TS} \frac{\partial}{\partial \Psi_T} + \cdots \tag{15.3.17}$$

Now the invariance condition for the Eq. (10) is given as follows:

$$Pr^{(3)}\tilde{V}\left[\Psi_T + A\Psi\Psi_X + B\Psi_{XXX} + C\Psi_{XYY} + C\Psi_{XZZ} + D\Psi_{XX}\right] = 0. \tag{15.3.18}$$

The infinitesimals $\xi^1, \xi^2, \xi^3, \xi^4$ and η^1 can be calculated by solving the over determined linear partial differential equations which are obtained from the above invariance condition (15.3.18). Since $\xi^1, \xi^2, \xi^3, \xi^4$ and η^1 are independent of the derivatives of Ψ, so the coefficients like derivatives of Ψ in (15.3.18) can be equated to obtained the following over determined system of equations

$$\eta^1_{\Psi\Psi} = 0, \eta^1_{X\Psi} = 0, \eta^1_{Y\Psi} = 0, \eta^1_{Z\Psi} = 0,$$
$$\xi^1_{\Psi} = 0, \xi^1_X = 0, \xi^1_Y = 0, \xi^1_Z = 0,$$
$$\xi^2_{\Psi} = 0, \xi^2_X = 0, \xi^2_Y = 0, \xi^2_{ZZ} = 0, \xi^2_T = 0$$
$$\xi^3_{\Psi} = 0, \xi^3_X = 0, \xi^3_Z = 0, \ \xi^3_{YY} = 0, \xi^3_T = 0$$
$$\xi^4_{\Psi} = 0, \xi^4_X = 0, \xi^4_Y = 0, \xi^4_Z = 0,$$
$$A\eta^1 - \xi^1_T = 0$$

$$A\Psi\,\eta^1_X + C\eta^1_{XZZ} + D\eta^1_{XX} + C\eta^1_{XYY} + B\eta^1_{XXX} + \eta^1_T = 0,$$

solving the above system, one can get

$$\eta^1 = c_1, \xi^1 = Ac_1T + c_6, \xi^2 = c_3Z + c_4, \xi^3 = -c_3Y + c_5, \xi^4 = c_2, \tag{15.3.19}$$

where c_1, c_2, c_3, c_4, c_5 and c_6 are arbitrary constants.

Hence the infinitesimal symmetries of Eq. (15.3.1) form the Lie algebra generated by following six infinitesimal operators:

$$V_1 = \frac{\partial}{\partial T}, \tag{15.3.20}$$

$$V_2 = \frac{\partial}{\partial X}, \tag{15.3.21}$$

$$V_3 = \frac{\partial}{\partial Y}, \tag{15.3.22}$$

$$V_4 = \frac{\partial}{\partial Z}, \quad (15.3.23)$$

$$V_5 = AT\frac{\partial}{\partial X} + \frac{\partial}{\partial \Psi}, \quad (15.3.24)$$

$$V_6 = -Y\frac{\partial}{\partial Z} + Z\frac{\partial}{\partial Y}. \quad (15.3.25)$$

Table 15.1 Commutation table for Lie algebra

$[V_i, V_j]$	V_1	V_2	V_3	V_4	V_5	V_6
V_1	0	0	0	0	AV_2	0
V_2	0	0	0	0	0	0
V_3	0	0	0	0	0	$-V_4$
V_4	0	0	0	0	0	V_3
V_5	$-AV_2$	0	0	0	0	0
V_6	0	0	V_4	$-V_3$	0	0

Table 15.1 displays the Lie algebraic commutation relations caused by infinitesimal operators V_1, V_2, V_3, V_4, V_5, and V_6. The $(i, j)th$ entry of the Table (15.1) can be found using the Lie bracket formula $[V_i, V_j] = V_iV_j - V_jV_i$. According to the table, $[V_1, V_2, V_3, V_4, V_5, V_6]$ is closed with respect to the Lie bracket.

15.4 Symmetry group of (3+1)-dimensional Zakharov-Kuznetsov Burgers Equation

In this section, we will obtain the symmetry group of transformations $G_i : (\Psi, T, X, Y, Z) \to (\bar{\Psi}, \bar{T}, \bar{X}, \bar{Y}, \bar{Z})$ which is derived by the infinitesimal operator V_i, $(i = 1, 2, 3, 4, 5, 6)$. Before that we have to solve the following system of ordinary differential equations with initial conditions

$$\frac{d\bar{X}}{d\varepsilon} = \xi^1(\bar{\Psi},\bar{T},\bar{X},\bar{Y},\bar{Z}), \quad \bar{X}|_{\varepsilon=0} = X, \tag{15.4.1}$$

$$\frac{d\bar{Y}}{d\varepsilon} = \xi^2(\bar{\Psi},\bar{T},\bar{X},\bar{Y},\bar{Z}), \quad \bar{Y}|_{\varepsilon=0} = Y, \tag{15.4.2}$$

$$\frac{d\bar{Z}}{d\varepsilon} = \xi^3(\bar{\Psi},\bar{T},\bar{X},\bar{Y},\bar{Z}), \quad \bar{Z}|_{\varepsilon=0} = Z, \tag{15.4.3}$$

$$\frac{d\bar{T}}{d\varepsilon} = \xi^4(\bar{\Psi},\bar{T},\bar{X},\bar{Y},\bar{Z}), \quad \bar{T}|_{\varepsilon=0} = T, \tag{15.4.4}$$

$$\frac{d\bar{\Psi}}{d\varepsilon} = \eta^1(\bar{\Psi},\bar{T},\bar{X},\bar{Y},\bar{Z}), \quad \bar{\Psi}|_{\varepsilon=0} = \Psi. \tag{15.4.5}$$

Now the one parameter Lie group G_i generated by V_i is given by as follows:

$$G_1:(\Psi,T,X,Y,Z)\to(\Psi,T+\varepsilon,X,Y,Z), \tag{15.4.6}$$

$$G_2:(\Psi,T,X,Y,Z)\to(\Psi,T,X+\varepsilon,Y,Z), \tag{15.4.7}$$

$$G_3:(\Psi,T,X,Y,Z)\to(\Psi,T,X,Y+\varepsilon,Z), \tag{15.4.8}$$

$$G_4:(\Psi,T,X,Y,Z)\to(\Psi,X,Y,Z+\varepsilon), \tag{15.4.9}$$

$$G_5:(\Psi,T,X,Y,Z)\to(\Psi+\varepsilon,T,X+\varepsilon AT,Y,Z,T), \tag{15.4.10}$$

$$G_6:(\Psi,T,X,Y,Z)\to(\Psi,T,X,Y+\varepsilon Z,Z-\varepsilon Y). \tag{15.4.11}$$

Now suppose $\Psi = f(T,X,Y,Z)$ is a known solution of given Eq. (15.3.1), then from the above Lie symmetry groups G_i the corresponding new solutions Ψ_i can be derive as follows:

$$\Psi_1 = f(T-\varepsilon,X,Y,Z), \tag{15.4.12}$$

$$\Psi_2 = f(T,X-\varepsilon,Y,Z), \tag{15.4.13}$$

$$\Psi_3 = f(T,X,Y-\varepsilon,Z), \tag{15.4.14}$$

$$\Psi_4 = f(T,X,Y,Z-\varepsilon), \tag{15.4.15}$$

$$\Psi_5 = f(T,X-\varepsilon AT,Y,Z)-\varepsilon, \tag{15.4.16}$$

$$\Psi_6 = f(T,X,Y-\varepsilon Z,Z-\varepsilon Y). \tag{15.4.17}$$

15.5 Symmetry reduction and invariant solutions of (3+1)-dimensional Zakharov-Kuznetsov Burgers Equation

In this section, we will acquire similarity variables and their reduction equations, and then solve the reduction equations to achieve the similarity solution.

Case-I First consider the infinitesimal operator $V_1 = \frac{\partial}{\partial T}$, the characteristic (Lagrange's) equation is

$$\frac{dX}{0} = \frac{dY}{0} = \frac{dZ}{0} = \frac{dT}{1} = \frac{d\Psi}{0} \tag{15.5.1}$$

The similar(invariant) solution of Eq. (15.3.1) is

$$\Psi(T, X, Y, Z) = f(x, y, z) \tag{15.5.2}$$

with similarity variables

$$x = X, y = Y, z = Z. \tag{15.5.3}$$

Now by inserting the value of Ψ from Eq. (15.5.2) into Eq. (15.3.1), we will get

$$Aff_x + Bf_{xxx} + Cf_{xyy} + Cf_{xzz} + Df_{xx} = 0. \tag{15.5.4}$$

Now by applying the same process of similarity transformations discussed in Section (2) on Eq. (15.5.4) until we get converted ordinary differential equation. After solving converted ODE by tanh method [2], we will get the following similarity(invariant) solution

$$\Psi(T, X, Y, Z) = \frac{2mD}{A} \tanh\left(\frac{mCX + nCY + lC - \sqrt{-m^2BC - C^2l^2}\,Z}{C}\right), \tag{15.5.5}$$

where l, m, n are arbitrary constants.

Case-II By considering the infinitesimal operator $V_2 = \frac{\partial}{\partial X}$, the

Lagrange's equation is

$$\frac{dX}{1}=\frac{dY}{0}=\frac{dZ}{0}=\frac{dT}{0}=\frac{d\Psi}{0} \tag{15.5.6}$$

Then the new form of the solution of Eq. (15.3.1) is

$$\Psi(T,X,Y,Z)=f(y,z,t) \tag{15.5.7}$$

with similarity variables

$$y=Y, z=Z, t=T. \tag{15.5.8}$$

Now by inserting the value of Ψ from Eq. (15.5.7) into Eq. (15.3.1), we will get

$$f_t=0. \tag{15.5.9}$$

So in this case, we obtain the new similar(invariant) solution as $\Psi=f(Y,Z)$.

Case-III For another operator $V_3=\frac{\partial}{\partial Y}$, the characteristic equation is

$$\frac{dX}{0}=\frac{dY}{1}=\frac{dZ}{0}=\frac{dT}{0}=\frac{d\Psi}{0} \tag{15.5.10}$$

The another form of the solution of Eq. (15.3.1) is

$$\Psi(T,X,Y,Z)=f(x,z,t) \tag{15.5.11}$$

with similarity variables

$$x=X, z=Z, t=T. \tag{15.5.12}$$

Now by inserting the value of Ψ from Eq. (15.5.11) into Eq. (10), we will get

$$f_t + Aff_x + Bf_{xx} + Cf_{xzz} + Df_{xx} = 0. \tag{15.5.13}$$

Now by applying the same process of similarity transformations discussed in Section (2) on Eq. (15.5.13) until we get converted ordinary differential equation. After solving converted ODE by tanh method, we will get the following similarity(invariant) solution

$$\Psi(T,X,Y,Z) = \frac{2mD}{A}\tanh\left(\frac{mCX + nCT + lC - m\sqrt{-BC}Z + lC}{C}\right) - \frac{n}{mA}. \tag{15.5.14}$$

where l, m, n are arbitrary constants.

Case-IV For the operator $V_4 = \frac{\partial}{\partial Z}$, the characteristic (Lagrange's) equation is

$$\frac{dX}{0} = \frac{dY}{0} = \frac{dZ}{1} = \frac{dT}{0} = \frac{d\Psi}{0} \tag{15.5.15}$$

The similar form of the solution of Eq. (15.3.1) is

$$\Psi(T,X,Y,Z) = f(x,y,t) \tag{15.5.16}$$

with similarity variables

$$x = X,\ y = Y,\ t = T. \tag{15.5.17}$$

Now by inserting the value of Ψ from Eq. (15.5.16) into Eq. (10), we will get

$$f_t + Aff_x + Bf_{xx} + Cf_{xxx} + Cf_{xyy} + Df_{xx} = 0. \tag{15.5.18}$$

Similarly, as above we will get the following similarity(invariant) solution for Eq. (15.5.18)

$$\Psi(T,X,Y,Z) = \frac{2mD}{A}\tanh\left(\frac{mCX + nCT + lC - m\sqrt{-BC}Y + lC}{C}\right) - \frac{n}{mA}.$$

(15.5.19)

where l, m, n are arbitrary constants.

Case-V For the infinitesimal operator $V_5 = AT\frac{\partial}{\partial X} + \frac{\partial}{\partial \Psi}$, the characteristic equation is

$$\frac{dX}{AT} = \frac{dY}{0} = \frac{dZ}{0} = \frac{dT}{0} = \frac{d\Psi}{1} \tag{15.5.20}$$

The new form of the solution of Eq. (15.3.1) is

$$\Psi(T, X, Y, Z) = f(y, z, t) + \frac{X}{AT}, \tag{15.5.21}$$

with similarity variables

$$y = Y, z = Z, t = T. \tag{15.5.22}$$

Now by inserting the value of Ψ from Eq. (15.5.21) into Eq. (15.3.1), we will get

$$tf_t + f = 0. \tag{15.5.23}$$

Similarly, as above we will get the following similarity(invariant) solution for Eq. (15.5.23)

$$\Psi(T, X, Y, Z) = \frac{f(Y, Z)}{T} + \frac{X}{AT}. \tag{15.5.24}$$

Case-VI For the operator $V_5 = -Y\frac{\partial}{\partial Z} + Z\frac{\partial}{\partial Y}$, the characteristic equation is

$$\frac{dX}{0} = \frac{dY}{Z} = \frac{dZ}{-Y} = \frac{dT}{0} = \frac{d\Psi}{0} \tag{15.5.25}$$

The another form of the solution of Eq. (15.3.1) is

$$\Psi(T, X, Y, Z) = f(x, t, \alpha) \tag{15.5.26}$$

with similarity variables

$$x = X, t = T, \alpha = Y^2 + Z^2. \tag{15.5.27}$$

Now by inserting the value of Ψ from Eq. (15.5.26) into Eq. (15.3.1), we will get

$$f_t + Aff_x + 4C\alpha f_{x\alpha\alpha} + Bf_{xxx} + 4Cf_{x\alpha} + Df_{xx} = 0. \tag{15.5.28}$$

Similarly, as above we will get the following similarity(invariant) solution for Eq. (15.5.28)

$$\Psi(T, X, Y, Z) = \frac{f(Y^2 + Z^2)}{T} + \frac{X}{AT}. \tag{15.5.29}$$

15.6 Conclusion

In this article, we use the Lie symmetry approach to find a similarity solution for the non-linear partial differential equations. We investigate new invariant solutions of (3+1)-dimensional Zakharov-Kuznetsov-Burgers equation. Using symmetry transformations, we can reduce the number of independent variable of the partial differential equations and also convert it into ODEs. After transforming into ODEs, we can solve these ODEs using tanh method and get a new invariant solution. The Lie symmetry approach is trustworthy and useful for getting new exact solutions to nonlinear differential equations that occur in physics and engineering.

References

[1] Robert Conte. Invariant Painlev′e analysis of partial differential equations. Physics Letters A, 140(7-8):383-390, 1989.

[2] MA Abdou. The extended tanh method and its applications for solving nonlinear physical models. Applied Mathematics and Computation, 190(1):988-996, 2007.

[3] Wen-Xiu Ma, Tingwen Huang, and Yi Zhang. A multiple exp-function method for nonlinear differential equations and its application. Physica Scripta, 82(6):065003, 2010.

[4] Xing L¨u, Fuhong Lin, and Fenghua Qi. Analytical study on a two-dimensional Korteweg–de Vries model with bilinear representation, B¨acklund transformation and soliton solutions. Applied Mathematical Modelling, 39(12):3221-3226, 2015.

[5] Sophus Lie. Theory of Transformation Groups I: General Properties of Continuous Transformation Groups. A Contemporary Approach and Translation. Springer, 2015.

[6] Peter E Hydon and Peter Ellsworth Hydon. Symmetry methods for differential equations: a beginner's guide. Number 22. Cambridge University Press, 2000.

[7] George Bluman and Stephen Anco. Symmetry and integration methods for differential equations, volume 154. Springer Science & Business Media, 2008.

[8] R Sahadevan and T Bakkyaraj. Invariant analysis of time fractional generalized Burgers and Korteweg-de Vries equations. Journal of Mathematical Analysis and Applications, 393(2):341-347, 2012.

[9] Evelyn Buckwar and Yuri Luchko. Invariance of a partial differential equation of fractional order under the Lie group of scaling transformations. Journal of Mathematical Analysis and Applications, 227(1):81-97, 1998.

[10] George W Bluman and Julian D Cole. The general similarity solution of the heat equation. Journal of Mathematics and Mechanics, 18(11):1025-1042, 1969.

[11] Peter J Olver. Applications of Lie groups to differential equations, volume 107. Springer Science & Business Media, 1993.

[12] Aly R Seadawy, Wang Jun, et al. Mathematical methods and solitary wave solutions of three-dimensional Zakharov-Kuznetsov-Burgers equation in dusty plasma and its applications. Results in Physics, 7:4269-4277, 2017.

[13] Aly R Seadawy. Nonlinear wave solutions of the three-dimensional Zakharov-Kuznetsov-Burgers equation in dusty plasma. Physica A: Statistical Mechanics and its Applications, 439:124-131, 2015.

[14] Waleed M Moslem and R Sabry. Zakharov-Kuznetsov-Burgers equation for dust ion acoustic waves. Chaos, Solitons & Fractals, 36(3):628-634, 2008.

Chapter 16

Parallel Computing of Beta, Incomplete Gamma and Meijer-G Functions using GPU Platform

Vishwesh A. Vyawahare, Navin G. Singhaniya, Divya K. Shah, Shrunoti Gulhane, Stephen Fernandes, Prasad Jadhav

Department of Electronics Engineering, Ramrao Adik Institute of Technology, D. Y. Patil Deemed to be University, Nerul-400706, Navi Mumbai, India

Abstract

This chapter presents novel results on the parallel computation of special functions used in fractional calculus, namely, the Beta function, Incomplete Gamma function and the Meijer-G Function. These functions have wide applications in mathematical modelling especially, in fractional-order modelling of many real world and engineering systems like population biology, ecological systems, heat conduction, nuclear reactor engineering, and many more. Due to their ubiquitous presence in the mathematical representations of physical systems, the reliable, accurate and fast computation of these functions is of utmost importance. The present literature on the computing of these functions is based on serial computing framework, which makes the computational task slower and may lead to inaccuracies. This work, probably for the first time, presents a systematic methodology for the parallel computing of these functions. The work employs series representations of these functions for computation. The parallel computing of these functions is carried out on a general purpose GPU (GPGPU) platform. Beta function Incomplete Gamma function Meijer G-function GPU Computing Parallel Computing

Keywords: Beta function, Incomplete Gamma function, Meijer G-function, GPU Computing, Parallel Computing

16.1 Introduction

Special functions are mathematical functions that have unique properties and applications in various branches of mathematics, physics, engineering, and other scientific disciplines [1]. They often arise as solutions to important differential equations or mathematical representations of certain phenomena. These functions play a fundamental role in modeling and understanding complex systems and have been extensively studied and applied throughout mathematics. The special functions play a crucial role in fractional calculus (FC) definitions and solution of fractional differential equations (FDEs), dealing with non-integer order derivatives and integrals, enabling the study and analysis of systems and phenomena that exhibit fractional dynamics [2]. Some of the important special functions are the Gamma function (extension of the factorial function to real and complex numbers), the Mittag-Leffler function (generalization of the exponential function), the Wright function, the Fox's H-function, the H-function, the Beta function, and the Meijer G-function [3,4].

A comprehensive summary of mathematical functions is given in the handbook [5]. It provides formulae, numerical tables, and algorithms that can be implemented on GPUs. Special functions and their application in the theory of fractional differential equations are given by Podlubney [6]. Amparo Gil et al. [7] focus on the numerical computation of special functions and present various algorithms and techniques. A comprehensive treatment of special functions and their applications in science and engineering is given by Bell et al. [8]. It provides a good resource for understanding the mathematical properties and computational methods of special functions. The hardware implementation of special functions is crucial in various fields, such as mathematics, signal processing, communication systems, scientific simulations, and artificial intelligence. Hardware implementation of special function on field programmable gate array (FPGA) with analysis is reported in [9].

A Graphics Processing Unit (GPU) is a specialized electronic circuit designed to handle and accelerate the processing of computer graphics and visual data [10]. It was initially developed for rendering images, videos, and animations in video games and other multimedia applications. GPUs have evolved to become powerful parallel processors capable of performing various tasks beyond graphics. Unlike the Central Processing Unit (CPU), which is a general-purpose processor responsible for executing various tasks in a computer system, a GPU is optimized for parallel processing and specifically designed to handle complex mathematical and geometric calculations required for rendering graphics [11]. This parallel architecture allows the GPU to process multiple tasks simultaneously, leading to significant speed and performance improvements compared to a CPU [12].

The power of GPUs lies in their ability to perform massive amounts of computations in parallel [13]. They consist of thousands of cores, each capable of executing multiple instructions simultaneously. This architecture makes GPUs highly efficient for tasks that can be parallelized, such as image and video processing, scientific simulations, and machine learning [14]. In addition to their role in graphics rendering, GPUs have found applications in various fields that require intensive computational tasks [15]. For example, in machine learning and artificial intelligence, GPUs have become an essential tool for training and running deep neural networks, as they can process vast amounts of data and perform complex calculations much faster than traditional CPUs [16]. GPU can be seen as a parallel array with many smaller processors acting in parallel. In this work, the parallel nature of the GPU is used to execute the special functions for speed-up and accuracy.

To utilize the computational power of a GPU, specialized programming frameworks and languages have been developed, such as Compute Unified Device Architecture (CUDA) by NVIDIA and Open Computing Language (OpenCL), which enable developers to write software that can leverage the parallel processing capabilities of GPUs. The book [17] is a collection of GPU computing techniques and applications. It covers many hands-on examples of NVIDIA's CUDA parallel computing architecture. Parallel algorithms to compute a few special functions are demonstrated in [18] using a graphic processing unit. This implementation showed parallel computation, which helps reduce the execution time to compute the functions. K. Ahnert et al. [19] describe solving ordinary differential equations on GPUs. The Parallel Computing Toolbox (PCT) is a MATLAB toolbox [20] provides high-level functions and tools for parallel and distributed computing. With the PCT, one can distribute the MATLAB computations across multiple processors, clusters, or grids to speed up the execution time of the code [21].

This work reports the computation of special mathematical functions using GPU. It involves developing and optimizing algorithms for special mathematical functions, such as Beta, Incomplete Gamma and the Meijer-G functions. These functions involve computationally intensive operations, often requiring repetitive calculations, making them suitable for parallelization on GPUs. The results presented demonstrate the effectiveness of GPU computing for the computation of these functions. The performance improvements achieved through GPU acceleration is measured in terms of speedup. Comparative analysis is carried out to evaluate the advantages of GPU-based computations over traditional CPU-based methods. The findings contribute to numerical computation by providing insights into the efficient utilization of GPUs and demonstrating the benefits of parallelization in mathematical function evaluation. The research gives way for further advancements in GPU-accelerated numerical computing and opens up new possibilities for high-performance scientific computing applications. The salient contributions of this work can be summarized as follows:

1. Analysis of the sequential computing algorithms for speed and accuracy.
2. Development of parallel computing algorithms/codes.
3. Benchmarking and analysis of the results.

The chapter is organized as follows. Section 2 discusses about special functions implemented in this chapter. GPU computing is explained in section 3. Section 4 presents results. Conclusion is given in section 5.

16.2 Special Functions

This section briefly introduces the special functions used in fractional calculus namely, the Gamma, the Beta, the Incomplete Gamma and the Meijer-G functions.

16.2.1 Gamma Function

The Gamma function is popularly known as the factorial of non-integers. Adrien-Marie Legendre introduced gamma function. It is defined in mathematics, as an extension of the factorial function. The integral definition of the gamma function is one of the ways to express and understand its properties. The integral definition of the gamma function is given as:

$$\Gamma(z)=\int_{0}^{\infty}s^{z-1}e^{-s}ds, \tag{16.2.1}$$

where, z is a complex number (with real part greater than 0 for convergence), and t is a positive real number. This integral representation defines the gamma function for complex numbers z with a positive real part. It is implemented by using the following infinite product definitions for the Gamma function, given by Karl Weierstrass [22]

$$\Gamma(z)=\frac{e^{-\gamma^{z}}}{z}\prod_{k=1}^{N}(1+\frac{z}{k})^{-1}e^{\frac{z}{k}}. \tag{16.2.2}$$

16.2.2 Beta Function

The Beta function, denoted by the symbol $B(x,y)$, is a mathematical function defined for positive real numbers x and y [23]. It is defined as:

$$B(x,y)=\int_{0}^{1}t^{x-1}(1-t)^{y-1}dt, \tag{16.2.3}$$

where t is the integration variable, and x and y are the parameters of the beta function. The relationship of the beta function with gamma function is an important property. The beta function is simplified using gamma function for easy implementation [24] as

$$B(x,y)=\frac{\Gamma(x)\Gamma(y)}{\Gamma(x+y)}. \tag{16.2.4}$$

The beta function has several important properties and relationships with other mathematical functions. Some of the key properties of the function include:

Symmetry:

$$B(x,y)=B(y,x). \tag{16.2.5}$$

Recurrence relation:

$$B(x,y)=\frac{(x-1)}{(x+y-2)}B(x-1,y). \tag{16.2.6}$$

The beta function finds applications in many areas of mathematics and science. Some notable applications include Probability and statistics, Integral calculus, and mathematical physics [25]. The properties and applications of the beta function make it a valuable tool in various mathematical and scientific disciplines, providing a way to generalize factorial-like calculations to non-integer values and solve diverse problems. The basic outline of a parallelizing approach for the computation of the beta function is given in algorithm 1.

16.2.3 Incomplete Gamma Function

The incomplete gamma function is a special function that arises in mathematics and physics, particularly in the study of gamma distributions and exponential integrals. It is denoted by the symbol $\Gamma(s,x)$, where s is a positive real number (the shape parameter) and x is a non-negative real number (the integration limit). The upper incomplete gamma function is defined as [7]:

$$\Gamma(s,x)=\int_x^{\infty} t^{s-1}e^{-t}\,dt, \tag{16.2.7}$$

where t is the integration variable. The incomplete gamma function is closely related to the gamma function. The lower incomplete gamma function is defined as:

$$\gamma(s,x)=\int_0^{x} t^{s-1}e^{-t}\,dt. \tag{16.2.8}$$

The incomplete gamma function can be thought of as a generalization of the gamma function, where the integration limit is not necessarily infinity but instead a specific value x. Repeated application of the recurrence relation for the lower incomplete gamma function leads to the power series expansion [26]:

$$\gamma(s,x)=\sum_{k=0}^{\infty}\frac{x^{s}e^{-x}x^{k}}{s(s+1)....(s+k)}=x^{s}\Gamma(s)e^{-x}\sum_{k=0}^{\infty}\frac{x^{k}}{\Gamma(s+k+1)}. \quad (16.2.9)$$

This function has various properties and can be expressed in terms of other special functions, such as the exponential integral or the complementary error function. It is commonly used in probability theory, statistics, and mathematical physics for calculating various quantities related to gamma distributions, including survival probabilities, cumulative distribution functions, and moments [27]. Different ways to compute or approximate the incomplete gamma function include numerical methods, series expansions, or specialized algorithms. Many mathematical software packages and programming languages provide built-in functions for computing the incomplete gamma function [28]. Parallel computation of the incomplete gamma function is described in algorithm 2.

16.2.4 Meijar-G Function

The Meijer-G function, named after its discoverer, Eugen Meijer, is a special function that generalizes several other special functions, including the hypergeometric function. It is denoted by $G(a_i,b_j;c_j;z)$, where a_i, b_j, c_j, and z are complex numbers, and the parameters a_i and b_j can be either real or complex. The Meijer-G function is defined as an integral representation [7]:

$$G_{p,q}^{m,n}\left(\begin{matrix}a_1,\dots,a_p\\ b_1,\dots,b_q\end{matrix}\,|\, z\right)=\frac{1}{2\pi i}\int_L\frac{\prod_{j=1}^{m}\Gamma(b_j-s)\prod_{j=1}^{n}\Gamma(1-a_j+s)}{\prod_{j=m+1}^{q}\Gamma(1-b_j+s)\prod_{j=n+1}^{p}\Gamma(a_j-s)}z^{s}\,ds,$$

(16.2.10)

where $\Gamma(x)$ represents the gamma function, z is a complex number with non-zero real part, and the integration is performed along a suitable contour L in the complex plane.

If the integral converges and is evaluated through the second path given above, and if no confluent poles appear among the $\Gamma(bj-s)$ for $j=1,2,3,.....,m$ then the Meijer G-function can be expressed as a sum of residues in terms of generalized hypergeometric functions ${}_pF_{q-1}$ (Slater's theorem):

$$G_{p,q}^{m,n}\left(\begin{matrix}a_p\\b_q\end{matrix}\,\middle|\, z\right)=\sum_{h=1}^{m}\frac{\prod_{j=1}^{m}\Gamma(b_j-b_h)^{*}\prod_{j=1}^{n}\Gamma(1+b_h-a_j)z^{b_h}}{\prod_{j=m+1}^{q}\Gamma(1+b_h-b_j)\prod_{j=n+1}^{p}\Gamma(a_j-b_h)}\times \tag{16.2.11}$$

$${}_pF_{q-1}\left(\begin{matrix}1+b_h-a_p\\(1+b_h-b_q)^{*}\end{matrix}\,\middle|\,(-1)^{p-m-n}z\right).$$

The Meijer-G function has several important properties and applications in mathematics and physics. It is instrumental in solving differential equations, integral transforms, and evaluating complex integrals. It can also be used to express various special functions, such as the exponential function, trigonometric functions, Bessel functions, and more [29]. Due to its complex nature, the Meijer-G function is typically computed using numerical methods or specialized software packages that provide routines for evaluating it accurately. Specific identities, transformations, and series representations are also available to simplify calculations involving Meijer-G functions. A general framework for parallel computation of the Meijer-G function is given in algorithm 3.

16.3 GPU Computing

This section gives brief introduction to the hardware description of GPU and the parallel computing toolbox (PCT) of MATLAB.

16.3.1 GPU Hardware

The Graphics Processing Unit (GPU) is widely employed for real-time graphics rendering in various applications, including high-definition gaming and computer-aided design software. GPUs were initially introduced in 1990 for rendering purposes. Rendering involves the generation of images from 2D or 3D models using computer programs. This process necessitates a significant number of calculations to be performed rapidly. GPUs are increasingly popular due to their affordability and parallel processing capabilities, similar to a supercomputer. The architecture of a GPU differs from that of a traditional CPU. While CPUs provide more resources to accelerate the execution of a single instruction, GPUs utilize numerous individual processing elements or cores to execute a single instruction on multiple data elements in parallel [30]. This parallel execution is achieved simultaneously by operating on many independent elements, often organized in arrays. However, it is essential to note that each element in GPU programming should be independent of one another, lacking direct communication. Consequently, GPU programs must be appropriately structured to effectively process multiple parallel elements simultaneously [31]. This parallel processing capability enables GPUs to handle the substantial computational requirements of real-time graphics rendering. Pictorial representation of CPU and GPU structures are shown in Figure 16.1.

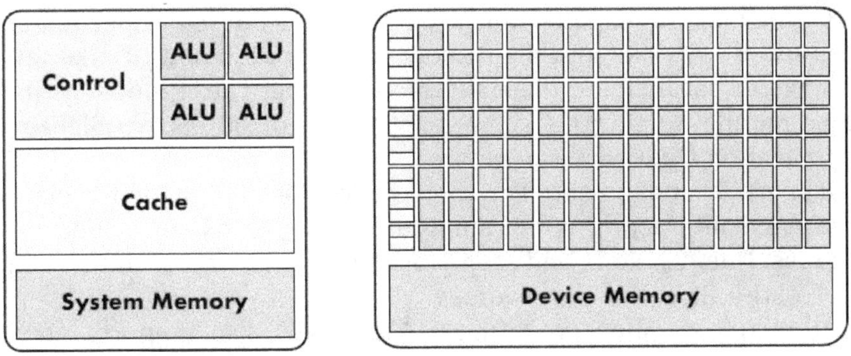

Figure 16.1 Pictorial representation of CPU and GPU structures [22].

16.3.2 MATLAB Parallel Computing Toolbox

MATLAB is widely used by engineers worldwide for various purposes, including system analysis and design. Its ability to operate on matrices and arrays makes it an efficient tool for expressing computational mathematics. The built-in graphics capabilities in MATLAB allow users to visualize and gain insights from data, making it easier to work. The desktop environment (DE) provided by MATLAB, offers a platform for experimentation, analysis, and discovery. The tools and features of MATLAB are designed and tested to work together, ensuring a seamless user experience [32].

The PCT in MATLAB enables users to solve computationally and data-intensive problems using multicore processors, such as GPUs

[20]. High-level constructs like parallel for loops and parallelized numerical algorithms allow for parallelization of MATLAB applications [13]. Some key features of the MATLAB PCT include:

• Parallel for-loops: parfor loop construct used to parallelize the execution of for-loops. MATLAB automatically partitions the loop iterations and distributes them across the available computational resources.
• Parallel computing constructs: The toolbox provides various constructs, such as parallel arrays (matrices distributed across multiple workers), distributed arrays (arrays that are partitioned across workers), and Single Program, Multiple Data (SPMD) blocks, which allow programmer to write code that executes simultaneously on multiple MATLAB workers.
• High-level parallel algorithms: MATLAB PCT offers a collection of high-level functions and algorithms that are designed to take advantage of parallel and distributed computing. These functions include parallel sorting, distributed array manipulations, parallelized optimization algorithms, and more.
• Resource management: The toolbox provides functions for managing and interacting with compute clusters, grids, and cloud resources. One can start and stop workers, monitor their status, and control the allocation of resources.
• Interactive parallel computing: MATLAB PCT also supports interactive parallel computing through the PCT Job Scheduler. It allows the user to create and submit batch jobs for parallel execution, manage job dependencies, and monitor job progress.

By utilizing the full processing power of multicore desktops, applications can be executed on local workers, thereby enhancing performance. To utilize the PCT, a NVIDIA CUDA-enabled device with a compute capability of 1.3 or greater is required. Some key functions used for GPU computing in MATLAB include:

1. gpuArray: This function transfers input data from the MATLAB workspace to the GPU. It converts the array in the MATLAB workspace into a gpuArray object, allowing computations to be performed on the GPU.

2. gather: This function is used to transfer data from the GPU back to the MATLAB workspace. It is typically used after performing all the necessary operations on the data using the GPU.

3. gpuDevice: This command is used to display the properties

of the currently selected GPU device. It provides information about the GPU device being used, such as its name, compute capability, memory, and other relevant details.

These functions and features of MATLAB's PCT provide users with the ability to leverage the computational power of GPUs for efficient and parallel processing of data and computations.

16.3.3 Parallel Computing using PCT

To execute a program in parallel mode using MATLAB PCT, steps are as follows:

1. Start with accurate and correct code that produces the desired output.
2. Optimize the code through vectorization, which involves using MATLAB's built-in vector and matrix operations instead of iterative loops whenever possible. Vectorization can significantly improve code performance.
3. Use the Run command or execute the code directly in the MATLAB command window or script editor to run the program. MATLAB will execute the code line by line and display any output or results.
4. The power of a GPU, can be exploited in total by transferring the variables to GPU memory using the command gpuArray.
5. Pose execution, transfer the outputs from GPU memory back to the CPU memory using the gather command. This step is necessary for accessing and analyzing the results in the MATLAB workspace.

Visual representation of converting MATLAB program on run on GPU is given in Fig. 16.2. These steps illustrate the complete procedure to effectively run MATLAB programs, optimize their performance, utilize GPU processing power, and transfer data between CPU and GPU memory as needed.

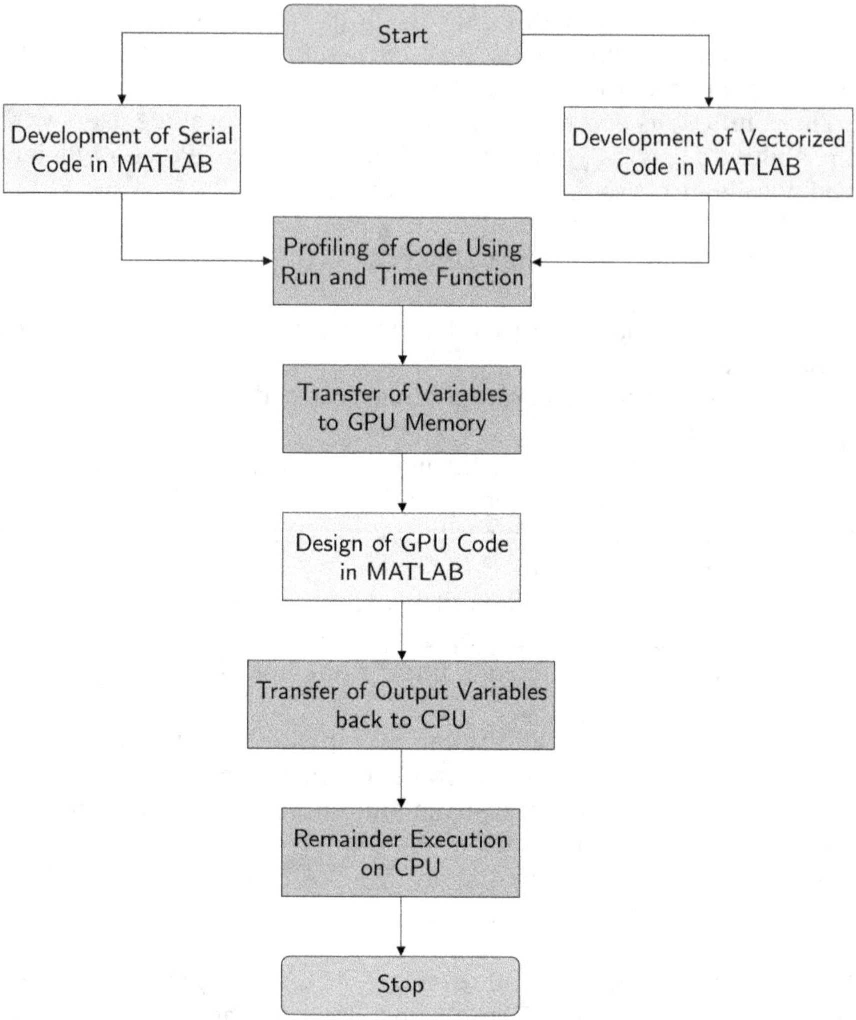

Figure 16.2 Visual Representation of MATLAB program on GPU.

MATLAB provides a built-in profiler tool that helps analyze the performance of your code and identify bottlenecks. The profiler gives you insights into the execution time and memory usage of different functions and lines of code, allowing you to optimize your MATLAB programs. To use the MATLAB profiler, steps are as follows:

1. Open the MATLAB script or function.
2. Click on the "Profile" tab in the MATLAB Editor.
3. Click on the "Run and Time" button in the "Profiling" sec-

tion of the toolbar. Alternatively, the "Profile" command in the Command Window can also be used.

4. MATLAB will execute the code while collecting profiling data.

5. After the execution is complete, the profiler window will open, displaying the results.

The profiler window gives insights into the code's performance. Which is helpful in focusing on the most time-consuming parts of the code. By analyzing the profiling data, performance bottlenecks, optimizing critical sections, and improve the overall efficiency of the MATLAB code can be identified. It is important to note that the MATLAB profiler provides valuable insights, but it requires careful analysis and understanding of code and algorithm [33]. The profiling of serial code for CPU and parallel code for GPU for the Meijer-G Function are shown in Figs. 16.3 and 16.4 respectively.

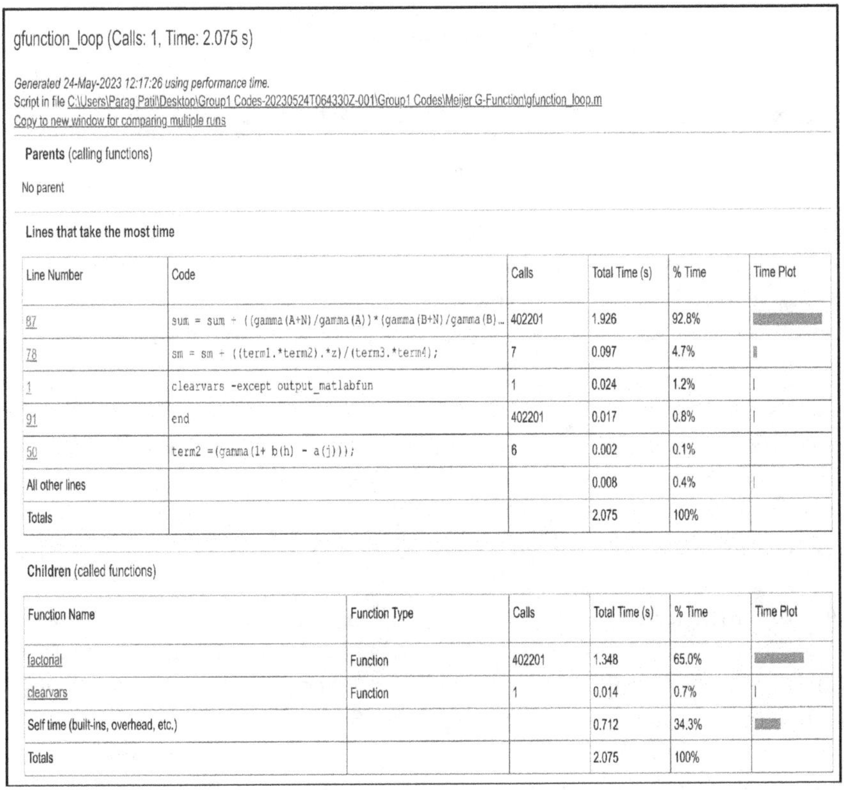

gfunction_loop (Calls: 1, Time: 2.075 s)

Generated 24-May-2023 12:17:26 using performance time.
Script in file C:\Users\Parag Patil\Desktop\Group1 Codes-20230524T064330Z-001\Group1 Codes\Meijer G-Function\gfunction_loop.m
Copy to new window for comparing multiple runs

Parents (calling functions)

No parent

Lines that take the most time

Line Number	Code	Calls	Total Time (s)	% Time	Time Plot
87	`sum = sum + ((gamma(A+N)/gamma(A))*(gamma(B+N)/gamma(B)...`	402201	1.926	92.8%	
78	`sm = sm + ((term1.*term2).*z)/(term3.*term4);`	7	0.097	4.7%	
1	`clearvars -except output_matlabfun`	1	0.024	1.2%	
91	`end`	402201	0.017	0.8%	
50	`term2 =(gamma(1+ b(h) - a(j)));`	6	0.002	0.1%	
All other lines			0.008	0.4%	
Totals			2.075	100%	

Children (called functions)

Function Name	Function Type	Calls	Total Time (s)	% Time	Time Plot
factorial	Function	402201	1.348	65.0%	
clearvars	Function	1	0.014	0.7%	
Self time (built-ins, overhead, etc.)			0.712	34.3%	
Totals			2.075	100%	

Figure 16.3 Profiling of Meijer-G Function when executed on CPU.

gfunction_GPU (Calls: 1, Time: 0.280 s)

Generated 24-May-2023 12:19:12 using performance time.
Script in file C:\Users\Parag Patil\Desktop\Group1 Codes-20230524T064330Z-001\Group1 Codes\Meijer G-Function\gfunction_GPU.m
Copy to new window for comparing multiple runs

Parents (calling functions)

No parent

Lines that take the most time

Line Number	Code	Calls	Total Time (s)	% Time	Time Plot
73	`h=legend('Output of CPU Execution');`	1	0.144	51.5%	
72	`plot(t,arr5d,'r','LineWidth',0.5);`	1	0.111	39.8%	
75	`grid on;`	1	0.007	2.3%	
57	`arr5f = gpuArray(factorial(N));`	1	0.003	1.0%	
74	`set(h,'FontSize',12);`	1	0.002	0.6%	
All other lines			0.014	4.8%	
Totals			0.280	100%	

Children (called functions)

Function Name	Function Type	Calls	Total Time (s)	% Time	Time Plot
legend	Function	1	0.144	51.5%	
newplotwrapper	Function	1	0.110	39.2%	
grid	Function	1	0.006	2.3%	
factorial	Function	1	0.002	0.6%	
Legend.Legend>Legend.set.FontSize	Class method	1	0.001	0.5%	
Self time (built-ins, overhead, etc.)			0.016	5.9%	
Totals			0.280	100%	

Figure 16.4 Profiling of Meijer-G Function when executed on GPU.

16.4 Results

The computations were carried out on a PC with Intel Xeon E5-2623 v4 CPU (10M Cache, 2.60 GHz) RAM 4 * 16.00 GB system type 64-bit operating system, x64-based processor for its CPU and NVIDIA Tesla K80 for its GPU (See Fig. 16.5).

Algorithm1: Beta Function

1. Set step size and input.
2. Fix the value of y.
3. Start Timer.
4. Start $for\ i = 1 \longrightarrow length(x)$ loop.
5. Summation Equation: $B(x,y) = \dfrac{\Gamma(x)\Gamma(y)}{\Gamma(x+y)}$
6. Calculate Sum of gamma of first input (x) and gamma of second input (y).
7. Divide the sum obtained by gamma of addition of first and second input ($x+y$).
8. Calculate gamma( ) function.
9. End 'for' loop.
10. Stop timer.
11. Output is obtained, recorded with time as sequential runtime (T_S) and plotted with step size.
12. Initialize steps 1 and 4 again.
13. Start timer.
14. Convert all the elements from step 5 to arrays. All terms will be converted to 2-D matrices.
15. Operations like addition, multiplication and division are performed and store it in a variable '$output\_cpu$'.
16. Variable '$output\_cpu$' is a 2-D matrix.
17. Stop timer.
18. Time taken for the output to be obtained is recorded as vectorized runtime (T_V) and plotted v/s step size.
19. Initialize steps 1 and 4 again.
20. Start timer.
21. Transfer all the terms in array to GPU using gpuArray() function.
22. After transfer perform operations like addition, multiplication and division and store it in a variable '$output\_gpu$'.
23. Variable '$output\_gpu$' is a 2-D matrix in the parallel computing platform.
24. Stop timer.
25. Time taken for the output to be obtained is recorded as GPU time (T_G) and plotted v/s step size.
26. All the 3 times for Beta function i.e. T_S, T_V and T_G are

compared.

27. Speedups are calculated as mentioned in results.

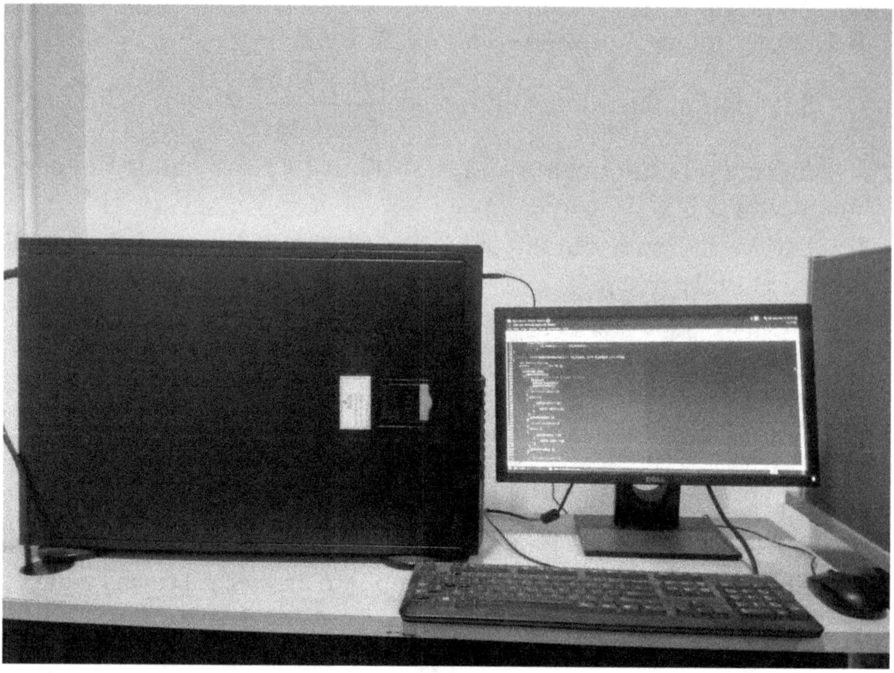

Figure 16.5 Workstation with GPU Tesla K80.

Algorithm2: Incomplete Gamma Function

1. Set step size and input.
2. Fix the value of x.
3. Start Timer.
4. Start $for\ i = 0 \longrightarrow N$ loop.
5. Summation Equation: $\gamma(s,x) = x^s\Gamma(s)e^{-x}\sum_{k=0}^{\infty}\frac{x^k}{\Gamma(s+k+1)}$
6. Calculate x^k for the current value.
7. Calculate gamma of $(s+k+1)$ by using inbuilt gamma function.
8. Divide step 6 with 7
9. Calculate previous three steps within the summation limit of 0 to k within the mentioned step size.

10. End '*for*' loop.

11. A constant F = $(x^s \times (s-1)! e^{(-x)})$ is multiplied to every output of for loop in Incomplete Gamma function.

12. Stop timer.

13. Output is obtained, recorded with time as sequential runtime T_S and plotted with step size.

14. Initialize steps 1 and 4 again.

15. Start timer.

16. Convert all the elements from step 5 to arrays. All terms will be converted to 2-D matrices.

17. Calculate array as in step 5.

18. Operations like multiplication, division and addition are performed and the final array is stored in a variable '*output_cpu*'.

19. Variable '*output_cpu*' is a 2-D matrix.

20. Stop timer.

21. Time taken for the output to be obtained is recorded as vectorized runtime T_V and plotted v/s step size.

22. Initialize steps 1 and 4 again.

23. Start timer.

24. Transfer all the terms of array to parallel computing platform using gpuArray().

25. After transfer perform operations like addition, multiplication and division and store it in a variable '*output_gpu*'.

26. Variable '*output_gpu*' is a 2-D matrix in the parallel computing platform.

27. Stop timer.

28. Time taken for the output to be obtained is recorded as GPU time (T_G) and plotted v/s step size. All the 3 times for Incomplete Gamma function i.e. T_S, T_V and T_G are compared.

29. Speedups are calculated as mentioned in results.

Table 16.1 Workstation specifications

Parameter	**Value**
CPU	Intel Xeon E5-2623 v4
Cores	4
Total RAM	4 * 16 GB
Base Clock	2.6 GHz
Total Cache	10 MB
GPU	NVIDIA Tesla K80
CUDA Cores	4992
GPU Memory	24 GB
Support for Unified Memory	Yes
CUDA Driver version	11.2

Algorithm 3: Meijer-G Function

1. Set step size and input.
2. Start Timer.
3. Derive summation equation
4. $$G_{p,q}^{m,n}\left(\begin{matrix}a_p\\b_q\end{matrix}\middle|\,z\right)=\sum_{h=1}^{m}\frac{\prod_{j=1}^{m}\Gamma(b_j-b_h)^{*}\prod_{j=1}^{n}\Gamma(1+b_h-a_j)z^{b_h}}{\prod_{j=m+1}^{q}\Gamma(1+b_h-b_j)\prod_{j=n+1}^{p}\Gamma(a_j-b_h)}\times{}_pF_{q-1}\left(\begin{matrix}1+b_h-a_p\\(1+b_h-b_q)^{*}\end{matrix}\middle|\,(-1)^{p-m-n}z\right)$$

Calculate Total. Total is equal to multiplication of two entities.

5. Entity one 'sm' is constructed in a '$for\ h=1 \longrightarrow n$' loop as $sm=(term1\times term2\times z)/(term3\times term4)$
6. term1: pi-product $for\ j=1 \longrightarrow n$
7. term2: pi-product $for\ j=1 \longrightarrow m$
8. term3: pi-product $for\ j=n+1 \longrightarrow p$
9. term4: pi-product $for\ j=m=1 \longrightarrow q$
10. The 'sm' is calculated within the summation limit of $h=1 \longrightarrow n$ within the mentioned step size.
11. Entity two 'Sum' is the hypergeometric part of the equation.
12. Stop timer.
13. Output is obtained, recorded with time as sequential runtime T_S and plotted with step size.

14. Initialize step 1 again and Start timer.
15. Convert all the elements from step 4 to arrays. Calculate the arrays as mentioned in step 4.
16. Operations like multiplication and division are performed and are stored in a variable '*output_cpu*'.
17. Variable '*output_cpu*' is a 2-D matrix.
18. Stop timer.
19. Time taken for the output to be obtained is recorded as vectorized runtime T_V and plotted v/s step size.
20. Initialize steps 1 again and Start timer.
21. Transfer all the terms from steps 7 and 11 to parallel computing platform.
22. After transfer perform operations like addition, multiplication and division and store it in a variable '*output_gpu*'.
23. Variable '*output_gpu*' is a 2-D matrix in the parallel computing platform.
24. Stop timer.
25. Time taken for the output to be obtained is recorded as GPU time T_G and plotted v/s step size.
26. All the 3 times for Meijer-G function i.e. T_S, T_V and T_G are compared.
27. Speedups are calculated as mentioned in results.

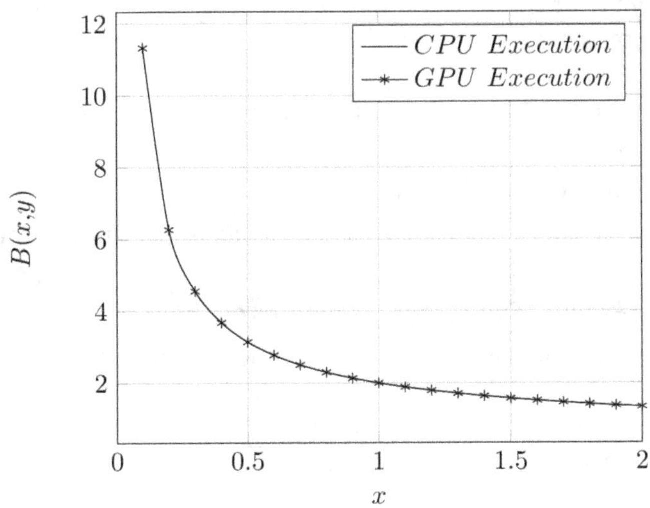

Figure 16.6 Incomplete gamma function.

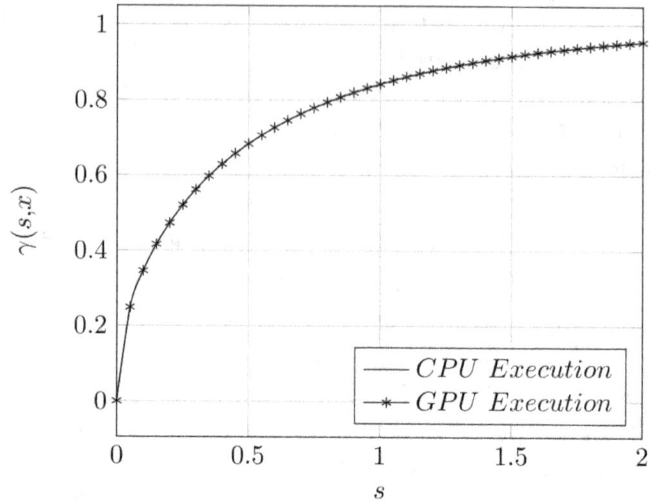

Figure 16.7 Meijer-G function.

Thc speedup value represents the factor by which the parallel execution is faster than the serial execution. If the speedup is greater than 1, it indicates that the parallel execution is faster than the serial execution. The larger the speedup value, the more efficient the parallelization. The speedup is computed by measuring the execution time of the serial version of the code and the parallel version of the code. The execution time can be measured using various techniques, such as timing functions or profiling tools. To calculate the speedup, steps are as follows:

1. Measure the execution time of the serial version of the code (Serial Execution Time denoted by T_S).
2. Implement and measure the execution time of the parallel version of the code (Parallel Execution Time denoted by T_G).
3. Calculate the speedup using the formula:

$$Speedup = \frac{Serial\ Execution\ Time}{Parallel\ Execution\ Time} = \frac{T_S}{T_G}. \tag{16.4.1}$$

It's important to note that achieving a high speedup depends on various factors, including the nature of the problem being solved, the efficiency of the parallel algorithm, the available hardware resources, and the scalability of the parallel implementation [34].

From the benchmarking plots (See Figs. 16.6 and 16.7), it has been observed that the output of CPU and GPU execution exactly overlaps for a given set of parameters. It suggests that both the CPU and GPU computation are producing the same results for the computation being performed. It's worth noting that in some cases, due to the inherent parallelism and architecture differences between CPUs and GPUs, it is common to see slight differences in output. However, the observed plots are exact overlap, indicating a high level of consistency between the CPU and GPU executions for the given parameters.

The results of Beta function, Incomplete gamma function, Meijer-G function are presented in Tables 16.3-16.4. The details are as follows:

- Parameters: Varying Step sizes and Inputs.
- Runtime: The runtime taken by a code for its complete execution is recorded as:

  (a) T_S shows the time taken by the sequential code for its complete execution.

  (b) T_V shows the time taken by the vectorized code for its complete execution.

  (c) T_P shows the time taken by the Parallel code for its complete execution.

- Speed Up: In this section the runtime of different types of codes are calculated as a ratio to calculate the speed of one type with respect to other type as:

  (a) T_S / T_V is the ratio of sequential runtime to vectorized runtime.

  (b) T_V / T_P is the ratio of vectorized runtime to parallel runtime.

  (c) T_S / T_P is the ratio of sequential runtime to parallel runtime.

• Average speed up is calculated in sets of three with one varying input only.

• Three of such sets are calculated and averaged to get the total average speed up achieved.

Table 16.2 GPU computation of beta function

Parameters		Runtime			Speed Up		
Input	Input	Sequential	Vectorized	GPU Time	Speed up	Speed Up	Speed Up
(x)	(y)	T_S	T_V	T_G	$\frac{T\_S}{T\_V}$	$\frac{T\_V}{T\_G}$	$\frac{T\_S}{T\_G}$
	0:0.01:100000;	174.5794	2.3034	2.0096	75.79	1.15	86.87
	0:0.01:10000;	15.4093	0.2648	0.1079	58.19	2.45	142.81
	0:0.01:1000;	1.9357	0.0569	0.0299	34.02	1.9	64.74
		Average Speedup			56	1.83	98.14
	0:0.01:100000;	211.8528	1.9106	1.5794	110.88	1.21	134.13
	0:0.01:10000;	30.2968	0.2176	0.1937	139.23	1.12	156.41
	0:0.01:1000;	1.4768	0.0335	0.0194	44.08	1.73	76.12
		Average Speedup			98.07	1.35	122.22
	0:0.01:100000;	230.8079	1.9916	1.6792	115.89	1.19	137.45
	0:0.01:10000;	20.712	0.2342	0.1313	88.44	1.78	157.75
	0:0.01:1000;	2.0722	0.0472	0.0154	43.9	3.06	134.56
		Average Speedup			82.74	2.01	143.25
		Total Avg Speed up			**78.94**	**1.73**	**121.21**

It is observed from the table 16.2 reveals that in the computation of Beta function a speedup of 78 times is yielded from just serialization to vectorization and it increases to 121 times from serialization to parallel computing platform.

Table 16.3 GPU Computation of incomplete gamma function

Parameters			**Runtime**			**Speedup**		
Step Size	**Input**	**Input**	**Sequen-tial**	**Vector-ized**	**GPU Time**	**Speed up**	**Speed Up**	**Speed Up**
(N)	**(s)**	**(x)**	**T_S**	**T_V**	**T_G**	$\frac{T\_S}{T\_V}$	$\frac{T\_V}{T\_G}$	$\frac{T\_S}{T\_G}$
	5	15	72.9736	0.2376	0.2156	307.13	1.1	338.47
	5	15	360.1299	1.7434	1.274	206.57	1.37	282.68
	5	15	586.8109	6.5118	5.2913	90.12	1.23	110.9
			Average Speedup			201.27	1.23	244.02
	5	45	22.7865	0.196	0.123	116.26	1.59	185.26
	5	45	215.19	1.7785	1.415	121	1.26	152.08
	5	45	595.4406	5.8168	5.314	102.37	1.09	112.05
			Average Speedup			113.21	1.32	149.8
	5	110	20.5516	0.3301	0.1858	62.26	1.78	110.61
	5	110	202.5839	2.0084	1.9886	100.87	1.01	101.87
	5	110	592.6492	6.7347	5.2177	88	1.29	113.58
			Average Speedup			83.71	1.36	108.69
			Total Avg Speedup			**132.73**	**1.3**	**167.5**

A glance at table 16.3 that in the computation of Incomplete Gamma function, a speedup of 132 times is achieved from just serialization to vectorization and it increases to 167 times from serialization to parallel computing platform.

Table 16.4 GPU computation of Meijer-G function

Parameters		**Runtime**			**Speedup**		
Step Size	**In-put**	**Sequen-tial**	**Vector-ized**	**GPU Time**	**Speed up**	**Speed Up**	**Speed Up**
(Z)	**(N)**	**T_S**	**T_V**	**T_G**	$\frac{T\_S}{T\_V}$	$\frac{T\_V}{T\_G}$	$\frac{T\_S}{T\_G}$
-1:1e-03:1	200	17.3352	0.45	0.32	38.52	1.41	54.17
-1:1e-03:1	300	43.3779	0.47	0.41	92.29	1.15	105.8
-1:1e-03:1	500	92.043	0.6996	1.03	131.57	0.68	89.36
		Average Speed up			87.46	1.08	83.11
-1:1e-04:1	200	248.8997	3.0012	1.496	82.93	2.01	166.38
-1:1e-04:1	300	395.9354	3.301	1.652	119.94	2	239.67
-1:1e-04:1	500	597.1667	4.9003	2.6009	121.86	1.88	229.6
		Average Speed up			108.25	1.96	211.88
-1:1e-05:1	200	1059.1	12.005	5.2449	88.22	2.29	201.93
-1:1e-05:1	300	1706.7	16.966	8.0269	100.6	2.11	212.62
-1:1e-05:1	500	2836.6	26.2685	12.6649	107.98	2.07	223.97

		Average Speed up	98.93	2.16	212.84
		Total Avg Speed up	**98.21**	**1.73**	**169.28**

It is observed from the table 16.4 that in the computation of Meijer-G function a speedup of 98 times is yielded from just serialization to vectorization and it increases to 169 times from serialization to parallel computing platform.

16.5 Conclusion

In this work, the acceleration of computation of special mathematical functions has been performed by harnessing the power of GPUs and their multiple processing cores. Initially, the functions were coded sequentially, and subsequently using the MATLAB Parallel Computing Toolbox vectorized version of the code is developed. The focus was on optimizing the sequential code section that consumed most of the execution time. The findings demonstrate the successful implementation of these special functions on GPUs, leveraging the parallel processing capabilities of graphics processors. As a result, significantly faster execution times than sequential implementation is achieved. Specifically, speedups ranging from 2x to 200x were observed. It is important to note that the obtained speedup is depended on the amount of data or the number of iterations involved in the computations. To further enhance the outcomes of this research, there is potential for optimizing the MATLAB codes and exploring more powerful platforms like NVIDIA's CUDA, which could offer additional opportunities for optimization and potentially achieve even more significant speedups.

References

[1] Kiryakova V. A guide to special functions in fractional calculus. Mathematics. 2021 Jan 5;9(1):106.

[2] Sun H, Zhang Y, Baleanu D, Chen W, Chen Y. A new collection of real world applications of fractional calculus in science and engineering. Communications in Nonlinear Science and Numerical Simulation. 2018 Nov 1;64:213-31.

[3] Gorenflo R, Kilbas AA, Mainardi F, Rogosin SV. Mit-

tag-Leffler functions, related topics and applications. Berlin: Springer; 2020 Oct 27.

[4] Kumar P, Agrawal OP. An approximate method for numerical solution of fractional differential equations. Signal processing. 2006 Oct 1;86(10):2602-10.

[5] Abramowitz M, Stegun IA, editors. Handbook of mathematical functions with formulas, graphs, and mathematical tables. US Government printing office; 1948.

[6] Podlubny I. An introduction to fractional derivatives, fractional differential equations, to methods of their solution and some of their applications. Math. Sci. Eng. 1999;198:340.

[7] Gil A, Segura J, Temme NM. Numerical methods for special functions. Society for Industrial and Applied Mathematics; 2007 Jan 1.

[8] Bell WW. Special functions for scientists and engineers. Courier Corporation; 2004.

[9] Shah DK, Chaurasiya RB, Vyawahare VA, Pichhode K, Patil MD. FPGA implementation of fractional-order chaotic systems. AEU-International Journal of Electronics and Communications. 2017 Aug 1;78:245-57.

[10] Brodtkorb AR, Hagen TR, Sætra ML. Graphics processing unit (GPU) programming strategies and trends in GPU computing. Journal of Parallel and Distributed Computing. 2013 Jan 1;73(1):4-13.

[11] Ryoo S, Rodrigues CI, Baghsorkhi SS, Stone SS, Kirk DB, Hwu WM. Optimization principles and application performance evaluation of a multithreaded GPU using CUDA. InProceedings of the 13th ACM SIGPLAN Symposium on Principles and practice of parallel programming 2008 Feb 20 (pp. 73-82).

[12] Jadhav AA, Kalamkar AD, Gaikwad PA, Vyawahare V, Singhaniya N. Analysis of GPU Computation of Parabolic, Bessel, Wright and Riemann Zeta Functions. InITM Web of Conferences 2021 (Vol. 40, p. 02005). EDP Sciences.

[13] Navarro CA, Hitschfeld-Kahler N, Mateu L. A survey on parallel computing and its applications in data-parallel problems using GPU architectures. Communications in Computational Physics. 2014 Feb;15(2):285-329.

[14] Pratx G, Xing L. GPU computing in medical physics: A review. Medical physics. 2011 May;38(5):2685-97.

[15] Rajasekaran S, Reif J, editors. Handbook of parallel computing: models, algorithms and applications. CRC press; 2007 Dec 20.

[16] Steinkraus D, Buck I, Simard PY. Using GPUs for machine learning algorithms. InEighth International Conference on Document Analysis and Recognition (ICDAR'05) 2005 Aug 31 (pp. 1115-1120). IEEE.

[17] Wen-Mei, W. Hwu, GPU computing gems emerald edition (Elsevier, Burlington, USA,

[18] Keluskar YC, Navada MM, Jage CS, Singhaniya NG. Implementation of Airy function using Graphics Processing Unit (GPU). InITM Web of Conferences 2020 (Vol. 32, p. 03052). EDP Sciences.

[19] Ahnert K, Demidov D, Mulansky M. Solving ordinary differential equations on GPUs. InNumerical Computations with GPUs 2014 Jun 9 (pp. 125-157). Cham: Springer International Publishing.

[20] The Mathworks, Parallel computing toolbox, Natick, Massachusetts, United State. http:www.mathworks.com/help/parallel-computing/. Accessed: 2023-05-30.

[21] Hijma P, Heldens S, Sclocco A, Van Werkhoven B, Bal HE. Optimization techniques for GPU programming. ACM Computing Surveys. 2023 Mar 16;55(11):1-81.

[22] Patil P, Singhaniya N, Jage C, Vyawahare VA, Patil MD, Nataraj PS. GPU Computing of Special Mathematical Functions used in Fractional Calculus. Frontiers in Fractional Calculus. 2018 Mar 21;199.

[23] Li A, Sun Z, Qin H. The algorithm and application of the beta function and its partial derivatives. Engineering Letters. 2015 Jul 1;23(3):140-4.

[24] Kiryakova V. The special functions of fractional calculus as generalized fractional calculus operators of some basic functions. Computers & mathematics with applications. 2010 Feb 1;59(3):1128-41.

[25] Olver FW, editor. NIST handbook of mathematical functions hardback and CD-ROM. Cambridge university press; 2010 May 17.

[26] Didonato AR, Morris Jr AH. Algorithm 708: Significant digit computation of the incomplete beta function ratios. ACM Transactions on Mathematical Software (TOMS). 1992 Sep 1;18(3):360-73.

[27] Gil A, Segura J, Temme NM. Efficient and accurate algorithms for the computation and inversion of the incomplete gamma function ratios. SIAM Journal on Scientific Computing. 2012;34(6):A2965-81.

[28] Singhaniya NG, Patil MD, Vyawahare VA. Implementation of special mathematical functions for fractional calculus using DSP

processor. In2015 International Conference on Information Processing (ICIP) 2015 Dec 16 (pp. 811-816). IEEE.

[29] Beals R, Szmigielski J. Meijer G-functions: a gentle introduction. Notices of the AMS. 2013 Aug;60(7):866-72.

[30] Messmer P, Mullowney PJ, Granger BE. GPULib: GPU computing in high-level languages. Computing in Science & Engineering. 2008 Sep 1;10(05):70-3.

[31] Zhang B, Xu S, Zhang F, Bi Y, Huang L. Accelerating matlab code using gpu: A review of tools and strategies. In2011 2nd International Conference on Artificial Intelligence, Management Science and Electronic Commerce (AIMSEC) 2011 Aug 8 (pp. 1875-1878). IEEE.

[32] Suh JW, Kim Y. Accelerating MATLAB with GPU computing: A primer with examples. Newnes; 2013 Nov 18.

[33] Best practices for MATLAB GPU coding, Cornell university center for advanced computing.https://www.cac.cornell.edu/matlab/TechDocs/Examples/BestPracticesGPU/.Accessed:2023-06-11

[34] McLaughlin A, Bader DA. Scalable and high performance betweenness centrality on the GPU. InSC'14: Proceedings of the International Conference for High Performance Computing, Networking, Storage and Analysis 2014 Nov 16 (pp. 572-583). IEEE.

Chapter 17

On the Second Kind Chebyshev-Collocation Scheme for Solving Nonlinear Fredholm Integro-Differential Equations with Weakly Singular Kernels

Youcef Henka[1], Samir Lemita[2], Mohamed Zine Aissaoui [3]
[1,3] Laboratoire de Mathématiques Appliquées et de Modélisation, Université de 08 Mai 1945 Guelma. B.P. 401 Guelma 24000. Algérie
[2] École Normale Supérieure - Ouargla, Cité Ennacer, Ouargla 30000 Algérie

Abstract

This paper presents a rigorous numerical investigation focusing on nonlinear Fredholm integro-differential equations with weakly singular kernels. To tackle these equations, we use an approximation scheme based on the utilization of Chebyshev polynomials of the second kind, which enables us to approximate the true solution effectively. Subsequently, we apply the renowned collocation technique to transform the main equation into a nonlinear algebraic system, thereby facilitating its solution. To showcase the effectiveness and efficiency of our approach, we provide a selection of illustrative examples that demonstrate the accuracy and reliability of the applied method.

Keywords: Fredholm integro-differential equations, Nonlinear equations, Orthogonal polynomials, Fredholm integro-differential equation.

Subject Classification: 45J05, 65M60, 65J15.

17.1 Introduction

Integral equations and integro-differential equations play a crucial role in modeling various phenomena in physics, chemistry, and other research domains. These equations often arise as reformulations of boundary value problems and initial value problems. However, finding analytical solutions to these equations is generally impractical, leading researchers to explore numerical approaches. Numerous authors have devoted significant efforts to studying and developing numerical methods for solving such equations. These methods aim to provide efficient and accurate solutions to these challenging mathematical problems. By employing numerical techniques. In the case of linear equations with regular kernels, the author in [2] employed the Legender Galerkin method to find numerical solutions for boundary value problems. Similarly, in [1], an approximate solution based on Bernoulli polynomials was presented for solving mth-order linear FIDEs. Another approach, as demonstrated in [7], involved applying the homotopy perturbation method to solve FIDEs with separable kernels. Furthermore, [3] discussed the numerical solution for FIDEs using the exponential spline method. For linear equations with weakly singular kernels, the researchers explored different methods, for instance in [5], weakly singular Fredholm integro-differential equations were numerically examined using Legendre multiwavelets. Additionally, in [6], the researcher developed a fully discrete version of a piecewise polynomial collocation method specifically designed for addressing initial or boundary value problems associated with linear Fredholm integro-differential equations with weakly singular kernels.

This research focuses on conducting a numerical study for nonlinear Fredholm integro-differential equations with weakly singular kernels. The equations under consideration are of the form specified below:

$$\begin{cases} \xi(\zeta) = f(\zeta) + \int_{-1}^{1} p(|\zeta-\tau|)\Psi(\zeta,\tau,\xi(\tau),\xi'(\tau))d\tau, \zeta \in [-1,1], \\ \xi(-1) = \xi_0, \end{cases} \tag{17.1.1}$$

where,

$$(H)\begin{cases} f \in C^1([-1,1],R), \quad \Psi, \dfrac{\partial\Psi}{\partial\zeta} \in C^0([-1,1]^2 \times R^2), \\ p(\tau) \in \mathcal{W}^{1,1}(0,2), \\ \lim\limits_{\tau\to 0} p'(\tau) = +\infty. \end{cases}$$

$$\mathcal{W}^{1,1}(0,2) = p \in \mathcal{L}^1(0,2): p' \in \mathcal{L}^1(0,2).$$

The numerical solution for equation (17.1.1) is given by Chebyshev polynomials of the second kind (CPSKs) and the operational matrix of integration.

The organization of this chapter is as follows: In section 2, we present a short brief of the second kind of Chebyshev polynomials, section 3 introduces the function approximation, the operational matrix of integration is described in section 4, while section 5 illustrates the proposed method. Finally, section 6 gives some numerical experiments.

17.2 Preliminaries

The Chebyshev polynomials of the second kind are defined by the following reccurence relation:

$$\begin{cases} \vartheta_{n+1}(\tau) = 2\tau\vartheta_n(\tau) - \vartheta_{n-1}(\tau), \\ \vartheta_1(\tau) = 2\tau, \quad \vartheta_0(\tau) = 1, n \geq 1. \end{cases}$$

The present polynomials are orthogonal concerning the weight function $w(\tau) = \sqrt{1-\tau^2}$ on the interval $[-1,1]$, then we have the following property:

$$\int_{-1}^{1} w(\tau)\vartheta_i(\tau)\vartheta_j(\tau)d\tau = \begin{cases} \pi/2, & i = j, \\ 0, & i \neq j. \end{cases}$$

The first six Chebyshev polynomials of the second kind are

$$\begin{aligned} \vartheta_0(\tau) &= 1 \\ \vartheta_1(\tau) &= 2\tau \\ \vartheta_2(\tau) &= 4\tau^2 - 1 \\ \vartheta_3(\tau) &= 8\tau^3 - 4\tau \\ \vartheta_4(\tau) &= 16\tau^4 - 12\tau^2 + 1 \\ \vartheta_5(\tau) &= 32\tau^5 - 32\tau^3 + 6\tau \end{aligned}$$

In the next parts of this chapter, we will use these notations:

$$C^T = \left[c^{(0)}, c^{(1)}, \cdots, c^{(n)}\right],$$

$$P(\tau) = [\vartheta_0(\tau), \vartheta_1(\tau), \cdots, \vartheta_n(\tau)],$$

$$Q(\tau) = [\vartheta_0(\tau), \vartheta_1(\tau), \cdots, \vartheta_{n+1}(\tau)].$$

17.3 Function Approximation

Let N be a positive integer number, consider the following grid points in the interval $[-1,1]$

$$\Delta_N = \left\{\tau_i = -1 + ih, h = \frac{2}{N+1}, i = 0,1, \dots, N\right\}.$$

Let's introduce the following lemma:

17.3.1 Lemma: The matrix

$$W_n = \begin{pmatrix} \vartheta_0(\tau_0) & \vartheta_1(\tau_0) & \cdots & \vartheta_n(\tau_0) \\ \vartheta_0(\tau_1) & \vartheta_1(\tau_1) & \cdots & \vartheta_n(\tau_1) \\ \vdots & \vdots & \ddots & \vdots \\ \vartheta_0(\tau_n) & \vartheta_1(\tau_n) & \cdots & \vartheta_n(\tau_n) \end{pmatrix}$$

is invertible for any n.

Proof. Let C_n be the matrix that contains the coefficients for the Chebyshev polynomials of the second kind of a degree less than or equals n.

$$C_n = \begin{pmatrix} 1 & 0 & \cdots & c_n \\ 0 & 2 & \cdots & \vdots \\ \vdots & \vdots & \ddots & \vdots \\ 0 & 0 & \cdots & 2^n \end{pmatrix}, \quad \text{where,} \quad \begin{cases} 0, & n \equiv 1[2] \\ 1, & n \equiv 0[4] \\ -1, & n \equiv 2[4] \end{cases}$$

and let V_n be Vandermonde matrix

$$V_n = \begin{pmatrix} 1 & \tau_0 & \tau_0^2 & \cdots & \tau_0^n \\ 1 & \tau_1 & \tau_1^2 & \cdots & \tau_1^n \\ \vdots & \vdots & \vdots & \ddots & \vdots \\ 1 & \tau_n & \tau_n^2 & \cdots & \tau_n^n \end{pmatrix}$$

It is well known that the matrices C_n and V_n are invertible. We have

$$W_n = V_n C_n.$$

Hence, the matrix W_n is invertible as well.

We approximate the function $\xi(\tau)$ as

$$\xi(\tau) \approx C^T P(\tau), \tag{17.3.1}$$

such that

$$\xi(\tau_i) = C^T P(\tau_i), \quad \forall \quad i = 0, \cdots n.$$

17.4 Operational Matrix of Integration

The operational matrix of integration for $n = 4$ is obtained as follows:

$$\int_{-1}^{\tau} \vartheta_0(t)dt = \begin{bmatrix} 1 & \frac{1}{2} & 0 & 0 & 0 & 0 \end{bmatrix} P(\tau),$$
$$\int_{-1}^{\tau} \vartheta_1(t)dt = \begin{bmatrix} \frac{-3}{4} & 0 & \frac{1}{4} & 0 & 0 & 0 \end{bmatrix} P(\tau),$$
$$\int_{-1}^{\tau} \vartheta_2(t)dt = \begin{bmatrix} \frac{1}{3} & -\frac{1}{6} & 0 & \frac{1}{6} & 0 & 0 \end{bmatrix} P(\tau),$$
$$\int_{-1}^{\tau} \vartheta_3(t)dt = \begin{bmatrix} \frac{-1}{4} & 0 & \frac{-1}{8} & 0 & \frac{1}{8} & 0 \end{bmatrix} P(\tau),$$
$$\int_{-1}^{\tau} \vartheta_4(t)dt = \begin{bmatrix} \frac{1}{5} & 0 & 0 & \frac{-1}{10} & 0 & \frac{1}{10} \end{bmatrix} P(\tau).$$

Therefore

$$\int_{-1}^{\tau} C^T P(t)dt = C^T M_{5\times 6} Q(\tau), \tag{17.4.1}$$

Where

$$M_{5\times 6} = \begin{pmatrix} 1 & \frac{1}{2} & 0 & 0 & 0 & 0 \\ -\frac{3}{4} & 0 & \frac{1}{4} & 0 & 0 & 0 \\ \frac{1}{3} & \frac{-1}{6} & 0 & \frac{1}{6} & 0 & 0 \\ -\frac{1}{4} & 0 & -\frac{1}{8} & 0 & \frac{1}{8} & 0 \\ \frac{1}{5} & 0 & 0 & -\frac{1}{10} & 0 & \frac{1}{10} \end{pmatrix}$$

Similarly, we can find the operational matrix of integration M for an arbitrary values n.

17.5 Description of method

Consider the following integro-differential equation:

$$\begin{cases} \xi(\tau) = f(\tau) + \int_{-1}^{1} p(|\tau-\zeta|)\Psi(\tau,\zeta,\xi(\zeta),\xi'(\zeta))d\zeta, \tau \in [-1,1], \\ \xi(-1) = \xi_0, \end{cases} \tag{17.5.1}$$

taking the derivative of (17.5.1), we have

$$\begin{aligned} \xi'(\tau) = f'(\tau) \quad &+ \int_{-1}^{1} sign(\tau-\zeta)p'(|\tau-\zeta|)\Psi(\tau,\zeta,\xi(\zeta),\xi'(\zeta))d\zeta \\ &+ \int_{-1}^{1} p(|\tau-\zeta|)\frac{\partial\Psi}{\partial\tau}(\tau,\zeta,\xi(\zeta),\xi'(\zeta))d\zeta, \end{aligned} \tag{17.5.2}$$

where

$$sign(\tau-\zeta) = \begin{cases} +1, & \tau > \zeta \\ 0, & \tau = \zeta \\ -1, & \tau < \zeta \end{cases}$$

We approximate the unknow function $\xi'(\tau)$ by using (17.3.1), then

$$\xi'(\tau) = C^T P(\tau), \tag{17.5.3}$$

Integrate (17.5.3) concerning τ from 0 to τ

$$\xi(\tau) = \xi_0 + C^T M Q(\tau). \tag{17.5.4}$$

Now, substitute (17.5.4) and (17.5.3) in (17.5.2), then we find the following equation:

$$\begin{aligned} C^T P(\tau) = f'(\tau) \quad &+ \int_{-1}^{1} sign(\tau-\zeta)p'(|\tau-\zeta|)\Psi(\tau,\zeta,\xi_0 + C^T MP(\zeta), C^T P(\zeta))d\zeta \\ &+ \int_{-1}^{1} p(|\tau-\zeta|)\frac{\partial\Psi}{\partial\tau}(\tau,\zeta,\xi_0 + C^T MP(\zeta), C^T P(\zeta))d\zeta \end{aligned} \tag{17.5.5}$$

Collocate the equation (17.5.5) by the grids points Δ_N leads to the nonlinear algebraic system

$$C^TP(\tau_i) = f'(\tau_i) \quad + \int_{-1}^{1} sign(\tau_i - \zeta)p'(|\tau_i - \zeta|)\Psi(\tau_i, \zeta, \xi_0 + C^TMP(\zeta), C^TP(\zeta))d\zeta$$
$$+ \int_{-1}^{1} p(|\tau_i - \zeta|)\frac{\partial\Psi}{\partial\tau}(\tau_i, \zeta, \xi_0 + C^TMP(\zeta), C^TP(\zeta))d\zeta \quad (17.5.6)$$

The system (17.5.6) can be written as

$$C^TW_n = D^T = [d_0(C^T), d_1(C^T), \cdots, d_n(C^T)], \quad (17.5.7)$$

such that W_n is described in the lemma above and for $i = 1, \dots, n$,

$$d_i(C^T) = f'(\tau_i) \quad + \int_{-1}^{1} sign(\tau_i - \zeta)p'(|\tau_i - \zeta|)\Psi(\tau_i, \zeta, \xi_0 + C^TMP(\zeta), C^TP(\zeta))d\zeta$$
$$+ \int_{-1}^{1} p(|\tau_i - \zeta|)\frac{\partial\Psi}{\partial\tau}(\tau_i, \zeta, \xi_0 + C^TMP(\zeta), C^TP(\zeta))d\zeta$$

Therefore, by lemma (3), then

$$C^T = D^TW_n^{-1}. \quad (17.5.8)$$

The nonlinear algebraic system (9) is solved by iterative process, let's consider the following system:

$$\begin{cases} C_{k+1}^T = \left[d_0(C_k^T), d_1(C_k^T), \cdots, d_n(C_k^T)\right]W_n^{-1}, \\ C_0^T = \left(f'(\tau_0), f'(\tau_1), \cdots, f'(\tau_n)\right), k \geq 1. \end{cases} \quad (17.5.9)$$

By solving the system (17.5.9) with an appropriate value of k, we can determine the coefficients C^T. Subsequently, we can compute the desired approximation for $\xi'(\tau)$ using equation (17.3.1).

17.6 Examples

In this part, some illustrative examples for equation (17.1.1) are solved by the proposed method and we show the effectiveness in the tables below. Let's define the error function as follow:

$$E(n,k) = \max_{i=1:n}|\xi(\tau_i) - \xi_n(\tau_i)|,$$

where, $\xi(\tau)$ is the exact solution, $\xi_n(\tau)$ is the approximate solution, n is the degree of the polynomial $\xi_n(\tau)$, and k is the order of the approximate solution of the algebraic system (17.5.9).

Example 17.6.1 Consider the following equation

$$\begin{cases} \xi(\tau) = f(\tau) + \int_{-1}^{1} \sqrt{|\tau - \zeta|} \cos[\tau + \xi(\zeta) + e^{\zeta} - \xi'(\zeta)] d\zeta \quad \forall \tau \in [-1,1], \\ u(-1) = -e^{-1}, \end{cases} \tag{17.6.1}$$

with $f(\tau) = e^{\tau} - \frac{2}{3}\left(\sqrt{(1-\tau)^3} + \sqrt{(1+\tau)^3}\right)\cos(\tau)$ and the exact solution is $\xi(\tau) = \tau e^{\tau}$.

Example 17.6.2 Let the Fredholm integro-differential equation

$$\begin{cases} \xi(\tau) = \sin(\tau) + \frac{8}{7}((1+\tau)^{\frac{7}{4}} + (1-\tau)^{\frac{7}{4}}) + \int_{-1}^{1} |\tau - \zeta|^{\frac{3}{4}} \sqrt{3 + \xi(\zeta)^2 + \xi'(\zeta)^2} d\zeta, \\ u(-1) = \sin(-1), \quad \forall \tau \in [-1,1], \end{cases}$$

such that the exact solution is $\xi(\tau) = \sin(\tau)$.

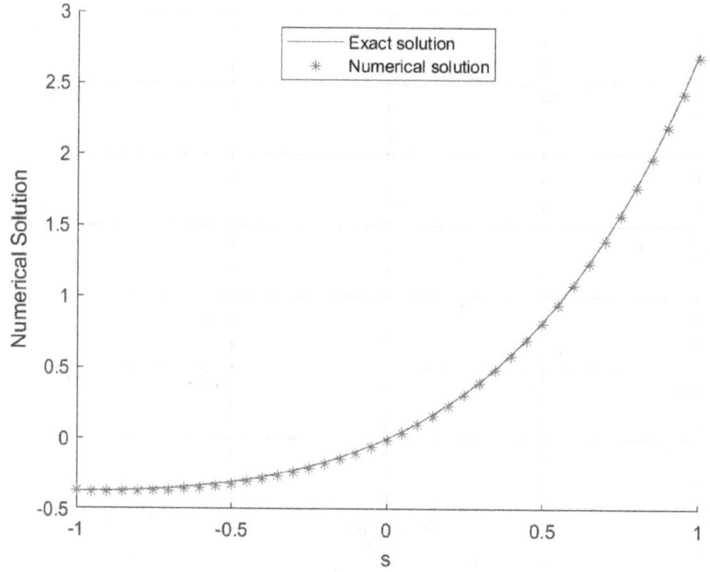

Figure 17.1 ξ versus ξ_n for Example 6.

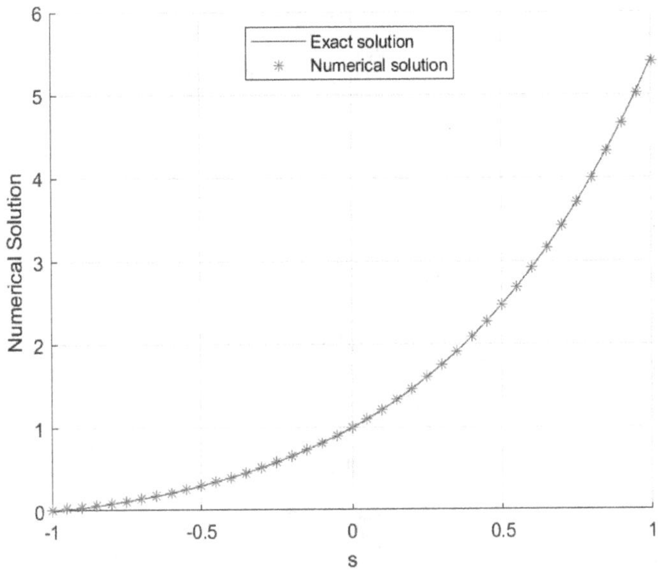

Figure 17.2 ξ' versus ξ_n' for Example 6, with $n = 4$.

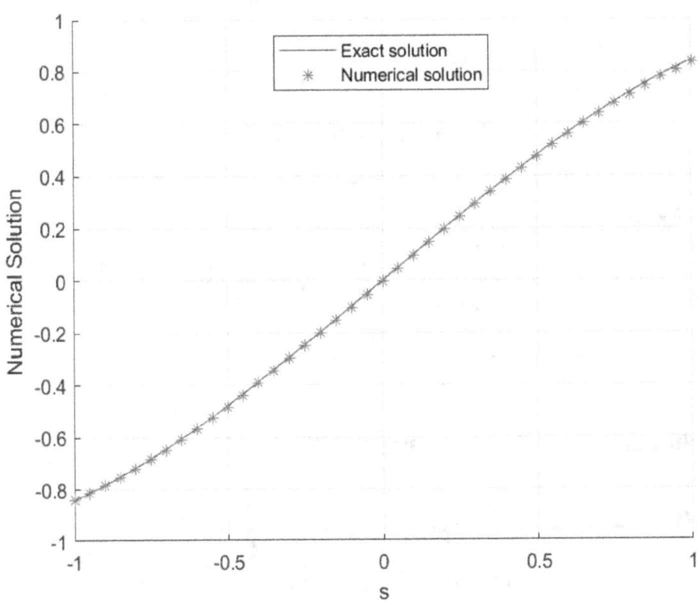

Figure 17.3 ξ versus ξ_n for Example 6 , with $n = 2$.

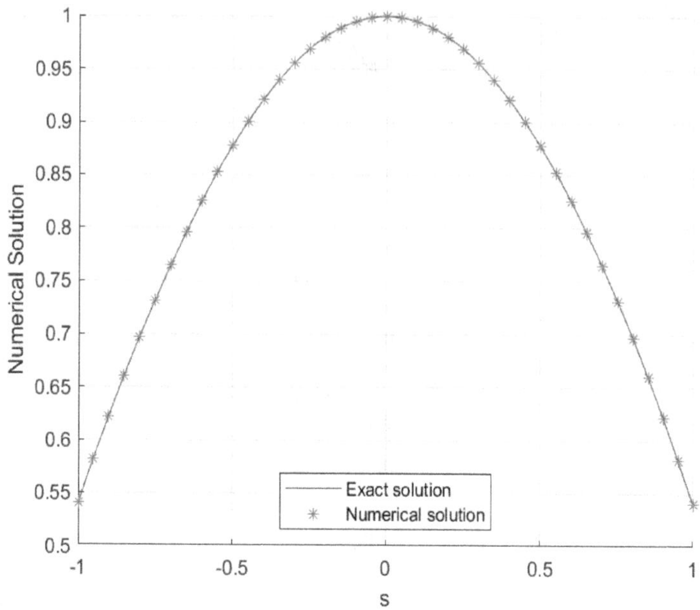

Figure 17.4 ξ' versus ξ'_n for Example 6, with $n = 4$.

Table 17.1 Numerical results for Example 17.6.1

n	$E(n, 4)$	CPU time
2	2.5634e-01	0.045137
4	1.5592e-02	0.031160
6	3.2462e-04	0.035663
8	3.3167e-05	0.049319

Table 17.2 Numerical results for Example 17.6.2

n	$E(n, 3)$	CPU time
2	1.1656e-01	0.029467
4	3.4925e-03	0.048081
6	4.7316e-05	0.077722
8	1.4533e-05	0.140198

17.7 Conclusion

In this work, we have applied the numerical method to solve a non-linear integro differential equation with a weakly singular kernel. The numerical solution is given by Chebyshev polynomials of the second kind and operational matrix of integration, thereafter, the collocation technique reduces the equation into a nonlinear algebraic system that is solved by the iterative process. By certain examples, we have shown the accuracy and efficiency of the used method.

References

[1] Biçer G., Öztürk Y. and Gülsu M. Numerical approach for solving linear Fredholm integro-diferential equation with piecewise intervals by Bernoulli polynomials. Int. J. Comp. Math. 95(10), 2100–2111, (2018).

[2] El-Gamel M., El-Azab M. S. Legendre-Galerkin method for the linear Fredholmintegro-diferential equations. Appl. Math. Comput. 243, 789–800, (2014).

[3] Jalilian R., Tahernezhad T. Exponential spline method for approximation solution of Fredholmintegro-diferential equation. Int. J. Comp. Math, (2019).

[4] Henka, Y., Lemita, S., Aissaoui, M. Z. Numerical study for a second order Fredholm integro-differential equation by applying Galerkin-Chebyshev-wavelets method. Journal of Applied Mathematics and Computational Mechanics, 21(4), (2022).

[5] Lakestani M., Saray B. N. and Dehghan, M. Numerical solution for the weakly singular Fredholm integro-differential equations using Legendre multiwavelets. Journal of Computational and Applied Mathematics, 235(11), 3291-3303. (2011).

[6] Pedas A., Tamme E. A discrete collocation method for Fredholm integro-differential equations with weakly singular kernels. Applied numerical mathematics, 61(6), 738-751, (2011), Lakestani, M., Saray, B. N., Dehghan, M. (2011).

[7] Yusufoglu E. Improved homotopy perturbation method for solving Fredholm type integro-diferential equations. Chaos Solitons Fractals 41, 28–37, (2009).

www.ingramcontent.com/pod-product-compliance
Lightning Source LLC
LaVergne TN
LVHW011527200926
848484LV00001B/175